T0199261

Recent Advances in AI-enabled Automated Medical Diagnosis

Editors

Richard Jiang
Lancaster University, UK

Li Zhang
Royal Holloway University of London, UK

Hua-Liang Wei
University of Sheffield, UK

Danny Crookes
Queen's University Belfast, UK

Paul Chazot
Durham University, UK

CRC Press
Taylor & Francis Group
Boca Raton London New York

CRC Press is an imprint of the
Taylor & Francis Group, an **informa** business
A SCIENCE PUBLISHERS BOOK

First edition published 2022
by CRC Press
6000 Broken Sound Parkway NW, Suite 300, Boca Raton, FL 33487-2742

and by CRC Press
4 Park Square, Milton Park, Abingdon, Oxon, OX14 4RN

CRC Press is an imprint of Taylor & Francis Group, LLC

Library of Congress Cataloging-in-Publication Data (applied for)

ISBN: 978-1-032-00843-1 (hbk)
ISBN: 978-1-032-00856-1 (pbk)
ISBN: 978-1-003-17612-1 (ebk)

DOI: 10.1201/9781003176121

Typeset in Times New Roman
by Innovative Processors

Preface

Recent years have witnessed significant advancement of machine learning (ML), artificial intelligence (AI) and other data-driven techniques, bringing medical and healthcare data analysis to a new era with exciting progress in many areas including biomedical informatics, health informatics, AI in medicine, AI of Medical Things (AIoMT), healthcare AI and smart homecare. It is an amazing age where opportunities and challenges exist equally in nearly all application areas for all these modern machine learning and data analytics techniques. We are lucky because we have now achieved many state-of-the-art results in medical diagnosis and healthcare services using different advanced machine learning techniques. For example, nowadays many shallow and deep learning frameworks and AI techniques have been successfully applied to medical diagnosis and healthcare services. However, there are still many challenges that need to be overcome to obtain satisfactory solutions when trying to resolve complicated problems. For example, there are some increasing concerns about data-driven AI and black-box deep neural networks from the viewpoints of reliability, ethics, explainability and interpretability, among others.

We have seen an increasing amount of applications of machine learning, artificial intelligence and other data analytics (e.g. data modelling, data mining) techniques in the fields of medical diagnosis and healthcare services, such as neurophysiological signal and neuro-imaging processing, cancer and disease diagnosis, healthcare (e.g. electronic health records) and wellbeing data analysis, pandemic diagnosis and prediction, just mention a few. A lot of new methods and algorithms have been developed, aiming to solve either the existing problems or emerging issues which are closely related to or significantly impact on our daily life.

It is right time to stimulate the dissemination of recent results and findings of applying AI and ML techniques to solving problems in the fields of medical diagnosis and healthcare services. This book aims to show the most recent advancement of data-driven and data-based ML and AI techniques, with an emphasis on the implementation of methods and algorithms, including signal processing and system identification, data mining, image processing and pattern recognition, and deep neural networks, together with their applications.

The book collects a total of 22 papers, which can be roughly categorized into 4 groups: 1) COVID-19 diagnosis and pandemic prediction (Chapters 1, 2, 3, 4, 5 and 15); 2) Neurophysiological and neurobiological data analysis and brain–computer interface (Chapters 17, 18, 19, 20, 21 and 22); 3) Biological and healthcare data analysis (Chapters 13, 14, 15 and 16); and 4) Disease and cancer diagnosis (Chapters 7, 8, 9, 10, 11 and 12). Some details are as follows.

The emergency of COVID pandemic has engrossed vast efforts across worldwide research communities. Chapter 1 developed AI-based COVID-19 diagnosis based on texture analysis. Chapter 2 proposed a novel transparent, interpretable, parsimonious, and simulatable (TIPS) modelling framework to reveal the COVID-19 pandemic dynamics. Chapter 3 combined audio with CT scan for COVID-19 diagnosis. Chapter 4 reported the COVID-19 forecasting in India with locally collected data. Chapter 5 carried out a survey on COVID diagnosis techniques.

Medical AI has also been widely exploited in many specific cases. Chapter 6 exploited AI-based age estimation models in the evaluation of human brains. Chapter 7 carried out a survey on the brain age estimation. Chapter 8 reviewed the social behavioural evaluation on animals. Chapter 9 reports a new AI model on Leukaemia Diagnosis. Chapter 10 investigated on AI-powered medical prediction on trauma patients. Chapter 11 exploited AI techniques for face-based emotion evaluation. Chapter 12 utilized blood vessel segmentation for retinal related eye analysis.

Time-series medical sensory data and signal can always play an important role in medical diagnosis. Chapter 13 analysed the dynamics of the walking gaits of Parkinson's patients. Chapter 14 examined neural dynamics via sparse model identification. Chapter 15 studied the correlation between weather conditions and the spread of COVID pandemic via contrastive learning with NARMAX models. Chapter 16 exploited NARX predictive model on body mass index.

Brain waves and heart signals are two particularly widely exploited time-series data in various applications. Chapter 17 utilized brain-computer interfaces in the therapy for motor rehabilitation. Chapter 18 proposed new time-series modelling ECG based heart anomaly detection. Chapter 19 carried out a comparative study on several deep neural networks for ECG based cardiologic diagnosis. Chapter 20 compared convolutional neural networks and long-sort term memory neural networks in EEG based emotion diagnosis. Chapter 21 proposed a novel approach on motor imagery EEG classification. Chapter 22 studied the EEG signal classification for brain-computer interface.

It is expected that scientists, academics, researchers, clinicians, medical services, policy makers, technologists, engineers and so on who work in the areas of medicine and healthcare, and those who are interested in some of the subjects in these areas, would be able to benefit from the book.

Richard Jiang
Li. Zhang
Hua-Liang Wei

Contents

Preface   iii

1. Enhancement of COVID-19 Diagnosis Using Machine Learning and
   Texture Analyses of Lung Imaging   1
   Bhuvan Mittal, JungHwan Oh

2. Modeling COVID-19 Pandemic Dynamics Using Transparent, Interpretable,
   Parsimonious and Simulatable (TIPS) Machine Learning Models: A Case
   Study from Systems Thinking and System Identification Perspectives   13
   Hua-Liang Wei, Stephen A. Billings

3. Deep Learning-based Respiratory Anomaly and COVID Diagnosis
   Using Audio and CT Scan Imagery   29
   Conor Wall, Chengyu Liu and Li Zhang

4. COVID-19 Forecasting in India through Deep Learning Models   41
   Arindam Chaudhuri and Soumya K. Ghosh

5. Deep Learning-based Techniques in Medical Imaging for COVID-19
   Diagnosis: A Survey   62
   *Fozia Mehboob, Abdul Rauf, Khalid M. Malik, Richard Jiang,
   Abdul K.J. Saudagar, Abdullah AlTameem and Mohammed AlKhathami*

6. Embedding Explainable Artificial Intelligence in Clinical Decision
   Support Systems: The Brain Age Prediction Case Study   81
   *Angela Lombardi, Domenico Diacono, Nicola Amoroso, Alfonso Monaco,
   Sabina Tangaro and Roberto Bellotti*

7. Machine Learning-based Biological Ageing Estimation
   Technologies: A Survey   96
   Zhaonian Zhang, Richard Jiang, Danny Crookes and Paul Chazot

8. Review on Social Behavior Analysis of Laboratory Animals: From
   Methodologies to Applications   110
   Ziping Jiang, Paul L. Chazot and Richard Jiang

9. Acute Lymphoblastic Leukemia Diagnosis Using Genetic Algorithm and
   Enhanced Clustering-based Feature Selection   123
   *Siew Chin Neoh, Srisukkham Worawut, Li Zhang and Md. Mostafa
   Kamal Sarker*

10. Artificial Intelligence-enabled Automated Medical Prediction and
    Diagnosis in Trauma Patients   135
    *Lianyong Li, Changqing Zhong, Gang Wang, Wei Wu, Yuzhu Guo, Zheng
    Zhang, Bo Yang, Xiaotong Lou, Ke Li and Heming Yang*

11. DCGAN-based Facial Expression Synthesis for Emotion Well-being
    Monitoring with Feature Extraction and Cluster Grouping   146
    Eaby Kollonoor Babu, Kamlesh Mistry and Li Zhang

12. A Hybrid-DE for Automatic Retinal Image-based Blood
    Vessel Segmentation   157
    Colin Paul Joy, Kamlesh Mistry, Gobind Pillai and Li Zhang

13. Artificial Intelligence for Accurate Detection and Analysis of Freezing
    of Gait in Parkinson's Disease   173
    *Wenting Yang, Simeng Li, Debin Huang, Hantao Li, Lipeng Wang,
    Wei Zhang and Yuzhu Guo*

14. Sparse Model Identification for Nonstationary and Nonlinear Neural
    Dynamics Based on Multiwavelet Basis Expansion   215
    Song Xu, Lina Wang and Jingjing Liu

15. How Weather Conditions Affect the Spread of COVID-19: Findings
    of a Study Using Contrastive Learning and NARMAX Models   238
    Yiming Sun and Hua-Liang Wei

16. New Measurement of the Body Mass Index with Bioimpedance Using
    a Novel Interpretable Takagi-Sugeno Fuzzy NARX Predictive Model   253
    Changjiang He, Yuanlin Gu, Hua-Liang Wei and Qinggang Meng

17. Training Therapy with BCI-based Neurofeedback Systems
    for Motor Rehabilitation   268
    Jingjing Luo, Qiying Chen, Hongbo Wang, Youhao Wang and Qiang Du

18. A Modified Dynamic Time Warping (MDTW) and Innovative
    Average Non-self Match Distance (ANSD) Method for Anomaly
    Detection in ECG Recordings   281
    Xinxin Yao and Hua-Liang Wei

19. An Investigation on ECG-based Cardiological Diagnosis via Deep
    Learning Models   304
    Alex Meehan, Zhaonian Zhang, Bryan Williams and Richard Jiang

20. EEG-based Deep Emotional Diagnosis: A Comparative Study   317
    *Geyi Liu, Zhaonian Zhang, Richard Jiang, Danny Crookes
    and Paul Chazot*

21. A Novel Motor Imagery EEG Classification Approach Based on Time-
    Frequency Analysis and Convolutional Neural Network   329
    Qinghua Wang, Lina Wang and Song Xu

22. Classification of EEG Signals for Brain-Computer Interfaces using
    a Bayesian-Fuzzy Extreme Learning Machine   347
    Adrian Rubio-Solis, Carlos Beltran-Perez and Hua-Liang Wei

Index   363

Enhancement of COVID-19 Diagnosis Using Machine Learning and Texture Analyses of Lung Imaging

Bhuvan Mittal[1], JungHwan Oh[2]

Department of Computer Science and Engineering, University of North Texas, Denton, TX 76203, U.S.A.

1. Introduction

The COVID-19 is a highly contagious and virulent disease caused by the Severe Acute Respiratory Syndrome – Coronavirus-2 (SARS-CoV-2). Over 186 million cases and 4.0 million deaths were reported worldwide as of 13 July, 2021 [17]. A multinational consensus from the Fleischner Society reported that Computerized Tomography (CT) can be utilized for an early diagnosis of CT-based COVID-19 [14]. CT also has a high sensitivity in the classification of the COVID-19 disease [8]. However, this binary classification task of identifying COVID-19 positive from negative patients from lung CT involves a significant amount of radiologists' time and effort. Thus, it is crucial to develop an automated analysis of CT images to save radiologists' time in overstretched healthcare environments.

In our previous work [12], we implemented the CoviNet, a 3D CNN-based model for COVID-19 classification of lung CT images. In this paper, we propose an enhancement of this model [12], CoviNet Enhanced, using 3D CNN and Leung-Malik (LM) texture features [15] additionally. CoviNet Enhanced has a novel conditional majority voting algorithm with an ensemble of 3D CNN and SVMs, which provide better classification sensitivity and specificity than using the 3D CNN alone [12]. Since the SVM algorithm is complementary, this hybrid approach using SVM models and 3D CNN combined via conditional majority voting shows superior performance.

Contributions to CoviNet Enhanced: To achieve a better classification sensitivity and specificity than those of CoviNet, we add: 1) Leung-Malik (LM) texture features [15], 2) an ensemble classifier comprising a 3D CNN and texture features-based SVMs, and 3) the final classification is done via conditional majority voting.

The remainder of this paper is organized as follows: related work is presented in Section 2; the proposed method is described in Section 3; in Section 4, we discuss our experimental setup and results; finally, Section 5 presents some concluding remarks.

2. Related Work

In our previous work [12], we classified the CT into COVID-19 positive or negative, based on the 3D CNN which comprises 3D filters in convolutional layers to train a deep 3D CNN from scratch. This model utilized the 3D CT scan volumes as opposed to individual slice-level CT imaging data to come up with the patient-level diagnosis. It has a network depth of 16 layers comprising four 3D convolutional layers, four 3D max-pooling layers, four 3D batch normalization layers, one global average 3D pooling layer, two fully connected dense layers, one dropout layer, and a final softmax layer. All four convolutional layers have a kernel size of $3\times3\times3$ but use different numbers of kernels at 64, 128, and 256. The 'ReLU' activation function is used. The four 3D-max-pooling layers take a $2\times2\times2$ sliding cube which subsamples the image length, width, and depth dimensions and has a stride of 2. To speed up the training of CoviNet, 3D batch normalization layers are included after the 3D pooling layers. Then, 3D global average pooling takes a 4D input of size length$\times$width$\times$depth$\times$channels ($=12\times12\times2\times256$) and outputs a one-dimensional output of size 256 channels. Next, the fully connected layer follows with a dimension of 512, followed by a dropout layer with a dropout factor of 0.3 which is introduced to make the model robust to noise. The final softmax layer with sigmoid activation outputs the predicted probability of being COVID-19 positive. Adam optimizer is used with an initial learning rate of 0.001 with an exponential decay rate of 0.96 over 100,000 decay steps.

In Wang *et al.*'s DeCovNet [16], a single 3D CNN classifier, taking only original CTs as input data, is implemented for COVID-19 classification into positive and negative classes. In DeCovNet, the 3D convolutional layers are followed by 3D batch normalization, 'ReLU' activation, and 3D pooling layers. There are six 3D convolutional layers. Training and testing are done on a proprietary dataset for 100 epochs having an Adam optimizer with a constant learning rate of 1×10^{-5}.

Imani *et al.* [11] first used morphological filters to extract the shape and structural features without and with CNN processing. Second, they used Gabor filters to extract textural features without and with the use of a trained CNN for feature extraction. Then they classified the morphological and Gabor features independently via two separate classifiers: support vector machine and random forest. For the morphological model, the random forest classifier without CNN performed the best, whereas, for the Gabor model, the random forest classifier with trained CNN performed the best.

Goncharov *et al.* [4] proposed a multitask approach to solving the identification and segmentation tasks together via a 2D U-Net model which outputs a common intermediate feature map that is aggregated into a feature vector via a pyramid pooling layer. The classification layers follow the high-resolution upper part of U-Net and comprise two fully-connected layers, followed by a softmax layer which outputs the probability of the input CT volume being COVID-19 positive.

2.1 Methodology

The proposed approach consists of four components:

2.1.1 3D CNN

A 3D CNN, having four three-dimensional convolutional kernel layers from our previous work [12], is trained from scratch on the augmented CT data. The augmentation details can be found in our previous work [12], and the 3D CNN architecture was explained in detail in Section 2. It is a deep model with 17 layers comprising four 3D convolutional layers, four 3D max-pooling layers, four 3D batch normalization layers, one global average 3D pooling layer, two fully connected dense layers, one dropout layer, and a final softmax layer. 3D batch normalization layers after the 3D pooling layers help the model learn faster, and the dropout layer contributes to its noise robustness. Early stopping is also invoked if the model loss does not reduce over the last five epochs.

2.1.2 Texture Feature Extraction

The LM features have textural, shape, and intensity-based features. LM texture features [15] are typically extracted for an image using the 48 LM filters (LM filter bank) which are convolved over the entire input image. The LM filter bank (Fig. 1) has a mix of edge, bar, and spot filters at multiple scales and orientations. It has a total of 48 filters, namely two Gaussian derivative filters at six orientations and three scales, eight Laplacian of Gaussian filters, and four Gaussian filters.

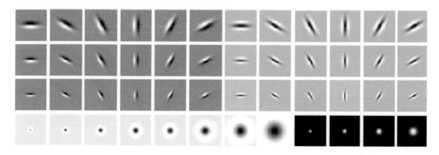

Fig. 1: LM filter bank with 48 filters [15]

When we apply the two Gaussian derivative filters at six orientations and three scales shown in the first three rows of Fig. 1, we have nearly all-black pixels as a result, since lungs do not have elongated objects. The features which can identify the ground glass opacities and consolidations will be suitable for our work. The results of convolving the 12 filters in the bottom row of Fig. 1 are shown in Fig. 2.

Intuitively, these are possibly informative features, which can be used for COVID-19 classification. We create 12 separate SVM models, taking one feature at a time for all the convolved outputs from the 12 filters. Then, the final feature selection is done by selecting the two top-performing LM filters. The selected top-performing filters are LM37 and LM41 and the results of convolving these filters with the input image are shown in Fig. 3.

2.1.3 SVM Models

The selected LM texture features, namely the LM37 and LM41, along with the

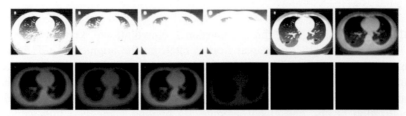

Fig. 2: Result of CT image convolving the 12 filters in the bottom row of Fig. 1 (LM 37 through LM48)

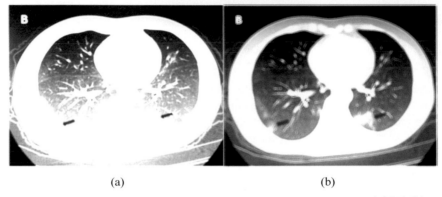

(a)   (b)

Fig. 3: Results of CT image convolving the two best-performing LM filters: (a) LM37, (b) LM41

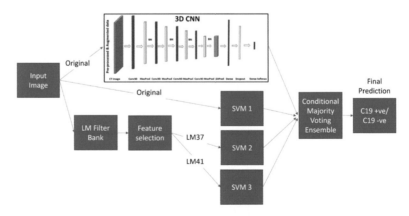

Fig. 4: Ensemble classifier of 3D CNNs and support vector machines

original image are classified via three separate SVM classifiers. We use the held-out test dataset approach [2] and 80:20 split is used for training versus testing. First, the original image datasets and the texture feature inputs are resized to 224×224. The COVID-19 CT images with COVID-19 positive or negative physician-provided labels are fed into the first SVM model. The Leung Malik feature LM41 is used to build the second SVM model. The Leung Malik feature LM37 is used to build the third SVM model. For the SVM, we perform a grid search for the parameter values of 'C' of 1, 10, 100, 1000, with a linear kernel for efficiency, and with gamma values of 0.001 and 0.0001 with the radial basis function kernel. We then get the predicted class for each image in the test dataset. This is repeated over the five folds.

2.1.4   Ensemble of 3D CNNs and SVMs

We found that the 3D CNN predictions have a positive class probability that varies for each input, ranging from 0.1% to 99.9%. For some inputs, our 3D CNN does have an uncertain prediction with COVID-19 positive class probability values (between 46% and 54%). In such cases, our 3D CNN is not very confident about the prediction. To improve the predictions for this range, we use conditional majority voting to combine the predicted outputs of the 3D CNN with the three SVMs.

When the original 3D CNN's prediction has a probability in a certain range, i.e. between the low threshold (Th_{Low}: 46% in our case) and the high threshold (Th_{High}: 54% in our case), then the model uses the LM texture-features' SVM model predictions along with 3D CNN via majority voting to make the final classification. When the original 3D CNN's prediction has a probability smaller than the low threshold (Th_{Low}: 46% in our case) or larger than the high threshold (Th_{High}: 54% in our case), only the 3D CNN's output makes the final classification. These probability threshold values were determined by experiments and fine-tuned on the validation dataset. This gives the final predicted classification of COVID-19 positive or COVID-19 negative. The model performance is evaluated, based on comparing this final prediction on held-out test data with the ground truth labels. The detailed architecture of CoviNet Enhanced is shown in Fig. 4. The conditional majority voting algorithm is showcased in Fig. 5.

Algorithm 1: Conditional Majority Voting Classification Rule
Inputs: 3D CNN's positive class probability (C), first SVM's positive class probability (S1), second SVM's positive class probability (S2), third SVM's positive class probability (S3)

if $0.46 < C < 0.54$ then
    'final predicted positive class probability' ← (C+S1+S2+S3)/4
else
    'final predicted positive class probability' ← C
end if

Fig. 5: Algorithm for conditional majority voting classification

3. Experiments and Results

CoviNet Enhanced was implemented in 'jupyter-notebook' using python's 'tensor-flow', 'keras', and other libraries [2, 5, 6, 9, 10]. The pre-processing of the data was done using the 'nibabel' library for medical image processing which can read the CT volume data provided in .nii format [7]. Data augmentation was performed on the entire dataset using the 'scipy' and 'ndimage' libraries. The augmentation details can be found in our previous work [12].

The dataset was split 80:20 for training and validation, and each patient's images are either entirely in the training dataset or entirely in the validation dataset. Five folds are used for cross-validation as discussed earlier. Adam optimizer with an exponentially decaying learning rate and an initial learning rate of 0.0001 with a decay rate of 0.96, and with 100,000 decay steps, was used. The training was done with an upper limit of 100 epochs with an early stopping criterion based on validation accuracy not improving over the next five epochs.

We trained and evaluated our CoviNet Enhanced model's performance on UCSD-AI4H [18], MosMed [13], and MosMed selected on lung CT datasets described in Tables 1, 2, and 3 respectively. Since the MosMed dataset in Table 2 is very large, we selected some of them for faster evaluations, which is shown in Table 3.

Table 1: UCSD-AI4H dataset [18]

Class	Training	Validation
COVID-19 Positive	172 patients 279 images	41 patients 70 images
COVID-19 Negative	140 patients 317 images	31 patients 80 images

Table 2: MosMed dataset

Class	Train	Validation	Test	Totals
COVID-19 Positive	582 patients 14,681 images	102 patients 2,573 images	172 patients 4,339 images	856 patients 21,593 images
COVID-19 Negative	173 patients 4,321 images	30 patients 749 images	51 patients 1,337 images	254 patients 6,407 images
Overall	**755 patients** **19,002 images**	132 patients 3,322 images	223 patients 5,676 images	1,110 patients 28,000 images

Table 3: MosMed selected dataset

Class	Train	Validation
COVID-19 Positive	138 patients 3,385 images	34 patients 860 images
COVID-19 Negative	203 patients 5,305 images	51 patients 1,337 images

The various performance metrics to evaluate our model's performance are accuracy, precision, recall, sensitivity, specificity, F-score, and Matthew's correlation coefficient (MCC) [3]. These metrics are defined as follows in equations (1), (2), (3), (4), (5) and (6):

$$\text{Accuracy} = \frac{(TP + TN)}{(TP + TN + FP + FN)} \tag{1}$$

$$\text{Sensitivity (a.k.a. Recall)} = \frac{(TP)}{(TP + FN)} \tag{2}$$

$$\text{Specificity} = \frac{(TN)}{(TN + FP)} \tag{3}$$

$$\text{Precision} = \frac{(TP)}{(TP + FP)} \tag{4}$$

$$F1 = \frac{\left(2 \times \text{Precision} \times \text{Recall}\right)}{\left(\text{Precision} + \text{Recall}\right)} \frac{\left(2 \times TP\right)}{\left(2 \times TP + FP + FN\right)} \tag{5}$$

$$MCC = \frac{(TP \times TN - FP \times FN)}{\sqrt{((TP + FP)(TP + FN)(TN + FP)(TN + FN))}} \tag{6}$$

where TP, TN, FP, and FN stand for COVID-19 positive patients predicted as COVID-19 positive, COVID-19 negative patients predicted as COVID-19 negative, COVID-19 negative patients predicted as COVID-19 positive, and COVID-19 positive patients predicted as COVID-19 negative, respectively. The field of medical diagnosis is intolerant to FN, i.e. type 2 errors and therefore, a high recall or sensitivity is a hard constraint. We aim to achieve the highest sensitivity possible, and then find the best specificity corresponding to that highest sensitivity. Although the MCC is not derived from sensitivity and specificity directly, in essence, the MCC is weighting sensitivity and specificity equally since it is weighting the FN and FP equally. However, in our case, sensitivity is more important than specificity, so we want to place more weight on sensitivity. Thus, we should evaluate the performance of models based, not just on the MCC, but on both the MCC and sensitivity.

To perform a fair comparison against previously published work, we also re-implemented Wang *et al.*'s DeCovNet [16] on the three datasets in this paper. We trained and evaluated Wang *et al.*'s DeCovNet [16] model's performance on UCSD-AI4H [18], MosMed [13], and MosMed selected on lung CT datasets described in Tables 1, 2 and 3 respectively.

On the UCSD-AI4H dataset, the performance comparison of CoviNet Enhanced with previously published works, namely, Wang *et al.*'s DeCovNet [16], Imani *et al.*'s Gabor, and Morphological models [11] is shown in Table 4. For Imani's Gabor model, the random forest classifier with CNN feature extraction was used, and for Imani's Morphological model, the random forest classifier alone was used. Imani *et al.*'s [11] also used the UCSD-AI4H dataset for their COVID-19 classification from CT, so we report their best-performing results (for Gabor and Morphological models) directly from their paper.

Among all the four models shown in Table 4, CoviNet Enhanced exhibited the highest F1-score of 0.930 and the MCC of 0.842 on the UCSD-AI4H dataset. Further, CoviNet Enhanced showed significantly superior performance than the next best-performing model, Wang *et al.*'s DeCovNet [16], which had an F1-score of 0.861, and an MCC of 0.717.

Table 4: Performance comparison between CoviNet enhanced and other published works on the UCSD-AI4H dataset (Table 1)

Model	Accuracy	Precision	Sensitivity	Specificity	F1-score	MCC
Wang's DeCovNet [16]	85.5%	91.9%	80.9%	91.2%	0.861	0.717
Imani's Gabor [11]	76.7%	-	-	-	0.743	-
Imani's Morph. [11]	75.3%	-	-	-	0.753	-
CoviNet Enhanced	92.2%	93.0%	93.0%	91.2%	0.930	0.842

On the MosMed dataset, the performance comparison of CoviNet Enhanced with Wang *et al.*'s DeCovNet [16] is shown in Table 5. Among the two models shown in Table 5, CoviNet Enhanced exhibited the highest F1-score of 0.774 and the MCC of 0.608 on the MosMed selected dataset. Further, CoviNet Enhanced showed significantly superior performance than the next best-performing model, Wang *et al.*'s DeCovNet [16], which had an F1-score of 0.652, and MCC of 0.353.

The MosMed full dataset has more variance, and it is more complex to learn all the different and varied data. Still, our model outperformed better than Wang's DeCovNet's 3D CNN [16] considerably. The reason for the superior performance of CoviNet Enhanced over DeCovNet [16] is the ensemble approach, which leverages the complementary texture feature-based SVM models when 3D CNN is less confident in its classification. However, DeCovNet [16] has only a single 3D CNN classifier, taking only original CT images as input data.

Table 5: Performance comparison between CoviNet enhanced and the other published work on MosMed dataset (Table 2)

Model	Accuracy	Precision	Sensitivity	Specificity	F1-score	MCC
Wang *et al.*'s DeCovNet [16]	66.8%	74.1%	58.3%	76.7%	0.652	0.353
CoviNet Enhanced	**80.8%**	**75.7%**	**79.1%**	**82.0%**	**0.774**	**0.608**

On the MosMed selected dataset, the performance comparison of CoviNet Enhanced with Goncharov *et al.*'s [4] model is shown in Table 6. Goncharov *et al.*'s [4] also used this same dataset, so the results are reported directly from their paper. CoviNet Enhanced exhibited the highest F1-score of 0.957 and the MCC of 0.926. Further, CoviNet Enhanced showed significantly superior performance than the next best model, Goncharov *et al.*'s multitasksp1 U-Net [4], which had an F1-score of 0.827, and MCC of 0.770. Goncharov *et al.*'s method involved multitask learning by jointly learning to classify and segment, and it was very slow to train versus our relatively more efficient CoviNet Enhanced 3D CNN classification.

The performance comparison of CoviNet [12] (our previous work) and CoviNet Enhanced is shown in Table 7. On the UCSD-AI4H dataset, CoviNet Enhanced outperforms the CoviNet by 1.3% on sensitivity, 32.9% on specificity, 0.144 on F1-score, and 0.312 on MCC. Notably, CoviNet Enhanced has the highest F1-score and MCC score at 0.930 and 0.842 respectively versus CoviNet's F1-score and MCC of 0.786 and 0.530. On the MosMed selected dataset, CoviNet Enhanced outperforms the CoviNet by 3.7% on specificity, 0.027 on F1-score, and 0.044 on MCC as shown in Table 7. Notably, CoviNet Enhanced has a higher F1-score and MCC score at 0.957 and 0.926 respectively versus CoviNet's F1-score and MCC of 0.930 and 0.882.

Table 6: Performance comparison between CoviNet enhanced and Goncharov *et al.* [4] on MosMed selected dataset (Table 3)

Model	Accuracy	Precision	Sensitivity	Specificity	F1-score	MCC
Goncharov's U-Net [4]	89.4%	72.1%	96.9%	86.8%	0.827	0.770
CoviNet Enh.	**96.4%**	**94.3%**	**97.1%**	**95.9%**	**0.957**	**0.926**

Table 7: Performance comparison between CoviNet and CoviNet enhanced on UCSD-AI4H (Table 1) and MosMed selected dataset (Table 3)

Model/Data	Accuracy	Precision	Sensitivity	Specificity	F1-score	MCC
CoviNet/USDAI4H	75.0%	68.7%	91.7%	58.3%	0.786	0.530
CoviNet Enhanced/ UCSD-AI4H	**92.2%**	**93.0%**	**93.0%**	**91.2%**	**0.930**	**0.842**
CoviNet/ MosMed	94.1%	89.2%	97.1%	92.2%	0.930	0.882
CoviNet Enhanced/ MosMed	**96.4%**	**94.3%**	**97.1%**	**95.9%**	**0.957**	**0.926**

We now showcase the result of feeding the entire 3D volume of a patient compares with utilizing each image in isolation, using the same 3D CNN CoviNet Enhanced approach. The proposed CoviNet Enhanced model's 3D CNN component takes only 3D CT volumes since the 3D CNN processes the entire 3D volume together and makes predictions at volume only. Note that the SVMs take image-level data and the SVM predictions are averaged for the images constituting a volume to get the final volume level predictions. To get image-level results from 3D CNN, we provided input as a single CT slice duplicated five times to make volume data for feeding to the 3D CNN. Table 8 shows the image-level and patient-level metrics of the CoviNet Enhanced model on the UCSD-AI4H dataset.

As shown in Table 8, CoviNet Enhanced patient-level outperforms the CoviNet Enhanced image-level by 33.0% on sensitivity, 0.180 on F1-score, and 0.283 on MCC although it does have an 8.8% lower specificity. The increased sensitivity of the CoviNet Enhanced patient-level model versus the image-level model is much higher than the lowered specificity. CoviNet Enhanced patient-level has a higher F1-score and MCC score at 0.930 and 0.842 versus the CoviNet Enhanced image-level model with an F1-score and MCC of 0.750 and 0.559, respectively.

The results show the power of learning from the sequential nature of the consecutive slices in a CT volume, so if the disease prediction is uncertain in one slice, the corresponding slices in the vicinity will help provide greater context and help improve the performance. Even noisy pixels occurring due to random noise will not occur in the same place in consecutive slices, so the model will become more robust to noise and be able to learn better. So, to summarize, learning from three-dimensional volumes is better than from images since the model learns not only from the spatial dimension, but also from the depth dimension.

Table 8: Image-level and patient-level metrics of CoviNet enhanced

CoviNet Enhanced	Accuracy	Precision	Sensitivity	Specificity	F1-score	MCC
Patient level (volume-level)	92.2%	93.0%	93.0%	91.2%	0.930	0.842
Image level	72.1%	100.0%	60.0%	100.0%	0.750	0.559

We now demonstrate how the 3D CoviNet Enhanced model with a comparable 2D model. For the CNN part of the CoviNet Enhanced model, the 3D input CT volumes, 3D convolutional, and 3D pooling layers of the CoviNet Enhanced are replaced by 2D CT input images, 2D convolutional, and 2D pooling layers. The 3D layers of the 3D CNN have 4 dimensions of size length×width×depth×channels, but the corresponding 2D layers of the 2D CNN have 3 dimensions of size length×width×channels only. The SVM portion of the CoviNet Enhanced model remains unchanged. Table 9 shows the CoviNet Enhanced's performance on UCSD-AI4H with 3D CNN versus 2D CNN in which 3D CNN is between 11% and 20% and is better on the various performance measures than 2D CNN.

Table 9: Performance of CoviNet enhanced on UCSD-AI4H with 3D versus 2D CNN

Model	Accuracy	Precision	Sensitivity	Specificity	F1-score	MCC
2D CNN	77.10%	76.67%	73.64%	80.13%	0.750	0.541
3D CNN	92.2%	93.0%	93.0%	91.2%	0.930	0.842

4. Concluding Remarks

Automated COVID-19 diagnosis via deep learning on lung CT is highly sensitive in detecting COVID-19 disease-induced pneumonic changes [4]. The proposed CoviNet Enhanced model with a hybrid approach is an excellent diagnostic model for COVID-19 diagnosis from lung CT as it exhibits not only high sensitivity, but also high specificity. The higher performance is because it utilizes both a deep 3D convolutional neural network and Leung-Malik texture features-based SVMs. It outperforms the CoviNet, our previous work [12] because of the hybrid approach and our novel conditional majority voting ensemble approach. In the instances where the 3D CNN is uncertain about its COVID-19 prediction, the texture features-based SVMs help to complement the 3D CNN. The SVM is a non-neural network method and hence, it serves as a complementary technique to the 3D CNN. This hybrid approach of deep learning with textural features and conditional majority voting helps to overcome the weakness of the 3D CNN alone. The experimental results show the proposed method is highly effective for COVID-19 detection.

Our approach will not only save the radiologist's time but also improve the diagnostic performance in terms of much higher sensitivity and similar specificity. Our model achieved its best sensitivity of 97.1% and a specificity of 95.9%. In comparison, a radiologists' performance study in differentiating COVID-19 from other viral pneumonia reported that the median values of sensitivity and specificity were 83% (ranging between 67% and 97%) and 96.5% (ranging between 7% and 100%), respectively [1].

For future work, to further improve the model performance, a larger and higher resolution dataset should be leveraged which will allow for training a deeper and more complex model to train reliably. A temporal study of how the COVID-19 disease manifestation changes would also be helpful.

References

1. Bai, H.X., Hsieh, B., Xiong, Z., Halsey, K., Choi, J.W., Tran, T.M.L., Pan, I., Shi, L.-B., Wang, D.-C., Mei, J., Jiang, X.-L., Zeng, Q.-H., Egglin, T.K., Hu, P.-F., Agarwal, S., Xie, F.-F., Li, S., Healey, T., Atalay, M.K. and Liao, W.-H. (2020). Performance of Radiologists in Differentiating COVID-19 from non-COVID-19 Viral Pneumonia at Chest CT, *Radiology*, 296(2): E46–E54. https://doi.org/10.1148/radiol.2020200823
2. Bernico, M. (2018). Deep Learning Quick Reference: Useful Hacks for Training and Optimizing Deep Neural Networks with Tensorflow and Keras, chap. 1, pp. 23.
3. Chicco, D. and Jurman, G. (Dec. 2020). The advantages of the Matthews correlation coefficient (MCC) over F1 score and accuracy in binary classification evaluation, *BMC Genomics*, 21(1): 6.
4. Goncharov, M., Pisov, M., Shevtsov, A., Shirokikh, B., Kurmukov, A., Blokhin, I., Chernina, V., Solovev, A., Gombolevskiy, V., Morozov, S. and Belyaev, M. (2021). CT-based COVID-19 Triage: Deep multitask learning improves joint identification and severity quantification, *Medical Image Analysis*, 102054.
5. Harris, C.R., Millman, K.J., van der Walt, Stéfan, J, Gommers, R., Virtanen, P., Cournapeau, D., Wieser, E., Taylor, J., Berg, S., Smith, N.J., Kern, R., Picus, M., Hoyer, S., van Kerkwijk, M.H., Brett, M., Haldane, A., Del Río, J.F., Wiebe, M., Peterson, P. and Oliphant, T.E. (2020). Array programming with NumPy, *Nature* (London), 585(7825): 357-362.
6. https://jupyter.org
7. https://libraries.io/pypi/nibabel
8. https://radiologyassistant.nl/chest/covid-19/covid19-imaging-findings
9. https://scikit-learn.org/stable/index.html
10. Hunter, J.D. (2007). Matplotlib: A 2D graphics environment, *Computing in Science & Engineering*, 9(3): 90-95.
11. Imani, M. (2021). Automatic diagnosis of coronavirus (COVID-19) using shape and texture characteristics extracted from X-Ray and CT-Scan images, *Biomedical Signal Processing and Control*, 68: 102602.

12. Mittal, B. and Oh, J. (2021). CoviNet: COVID-19 Diagnosis using Machine Learning Analyses for Computerized Tomography Images. The 13th International Conference on Digital Image Processing (ICDIP 2021), Singapore.

13. Morozov, S.P., Andreychenko, A.E., Blokhin, I.A., Gelezhe, P.B., Gonchar, A.P., Nikolaev, A.E., Pavlov, N.A., Chernina, V.Y. and Gombolevskiy, V.A. (2020). MosMedData: Data set of 1110 chest CT scans performed during the COVID-19 epidemic, *Digital Diagnostics*, 1(1): 4959.

14. Rubin, G.D., Ryerson, C.J., Haramati, L.B., Sverzellati, N., Kanne, J.P., Raoof, S., *et al.* (2020). The Role of Chest Imaging in Patient Management during the COVID-19 Pandemic: A Multinational Consensus Statement from the Fleischner Society, *Radiology*, 296(1): 172-180.

15. Varnousfaderani, E.S., Yousefi, S., Bowd, C., Belghith, A. and Goldbaum, M.H. (2015). Vessel delineation in retinal images using Leung-Malik filters and two levels hierarchical learning, *AMIA Annual Symposium Proceedings*, AMIA Symposium, 015: 1140-1147.

16. Wang, X., Deng, X., Fu, Q., Zhou, Q., Feng, J., Ma, H., Liu, W. and Zheng, C. (2020). A weakly-supervised framework for COVID-19 classification and lesion localization from chest CT, *IEEE Transactions on Medical Imaging*, 39(8): 2615-2625. https://doi.org/10.1109/TMI.2020.2995965

17. Weekly epidemiological update on COVID-19-13 July, 2021 (n.d.); retrieved 22 July, 2021, from https://www.who.int/publications/m/item/weekly-epidemiological-update-on-covid-19--13-july-2021

18. Yang, X., He, X., Zhao, J., Zhang, Y., Zhang, S. and Xie, P. (2020). Covid-CT-Dataset: A CT Scan Dataset about COVID-19. ArXiv:2003.13865 [Cs, Eess, Stat].

Modeling COVID-19 Pandemic Dynamics Using Transparent, Interpretable, Parsimonious and Simulatable (TIPS) Machine Learning Models: A Case Study from Systems Thinking and System Identification Perspectives

Hua-Liang Wei[1,2]**, Stephen A. Billings**[1,2]

[1] Department of Automatic Control and Systems Engineering, The University of Sheffield
[2] INSIGNEO Institute for in Silico Medicine, The University of Sheffield, Sheffield, S1 3JD

1. Introduction

The past 18 months have witnessed the devastating spread of the COVID-19, a disastrous global pandemic which has been and still is affecting almost every single person at each corner of the world. The attention paid to COVID-19 over the past 18 months categorically surpasses that to anything else. For example, when searching with the keyword 'COVID-19' and the scope of 'abstract' in the database of Web of Science, the number of published articles is 94,026. With the same keyword and scope, the number of published articles in the Elsevier's abstract and citation database, Scopus, is over 112,000. With the same keyword, but only search with the scope of 'in the title of the article', the number of articles given by Google Scholar is over 263,000, and if the scope is changed to 'anywhere in the article', the number of publications reaches over 4,340,000. Clearly, the numbers of publications on COVID-19 are categorically astronomically larger than those on any other single subject.

Presently, there are a huge number of publications describing the spread dynamics of the pandemic, most of which employ the well-known susceptible infected removed (SIR) [24] and susceptible exposed infected removed (SEIR) models [22], or their variants [14, 26] to simulate the spread of the coronavirus. SIR and SEIR are continuous-time models which are a class of initial value problems (IVPs) of ordinary differential equations (ODEs). Regression and machine learning methods have also been applied to analyze COVID-vf19 pandemic data (e.g. predicting infection cases), but most of these methods use simplified models involving a small number of input variables pre-selected, based on a priori knowledge, or use very complex models (e.g. deep learning) [5, 1, 31], merely focusing on the prediction purpose (e.g. positive case prediction) and paying little attention to the model interpretability. There have been relatively fewer studies focusing on the investigations of the inherent time-lagged or time-delayed relationships, e.g. between the reproduction number (R number), infection cases, and deaths, analyzing the pandemic spread from a systems thinking and dynamic perspective. The present study, for the first time, proposes using systems thinking and system identification approach to build transparent, interpretable, parsimonious, and simulatable (TIPS) dynamic machine learning models [32, 6], establishing links between the R number, the infection cases, and deaths caused by COVID-19. The TIPS models are developed, based on the well-known NARMAX (Nonlinear AutoRegressive Moving Average with eXogenous inputs) method, which can help better understand the COVID-19 pandemic dynamics. A case study on the COVID-19 data of the UK is carried out, and the findings are as follows: (1) the number of daily infection cases (DIC) is closely related to the R number but lags R number from 12 to 42 days, and (2) the number of daily deaths is highly dependent on R and DCIC but lags R from 14 to 41 days and lags DIC from 13 to 27 days. These new findings make significant contributions to better understanding of the spread dynamics of the pandemic.

The remaining of this chapter is as follows. Section 2 briefly depicted the research problem. Section 3 provides a brief description of the method used for calculating the R number. Section 4 introduces the NARMAX methods. In Section 5, three cases studies based on the UK COVID-19 data are presented. Finally, the work is concluded in Section 6.

2. Problem Representation

This work aims to build transparent, interpretable, parsimonious, and simulatable (TIPS) dynamic machine learning models, which are used to achieve two objectives: (1) reveal the quantitative relationships of reproduction number (R number), infection cases, and deaths, and (2) make prediction of daily infection cases (DIC) and the number of daily deaths (NDD) in advance of at least 12 days.

The objectives and procedure of the TIPS machine learning is graphically depicted in Fig. 1. In this study, the TIPS model training procedure needs the historical records of the numbers of daily infection cases and daily deaths. The reproduction number R will be derived from the number of daily infection cases. The predictive models are built using the well-known NARMAX method [7]. The calculation of R number and the framework of the NARMAX method are presented in Sections 3 and 4, respectively.

3. Reproduction Number

As mentioned in Section 2, the models to be built in this work involve three variables: the R number and the numbers of daily infection cases, and daily deaths. The values of the R number are estimated using the method proposed in Zingano et al. [39]. The method is briefly presented as follows:

Let $S(t)$, $E(t)$, $I(t)$, $R(t)$ and $D(t)$ (t can be understood in time unit of day) denote the numbers of susceptible individuals, exposed (infected but not yet infectious to transmit the disease) individuals, infectious or infected actively individuals, those recovered from the disease, and those who have died from the disease, respectively. From the mathematical theory of infectious diseases, the spread dynamics of COVID-19 can be characterized by the following SEIR model [39]:

$$\frac{dS}{dt} = -\beta I(t)\left(\frac{1}{N}\frac{dS(t)}{dt}\right)$$
$$\frac{dE}{dt} = \beta I(t)\left(\frac{1}{N}\frac{dS(t)}{dt}\right) - \delta E(t)$$
$$\frac{dI}{dt} = \delta E(t) - (r+\gamma)I(t)$$
$$\frac{dR}{dt} = \gamma I(t)$$
$$\frac{dD}{dt} = rI(t) \tag{1}$$

with initial values $S(t_0) = S_0$, $E(t_0) = E_0$, $I(t_0) = I_0$, $R(t_0) = R_0$, and $D(t_0) = D_0$, satisfying $S_0 + E_0 + I_0 + R_0 + D_0 = N$, where N is the full size of the susceptible population initially exposed at the initial time t_0. The parameters β (average transmission rate) and r (average lethality) change with time t (e.g. day), and the other two parameters δ and γ are assumed to be positive constants, which are defined as $\delta = 1/T_{ALP}$ and $\gamma = 1/T_{ATT}$, where T_{AL} and T_{AT} represent the average latent period and the average transmission time, respectively. Clinical study results show that for COVID-19, T_{AL} is around 5 days on average [29] and T_{AT} is around 14 days [36].

Note that the five variables, $S(t)$, $E(t)$, $I(t)$, $R(t)$, and $D(t)$, in the ordinary differential equation model (1) obey the following conservation law [39]: $S(t)+E(t)+I(t)+R(t)+D(t) = N$ (for any $t \geq t_0$).

Based on the SEIR model (1), Zingano et al. [39] developed the formula for calculating the reproduction number R_t as follows:

$$RN(t) = \frac{1}{N}\frac{\beta(t)S(t)}{r(t)+\gamma} \tag{2}$$

4. NARMAX Methods

NARMAX methods were initially developed for solving control systems engineering modelling problems, especially complex nonlinear system identification tasks, but gradually have been successfully applied to a wide range of multidisciplinary domains, including medical [8], neuroscience [15], social science [33, 16.], climate and weather [21], space weather [17], among others [36].

NARMAX methods employ discrete-time dynamic models [27 Specifically, assuming that a system output y is potentially driven by r input variables, designated by u_1, u_2, ..., u_r, the general form of the NARMAX model is:

$$y(t) = f[y(t-\tau_y),\cdots,y(t-n_y),u_1(t-\tau_u),\cdots,u_1(t-n_u),...,$$

$$u_r(t-\tau_u),...,u_r(t-n_u),e(t-1),\cdots,e(t-n_e)]+e(t) \qquad (3)$$

where $y(t)$, $u(t)$ and $e(t)$ are the measured system output, input and noise sequences respectively at time instant t; n_y, n_u, and n_e are the maximum lags for the system output, input and noise; τ_y and τ_u are a time delay between the input and output, and usually $\tau_y = \tau_u = 1$; $f[\bullet]$ is some non-linear function that needs to be estimated from measured or observed input and output data. Note that the noise $e(t)$ is unmeasurable but can be replaced by the model prediction error in system identification procedure. The noise terms are included to accommodate the effects of measurement noise, modelling errors, and/or unmeasured disturbances.

In practice, NARMAX models can be implemented using different approaches, such as recurrent neural networks [25], radial basis function (RBF) networks [11, 9], wavelet neural networks [34, 35], along with others [36]. More than often, the polynomial representation, due to its attractive interpretation properties, is employed to implement NARMAX models [2, 3, 4]. In this study, the power-form polynomial basis is considered.

NARMAX model identification usually starts from a specified dictionary or library, consisting of a sufficiently large number of candidate model elements (e.g. model terms or regressors), each of which is formed by the lagged system input and output variables, such as $y(t-1)$, $u_1(t-2)$, $u_2(t-1)$, $u_1(t-1)u_2(t-1)$. A model construction algorithm (or a combination or ensemble of a several algorithms) can then be performed on the dictionary, together with a training dataset, to construct sparse or parsimonious models. For example, for a single-input single-output (SISO) system, set $n_y = 1$, $n_u = 1$, $n_e = 0$, $\tau_y = \tau_u = 1$, the candidate dictionary with the nonlinear degree $l = 3$ is:

$$D = \begin{cases} y(t-1), & u(t-1), & y^2(t-1), & y(t-1)u(t-1), & u^2(t-1) \\ y^3(t-1), & y^2(t-1)u(t-1), & y(t-1)u^2(t-1), & u^3(t-1) \end{cases} \qquad (4)$$

The final identified model may be in the form:

$$y(t) = ay(t-1) + by(t-1)u(t-1) + cu^2(t-1) \qquad (5)$$

An efficient and commonly used algorithm for NARMAX model construction is the well-known forward regression with orthogonal least squares (FROLS) [36, 12] and its variants, e.g. minimization error [28], iFROLS (iterative FROLS) [19], uFROLS (ultra-FROLS) [20]. Recently, the LASSO algorithm [30] has also been applied to system identification [37] and feature selection for classification tasks [13], but it turned out that LASSO did not outperform FROLS, because 'the LASSO is not a very satisfactory variable selection method in the $p \gg n$ case' where n is the number of observations and p is the number of predictors [40]. However, $p \gg n$ is a common case in many real applications. It was theoretically proved by Johnson *et al.* [23] that the 'optimal' L_1–norm solutions are often inferior to L_0–norm solutions found using stepwise regression; they also compared algorithms for solving these two problems and showed that although L_1–norm solutions can be efficient, the 'optimal' L_1–norm solutions are often inferior to L_0–norm solutions found using greedy classic stepwise regression.

The FROLS algorithm uses a simple but efficient index, called error reduction ratio (ERR) [36, 12], to measure the importance or significance of each candidate model term (element) included in the specified dictionary or library, and determine which ones should be included in the model in order of their importance, e.g. in terms of the contributions they can make to explaining the variation of the target signal (system output). The model construction procedure usually leads to transparent, interpretable, parsimonious, and simple/simulatable (TIPS) models. A rigid model validation approach [10, 38] can guarantee the validity of the final identified model to sufficiently represent the input-output relationship hidden in the data.

Taking the case study, on the UK Understanding Society (UKUS) data, presented in [16] as an example, the identified models using the NARMAX methods show that, on the collective national level of the UK, the factors that appear to have significantly positive impact on happiness are: 'income (living comfortably)', 'income (doing alright)', 'income (just about getting by)', 'retired', 'health (excellent)', 'health (every good)'; the combination of the two variables of 'retirement' and 'above 65' is also an important model term, which can be explained that the retired people who are over 65 years old are more likely to be happy. The model also revealed that marriage could enhance the positive relationship between good health status and happiness, while smoke could enhance the negative effect of low income on happiness.

More detailed descriptions of NARMAX methods can be found in [36]. This work uses a 10fold cross validation scheme and the FROLS algorithm to build TIPS models. In doing so, the entire dataset is first split into two parts (for training and testing, respectively), and the training data are then split into two parts (around 70% : 30%) which are used for model structure selection and performance validation. The standard version of the FROLS algorithm is described in detail in the Appendix.

5. Case Studies

The section provides case studies on the UK COVID-19 data modelling and analysis using the TIPS modelling approach based on the NARMAX methods. The two main objectives are: (1) To establish the relationships between the R number, the daily infection cases and deaths; (2) To make predictions of the daily infection cases and deaths in advance of more than 10 days.

The data were extracted from the Johns Hopkins Coronavirus Resource Center (https://coronavirus.jhu.edu/about/how-to-use-our-data). For all the case studies, a total of 529 daily data (infection cases and deaths), from 4 March, 2020-15 August, 2021, are considered for the analysis purpose here, of which the first 361 data (4 March, 2020-28 February, 2021) are used for model training, estimation and validation, and the remaining 168 data (1 March, 2021-15 August, 2021) are used for model performance testing.

5.1 The Impact of the R Number on Daily Infection Cases

To examine and investigate the impact of the daily R number values on the daily infection cases in later days, the daily R number is treated as an input and the daily infection cases as the output (response), and the settings for the NARMAX model (1) are: n_u = 42 (maximum time lag in input), n_y = 0 (without including lagged autoregressive term), n_e = 0 (without including noise term), τ_u = 1 (time delay). The identified model is:

$$y(t) = 3.5551 \times 10^4 u(t-12) - 6.2117 \times 10^3 u(t-40) - 1.17395 \times 10^4 \qquad (6)$$

where y = 'daily infection cases' and u = 'daily R number'. The model was simulated driven by the 513 daily data. A comparison between the model predicted values and the real data, over the 361 training data points (4 March, 2020-28 Feb., 2021) and 168 test data (1 March, 2021-15 August, 2021), is shown in Fig. 2.

Model (6) indicates that the R number value of the current day may impact the daily infection cases of 12 days later, lasting until 40 days. This is also reflected in Fig. 2. This important finding has not been noticed in any previous study.

5.2 Predicting Daily Infection Cases Using R Number

Using the information given in by model (6) setting n_u = 42 (maximum time lag in input), n_y = 42 (maximum lag in output), n_e = 0 (without including noise term), $\tau_y = \tau_u$ = 12 (time delay) and the nonlinear degree ℓ = 2, a best NARMAX model was identified, which is shown in Table 2.

The model was simulated over the whole 513 data (4 March, 2020 to 30 July, 2021). The values of R-square (the coefficient of determination) on the 361 training data points (4 March, 2020 – 28 Feb., 2021) and 168 test data points (1 March, 2021 – 15 August, 2021) are 0.8991 and 0.8544, respectively. A comparison between the model predicted values and the corresponding records, over the training and test data, is shown in Fig. 3.

Table 1: The model for daily infection cases, where y = 'daily infection cases' and u = 'daily R number'

Index	Model Term	Parameter	ERR (100%)	P-value
1	$u(t-12) \times y(t-12)$	1.1674e+01	81.9265	6.4150e-11
2	$y(t-12) \times y(t-18)$	8.9681e-06	6.4553	1.5963e-02
3	$u(t-13) \times y(t-12)$	-1.1246e+01	1.0121	6.3366e-10
4	$u(t-12) \times y(t-18)$	1.0908e+00	1.8987	5.1187e-09
5	$u(t-27) \times y(t-18)$	-1.1120e+00	1.1427	9.9920e-15
6	$y(t-14)$	1.4192e+00	0.5034	0
7	$y(t-12) \times y(t-14)$	-2.6652e-05	0.5831	2.5684e-12
8	$u(t-42) \times y(t-36)$	-1.7229e-01	0.7061	2.1496e-09

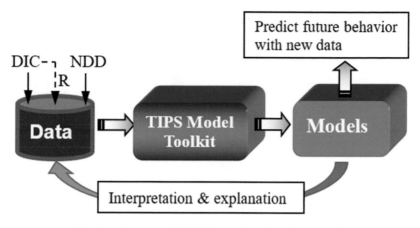

Fig. 1: A diagram of the TIPS machine learning procedure

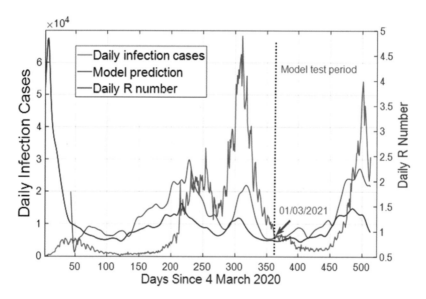

Fig. 2: A comparison between the model predicted values of daily infection cases and R number from model (6) and the real data.

Both the quantitative results (e.g. the R-square values) and graphical illustration show that the identified model show excellent prediction results. More importantly, it can be seen from Table 1 how the values of the R number and the infection cases can potential affect the pandemic spread after 12 days lasting until 42 days. For example, the combination of the two quantities of R number and infection cases of current day can potentially have a very high impact on the infection cases 12 days later, as the cross-product term of the two variables has a very high ERR value, showing that it can explain nearly 90% of the variance of the daily infection cases after 12 days.

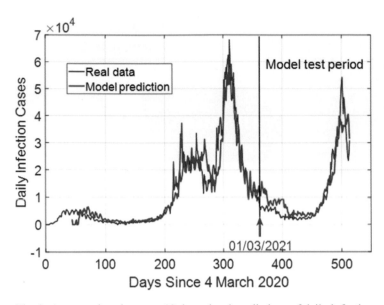

Fig. 3: A comparison between 12 days ahead predictions of daily infection cases and the real data

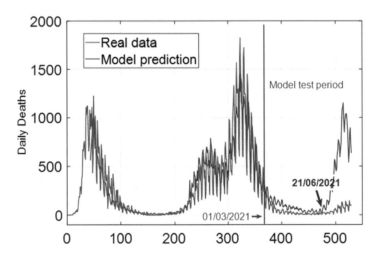

Fig. 4: A comparison between 12 days ahead predictions of daily deaths and the real data

5.3 Predicting Daily Mortality Using Daily Infection Cases and R Number

In this third case, the settings as follows. The R number and daily infection cases are used as two inputs, and the number of daily deaths is considered to be the output. The other coefficients of the NARMAX (1) are chosen as: $n_u = 42$ (for both inputs), $n_y = 0$ (no autoregressive variable is included in the model), $n_e = 0$ (without including noise term), τ_u =12 (time delay) and the nonlinear degree $\ell = 2$. The identified NARMAX model is shown in Table 2. Note that the model identification algorithm did not find any nonlinear models that outperforms the linear model shown in Table 1, this suggests or implies that there is no or very weak nonlinear relation along the input and output variables; the relationship is dominated by linearity.

Table 2: The model for daily infection cases, where y = 'number of daily deaths', u_1 = 'daily R number', and u_2 = 'daily infection cases'

Index	Model Term	Parameter	ERR (100%)	P-value
1	$u_2(t\text{-}13)$	1.7498e-02	75.9261	0
2	$u_1(t\text{-}41)$	2.4864e+02	7.3788	0
3	$u_1(t\text{-}12)$	-2.3836e+02	4.0250	0
4	$u_2(t\text{-}21)$	1.1513e-02	0.3910	3.6707e-04
5	$u_2(t\text{-}24)$	-1.3255e-02	0.5640	4.6917e-06
6	$u_2(t\text{-}27)$	6.3630e-03	0.2339	1.2048e-02

It can be observed from Table 2 that the values of the R number (input u_1) can potential affect the mortality after 12 days lasting until 41 days, whereas the daily infection cases (u_2) can potential affect the mortality after 13 days lasting until 27 days.

From the Table it can be noticed that the daily infection cases of current day (with ERR = 75.9%) potentially highly affect the number of deaths 13 days later.

The model was simulated over the entire data (4 March, 2020 to 15 August, 2021). A comparison between the model predicted values and the corresponding records, over the training and test data, is shown in Fig. 3, where it can be seen that the model performs excellent on the training data and most part of the test data (e.g. until 21 June, 2021), with the value of R-square is close to 0.8).

However, it can be clearly seen that the model fails to predict the number of daily deaths in most recent days (e.g. around 21 June, 2021 and onward), although the daily infection cases is still very high as shown in Fig. 3. This is reasonable and may probably be explained that more and more people have received a second vaccine which has helped significantly reduce the death rate, and this confirms the effectiveness of the UK government's vaccination policy.

It is worth mentioning the following two points: (1) Under the settings of $n_u = 42$, $n_y = 0$, $n_e = 0$, l =2, the best variables that the model training algorithm identified are $u_2(t\text{-}13)$, $u_1(t\text{-}14)$, etc. This suggests that the inclusion of any other variables cannot enhance the explanation of the variation of the daily death cases using the daily infection cases and the daily value of R number, and therefore cannot help improve the prediction performance. (2) It is straightforward to detect the periodical change of the daily death cases (with a period of 7 days), the correlation between the lagged variable $y(t-7)$ and the original signal is as high as 0.93. The main purpose of this section is to reveal the inherent dynamics that project daily infection cases and R-number to daily death cases many days later. To avoid the impact of the strong autocorrelation on the analysis of the underlying dynamics between the input and output variables, the lagged autoregressive variables, such as $y(t–1)$, $y(t–2)$, etc. were deliberately not considered when constructing the model in this case study.

6. Conclusion

The prediction of the COVID-19 pandemic is important and challenging. However, a complicated (black-box) model that lacks interpretability (e.g. without explicitly providing information on the inherent dynamics) may become less useful or powerful for applications where there is a need to know the relationship of the inherent dynamics. The main attention of this work was paid to developing a glass-box modelling approach. The main contributions of the work can be summarized as follows:

Firstly, it proposed a TIPS-ML framework based on the NARMAX methods, and applied the proposed approach to modelling the spread dynamics based on the UK COVID-19 data. In comparison with other complicated machine learning methods, the proposed method has several highly attractive properties, such as transparency, interpretability, parsimony, and simplicity/simulatability. These properties are very important for investigating and understanding the spread dynamics of the pandemic, which may not be able to obtained by using other machine learning methods (e.g. those complicated black-box neural network models).

Secondly, some important new findings have been obtained from the identified TIPS models. For example, the R number of the current day may significantly impact the daily infection cases (12 days) later and last as long as 42 days; the combinational effect of R number and infection cases of current day can be potentially very high on the infection cases 12 days later; the number of daily deaths is highly dependent on R and daily infection cases (DIC) but lags R from 14 to 41 days and lags DIC from 13 to 27 days. These new findings, which have not been observed before, are useful for better understanding the spread dynamics of the pandemic.

The case studies carried out in this work focused on the UK COVID-19 data. In future, more data of different countries will be considered and analyzed using the proposed method to investigate and compare the pandemic spread dynamic patterns, from which to acquired information that may be useful for healthcare and infectious disease studies.

Acknowledgements

This work was supported in part by the Natural Environment Research Council (NERC) under the Grant NE/V001787 and Grant NE/V002511, the Engineering and Physical Sciences Research Council (EPSRC) under Grant EP/I011056/1, the EPSRC Platform Grant EP/H00453X/1, and European Research Council under the Grant FP7-IDEAS-ERC (Grant agreement ID: 226037).

Appendix: The Forward Regression with Orthogonal Least Squares (FROLS) Algorithm

The TIPS models are built using a 10-fold-cross validation scheme and based on the FROLS algorithm. Taking a single-input, single-output as an example, an initial TIPS model can easily be converted into a linear-in-the-parameters form:

$$y(t) = \sum_{m=1}^{M} \theta_m \psi_m(t) + e(t) \tag{A1}$$

Where $\psi_m(t) = \psi_m(\mathbf{x}(t))$ are the model regressors, $\mathbf{x}(t) = [x_1(t), x_2(t), \cdots, x_n(t)]^T$ is a vector of model 'input' variables, each element $x_i(t)$ is either one of the n lagged variable such as $y(t-1), y(t-2), \ldots, y(t-n_y), u(t-1), u(t-2), \ldots, u(t-n_u)$ $(n = n_y + n_u)$, or cross-product of these lagged variables such as $y(t-12)u(t-12)$ and $y(t-12)u(t-18)$; θ_m are the model parameters, and M is the total number of candidate regressors.

The initial regression model (A1) often involves a large number of candidate model terms. Experience suggests that most of the candidate model terms can be removed from the model, and that only a small number of significant model terms are needed to provide a satisfactory representation for most nonlinear dynamical systems. The FROLS algorithms (10) can be used to select significant model terms.

Consider the term selection problem for the linear-in-the-parameters model (A1). Let $\{(\mathbf{x}(t), y(t)) : \mathbf{x} \in \mathbf{R}^n, y \in \mathbf{R}\}_{t=1}^{N}$ be a given training data set and $\mathbf{y} = [y(1), \cdots, y(N)]^T$ be the vector of the output. Let $I = \{1, 2, \cdots, M\}$, and denote by $\Omega = \{\psi_m : k \in I\}$ the dictionary of candidate model terms in an initially chosen candidate regression model similar to (9). The dictionary Ω can be used to form a variant vector dictionary $\mathcal{D} = \{\varphi_m : m \in I\}$, where the kth candidate basis vector φ_m is formed by the kth candidate model term $\psi_m \subset \Omega$, in the sense that $\varphi_m = [\psi_m(\mathbf{x}(1)), \cdots, \psi_m(\mathbf{x}(N))]^T$. The model term selection problem is equivalent to finding, from I, a subset of indices, $I_n = \{i_m : m = 1, 2, \cdots, n, i_m \in I\}$ where $n \leq M$, so that $\mathbf{y}$ can be approximated using a linear combination of $\alpha_{i_1}, \alpha_{i_2}, \cdots, \alpha_{i_n}$.

A.1 The Forward Orthogonal Regression Procedure

A non-centralised squared correlation coefficient will be used to measure the dependency between two associated random vectors. The non-centralised squared correlation coefficient between two vectors $\mathbf{x}$ and $\mathbf{y}$ of size N is defined as

$$C(\mathbf{x}, \mathbf{y}) = \frac{(\mathbf{x}^T \mathbf{y})^2}{\|\mathbf{x}\|^2 \|\mathbf{y}\|^2} = \frac{(\mathbf{x}^T \mathbf{y})^2}{(\mathbf{x}^T \mathbf{x})(\mathbf{y}^T \mathbf{y})} = \frac{\left(\sum_{i=1}^{N} x_i y_i\right)^2}{\sum_{i=1}^{N} x_i^2 \sum_{i=1}^{N} y_i^2} \tag{A2}$$

The squared correlation coefficient is closely related to the error reduction ratio (ERR) criterion (a very useful index in respect to the significance of model terms), defined in the standard orthogonal least squares (OLS) algorithm for model structure selection (28, 10).

The model structure selection procedure starts from equation (A1). Let $r_0 = y$, and

$$\ell_1 = \arg \max_{1 \le j \le M} \{C(\mathbf{y}, \boldsymbol{\varphi}_j)\} \tag{A3}$$

where the function $C(\because)$ is the correlation coefficient defined by (A2). The first significant basis can thus be selected as $\boldsymbol{\alpha}_1 = \boldsymbol{\varphi}_{\ell_1}$, and the first associated orthogonal basis can be chosen as $\mathbf{q}_1 = \boldsymbol{\varphi}_{\ell_1}$. The model residual, related to the first step search, is given as

$$\mathbf{r}_1 = \mathbf{r}_0 - \frac{\mathbf{y}^T \mathbf{q}_1}{\mathbf{q}_1^T \mathbf{q}_1} \mathbf{q}_1 \tag{A4}$$

In general, the kth significant model term can be chosen as follows. Assume that at the $(m-1)$th step, a subset $\mathcal{D}_{k-1}$, consisting of $(m-1)$ significant bases, $\boldsymbol{\alpha}_1, \boldsymbol{\alpha}_2, ..., \boldsymbol{\alpha}_{m-1}$, has been determined, and the $(m-1)$ selected bases have been transformed into a new group of orthogonal bases $\mathbf{q}_1, \mathbf{q}_2, \cdots, \mathbf{q}_{m-1}$ via some orthogonal transformation. Let

$$\mathbf{q}_j^{(m)} = \boldsymbol{\varphi}_j - \sum_{k=1}^{m-1} \frac{\boldsymbol{\varphi}_j^T \mathbf{q}_k}{\mathbf{q}_k^T \mathbf{q}_k} \mathbf{q}_k \tag{A5}$$

$$\ell_k = \arg \max_{j \ne \ell_i, 1 \le i \le k-1} \{C(\mathbf{y}, \mathbf{q}_j^{(k)})\} \tag{A6}$$

where $\boldsymbol{\varphi}_j \in \mathcal{D} - \mathcal{D}_{m-1}$, and $\mathbf{r}_{m-1}$ is the residual vector obtained in the $(m-1)$th step. The mth significant basis can then be chosen as $\boldsymbol{\alpha}_m = \boldsymbol{\varphi}_{\ell_m}$ and the mth associated orthogonal basis can be chosen as $\mathbf{q}_m = \mathbf{q}_{\ell_m}^{(m)}$. The residual vector $\mathbf{r}_m$ at the mth step is given by

$$\mathbf{r}_m = \mathbf{r}_{m-1} - \frac{\mathbf{y}^T \mathbf{q}_m}{\mathbf{q}_m^T \mathbf{q}_m} \mathbf{q}_m \tag{A7}$$

Subsequent significant bases can be selected in the same way step by step. From (A7), the vectors $\mathbf{r}_m$ and $\mathbf{q}_m$ are orthogonal, thus

$$\| \mathbf{r}_m \|^2 = \| \mathbf{r}_{m-1} \|^2 - \frac{(\mathbf{y}^T \mathbf{q}_m)^2}{\mathbf{q}_m^T \mathbf{q}_m} \tag{A8}$$

By respectively summing (A7) and (A8) for m from 1 to n, yields

$$\mathbf{y} = \sum_{m=1}^{n} \frac{\mathbf{y}^T \mathbf{q}_m}{\mathbf{q}_m^T \mathbf{q}_m} \mathbf{q}_m + \mathbf{r}_n \tag{A9}$$

$$\| \mathbf{r}_n \|^2 = \| \mathbf{y} \|^2 - \sum_{m=1}^{n} \frac{(\mathbf{y}^T \mathbf{q}_m)^2}{\mathbf{q}_m^T \mathbf{q}_m} \tag{10}$$

In general, the mth significant model term can be chosen as follows. Assume that at the $(m-1)$th step, a subset $\mathcal{D}_{m-1}$, consisting of $(m-1)$ significant bases, $\boldsymbol{\alpha}_1, \boldsymbol{\alpha}_2, \cdots, \boldsymbol{\alpha}_{m-1}$, has been determined, and the (m-1) selected bases have been transformed into a new group of orthogonal bases $\mathbf{q}_1, \mathbf{q}_2, \ldots, \mathbf{q}_{m-1}$ via some orthogonal transformation. Let

$$\mathbf{q}_j^{(m)} = \boldsymbol{\varphi}_j - \sum_{k=1}^{m-1} \frac{\boldsymbol{\varphi}_j^T \mathbf{q}_k}{\mathbf{q}_k^T \mathbf{q}_k} \mathbf{q}_k \tag{A11}$$

$$\ell_m = \arg \max_{j \neq \ell_k, 1 \leq k \leq m-1} \{C(\mathbf{y}, \mathbf{q}_j^{(m)})\} \tag{A12}$$

where $\boldsymbol{\varphi}_j \in \mathcal{D} - \mathcal{D}_{m-1}$, and $\mathbf{r}_{m-1}$ is the residual vector obtained in the (m-1)th step. The mth significant basis can then be chosen as $\boldsymbol{\alpha}_m = \boldsymbol{\varphi}_{\ell_m}$ and the mth associated orthogonal basis can be chosen as $\mathbf{q}_m = \mathbf{q}_{\ell_m}^{(m)}$. The residual vector $\mathbf{r}_m$ at the mth step is given by

$$\mathbf{r}_m = \mathbf{r}_{m-1} - \frac{\mathbf{y}^T \mathbf{q}_m}{\mathbf{q}_m^T \mathbf{q}_m} \mathbf{q}_m \tag{A13}$$

Subsequent significant bases can be selected in the same way step by step. From (A13), the vectors $\mathbf{r}_m$ and $\mathbf{q}_m$ are orthogonal, thus

$$\| \mathbf{r}_m \|^2 = \| \mathbf{r}_{m-1} \|^2 - \frac{(\mathbf{y}^T \mathbf{q}_m)^2}{\mathbf{q}_m^T \mathbf{q}_m} \tag{A14}$$

By respectively summing (A13) and (A14) for m from 1 to n, yields

$$\mathbf{y} = \sum_{m=1}^{n} \frac{\mathbf{y}^T \mathbf{q}_m}{\mathbf{q}_m^T \mathbf{q}_m} \mathbf{q}_m + \mathbf{r}_n \tag{A15}$$

$$\| \mathbf{r}_n \|^2 = \| \mathbf{y} \|^2 - \sum_{m=1}^{n} \frac{(\mathbf{y}^T \mathbf{q}_m)^2}{\mathbf{q}_m^T \mathbf{q}_m} \tag{A16}$$

The model residual $\mathbf{r}_n$ will be used to form a criterion for model selection, and the search procedure will be terminated when the norm $\| \mathbf{r}_n \|^2$ satisfies some specified conditions. Note that the quantity $\mathrm{ERR}_m = C(\mathbf{y}, \mathbf{q}_m)$ is just equal to the mth error reduction ratio (Chen *et al.*, 1989; Billings, 2013), brought by including the mth basis vector $\boldsymbol{\alpha}_m = \boldsymbol{\varphi}_{\ell_m}$ into the model, and that $\sum_{m=1}^{n} C(\mathbf{y}, \mathbf{q}_m)$ is the increment or total percentage that the desired output variance can be explained by $\boldsymbol{\alpha}_1, \boldsymbol{\alpha}_2, \cdots, \boldsymbol{\alpha}_n$.

Finally, a mean square error (MSE) based algorithm, e.g. Akaine's information criterion (AIC), Bayesian information criterion, generalized cross-validation (GCV) and adjustable prediction error sum of squares (APRESS) can be used to determine the model size [18].

A.2 Parameter Estimation

It is easy to verify that the relationship between the selected original bases $\boldsymbol{\alpha}_1, \cdots, \boldsymbol{\alpha}_n$, and the associated orthogonal bases $\mathbf{q}_1, \ldots, \mathbf{q}_n$, is given by

$$\mathbf{A}_n = \mathbf{Q}_n \mathbf{R}_n \tag{A17}$$

where $\mathbf{A}_n = [\alpha_1, \cdots, \alpha_n]$, $\mathbf{Q}_n$ is an $N \times n$ matrix with orthogonal columns $\mathbf{q}_1$, $\mathbf{q}_2$, ..., $\mathbf{q}_n$, and $\mathbf{R}_n$ is an $n \times n$ unit upper triangular matrix whose entries u_{ij} ($1 \le i \le j \le n$) are calculated during the orthogonalization procedure. The unknown parameter vector, denoted by $\theta_n = [\theta_1, \cdots, \theta_n]^T$, for the model with respect to the original bases, can be calculated from the triangular equation $\mathbf{R}_n \theta_n = \mathbf{g}_n$ with $\mathbf{g}_n = [g_1, g_2, \cdots, g_n]^T$, where $g_k = (\mathbf{y}^T \mathbf{q}_k) / (\mathbf{q}_k^T \mathbf{q}_k)$ for $k = 1, 2, ..., n$.

The model parameters reported in Tables 1 and 2 in Section 5 are the estimated values of $[\theta_1, \theta_2, \cdots, \theta_n]^T$.

References

1. Abbasimehr, H. and Paki, R. (2021). Prediction of COVID-19 confirmed cases combining deep learning methods and Bayesian optimization, *Chaos Solutions & Fractals*, 142: 110511; https://doi.org/10.1016/j.chaos.2020.110511
2. Aguirre, L.A. and Billings, S.A. (1994). Improved structure selection for nonlinear models based on term clustering, *International Journal Control*, 62: 569-587; https://doi.org/10.1080/00207179508921557
3. Aguirre, L.A. and Billings, S.A. (1995). Retrieving dynamical invariants from chaotic data using NARMAX models, *International Journal of Bifurcation and Chaos*, 5(2): 449-474; https://doi.org/10.1142/S0218127495000363
4. Aguirre, L.A. (2019). A Bird's Eyeview of Nonlinear System Identification, arXiv preprint arXiv:1907.06803; https://arxiv.org/pdf/1907.06803.pdf
5. Ayoobi, N., Sharifrazi, D., Alizadehsani, R., Shoeibi, A., Gorriz, J.M., Moosaei, H., *et al.* (2021). Time series forecasting of new cases and new deaths rate for COVID-19 using deep learning methods, arXiv:2104.15007; https://arxiv.org/ftp/arxiv/papers/2104/2104.15007.pdf
6. Billings, S.A. (2013). *Nonlinear System Identification: NARMAX Methods in the Time, Frequency, and Spatio-Temporal Domains*, John Wiley & Sons.
7. Billings, S.A., Wei, H.L. and Balikhin, M.A. (2007). Generalized multiscale radial basis function networks, *Neural Networks*, 20(10): 1081-1094.
8. Billings, S.A. and Wei, H.-L. (2019). NARMAX model as a sparse, interpretable and transparent machine learning approach for big medical and healthcare data analysis. *In:* 2019 IEEE 21st International Conference on High Performance Computing and Communications; IEEE 17th International Conference on Smart City; IEEE 5th International Conference on Data Science and Systems (HPCC/SmartCity/DSS), IEEE, pp. 2743-2750; doi: 10.1109/HPCC/SmartCity/DSS.2019.00385
9. Billings, C.G., Wei, H.-L., Thomas, P., Linnane, S.J. and Hope-Gill, B.D.M. (2013). The prediction of in-flight hypoxaemia using nonlinear equations, *Respiratory Medicine*, 107: 841-47; https://doi.org/10.1016/j.rmed.2013.02.016
10. Billings, S.A. and Zhu, Q.M. (1994). Nonlinear model validation using correlation tests, *International Journal of Control*, 60: 1107–1120; https://doi.org/10.1080/00207179408921513
11. Chen, S., Billings, S.A. and Luo, W. (1989). Orthogonal least squares methods and their application to non-linear system identification, *International Journal of Control*, 50(5): 1873-1896; https://doi.org/10.1080/00207178908953472

12. Chen, S., Hong, X. and Harris, C.J. (2005). Orthogonal forward selection for constructing the radial basis function network with tunable nodes, *Lecture Notes in Computer Science*, vol. 3644, Springer, Berlin, Heidelberg; https://doi.org/10.1007/11538059_81

13. Cui, M. and Prasad, S. (2016). Sparse representation-based classification: Orthogonal least squares or orthogonal matching pursuit? *Pattern Recognition Letters*, 84: 120-126; https://doi.org/10.1016/j.patrec.2016.08.017

14. de Camino-Beck, T. (2020). A modified SEIR model with confinement and lockdown of COVID-19 for Costa Rica; medRxiv; doi: https://doi.org/10.1101/2020.05.19.20106492

15. Gu, Y., Yang, Y., Dewald, J.P., Van der Helm, F.C., Schouten, A.C. and Wei, H.-L. (2020). Nonlinear modeling of cortical responses to mechanical wrist perturbations using the NARMAX method, *IEEE Transactions on Biomedical Engineering*, 68(3): 948-958; doi: 10.1109/TBME.2020.3013545

16. Gu, Y. and Wei, H.-L. (2018) Significant indicators and determinants of happiness: Evidence from a UK survey and revealed by a data-driven systems modeling approach, *Social Sciences*, **7**(4), art. 53; https://doi.org/10.3390/socsci7040053

17. Gu, Y., Wei, H.-L., Boynton, R.J., Walker, S.N. and Balikhin, M.A. (2019). System identification and data-driven forecasting of AE index and prediction uncertainty analysis using a new cloud-NARX model, *Journal of Geophysical Research – Space Physics*, 124: 248-263; https://doi.org/10.1029/2018JA025957

18. Gu, Y., Wei, H.-L. and Balikhin, M.M. (2018). Nonlinear predictive model selection and model averaging using information criteria, *Systems Science & Control Engineering*, 6(1): 319-328; https://doi.org/10.1080/21642583.2018.1496042

19. Guo, Y., *et al*. (2015). An iterative orthogonal forward regression algorithm, *International Journal of Systems Science*, 46: 776-789; https://doi.org/10.1080/00207721.2014.981237

20. Guo, Y., *et al*. (2016). Ultra-orthogonal forward regression algorithms for the identification of non-linear dynamic systems, *Neurocomputing*, 173: 715-723; https://doi.org/10.1016/j.neucom.2015.08.022

21. Hall, R.J., Wei, H.-L. and Hanna, E. (2019). Complex systems modeling for statistical forecasting of winter North Atlantic atmospheric variability: A new approach to North Atlantic seasonal forecasting, *Quarterly Journal of the Royal Meteorological Society*, 145: 2568-2585; https://doi.org/10.1002/qj.3579

22. He, S., Peng, Y. and Sun, K. (2020). SEIR modeling of the COVID-19 and its dynamics, *Nonlinear Dyn.*, 101: 1667-1680; https://doi.org/10.1007/s11071-020-05743-y

23. Johnson, K.D., Lin, D., Ungar, L.H., Foster, D.P. and Stine, R.A. (2015). A risk ratio comparison of L0 and L1 penalized regression; arXiv:1510.06319; https://arxiv.org/pdf/1510.06319.pdf

24. Kudryashov, N.A., Chmykhov, M.A. and Vigdorowitsch, M. (2021). Analytical features of the SIR model and their applications to COVID-19, *Applied Mathematical Modeling*, 90: 466-73; https://doi.org/10.1016/j.apm.2020.08.057

25. Liu, G.P. (2001). *Nonlinear Identification and Control: A Neural Network Approach*, Springer-Verlag: London.

26. Lopez, L.R. and Rodó, X. (2021). A modified SEIR model to predict the COVID-19 outbreak in Spain and Italy: Simulating control scenarios and multi-scale epidemics, Preprint at medRxiv; https://doi.org/10.1101/2020.03.27.20045005

27. Pearson, R.K. (1999). *Discrete Time Dynamic Models*, Oxford University Press.

28. Piroddi, L. and Spinelli, W. (2003). An identification algorithm for polynomial NARX models based on simulation error minimization, *International Journal of Control*, 76(17): 1767-1781; https://doi.org/10.1080/00207170310001635419

29. The Royal Society (2020). Reproduction number (R) and growth rate (r) of the COVID-19 epidemic in the UK: Methods of estimation, data sources, causes of heterogeneity, and

use as a guide in policy formulation, [Online] https://royalsociety.org/-/media/policy/projects/set-c/set-covid-19-R-estimates.pdf

30. Tibshirani, R. (1996). Regression shrinkage and selection via the lasso, *J.R. Statist. Soc. B*, 58: 267-288; https://doi.org/10.1111/j.2517-6161.1996.tb02080.x

31. Wang, P.O., Zheng, X., Ai, G., Liu, D. and Zhu, B. (2020). Time series prediction for the epidemic trends of COVID-19 using the improved LSTM deep learning method: Case studies in Russia, Peru and Iran. *Chaos, Solitons & Fractals*, 140: 110214; https://doi.org/10.1016/j.chaos.2020.110214

32. Wei, H.-L. (2019). Boosting wavelet neural networks using evolutionary algorithms for short-term wind speed time series forecasting. *In:* Rojas, I., Joya, G., Catala, A. (Eds.). *Advances in Computational Intelligence. IWANN 2019. Lecture Notes in Computer Science*, vol. 11506, Springer, Cham; https://doi.org/10.1007/978-3-030-20521-8_2

33. Wei, H.-L., Balikhin, M.A. and Walker, S.N. (2015). A new ridge basis function neural network for data-driven modeling and prediction. *In. Proceedings of the 10th International Conference on Computer Science and Education (ICCSE)*, pp. 125-130, Cambridge University, IEEE, U.K.; doi: 10.1109/ICCSE.2015.7250229

34. Wei, H.-L. and Bigg, G.R. (2017). The dominance of food supply in changing demographic factors across Africa: A model using a system identification approach, *Social Science*, 6(4): art. 122. https://doi.org/10.3390/socsci6040122

35. Wei, H.-L. (2019). Sparse, interpretable and transparent predictive model identification for healthcare data analysis. *In:* Rojas, I., Joya, G., Catala, A. (Eds.). *Advances in Computational Intelligence*, IWANN 2019, Lecture Notes in Computer Science, vol. 11506. Springer, Cham; https://doi.org/10.1007/978-3-030-20521-8_9

36. WHO (2020). Transmission of SARS-CoV-2: Implications for infection prevention precautions, *Scientific Brief.* [Online] https://www.who.int/news-room/commentaries/detail/transmission-of-sars-cov-2-implications-for-infection-prevention-precautions

37. Zhang, L., Li, K. and Bai, E.-W. (2013). A novel LOO based two-stage method for automatic model identification of a class of nonlinear dynamic systems. *In: Proc. IEEE 52nd Annual Conference on Decision and Control (CDC)*, pp. 4290-4295; doi: 10.1109/CDC.2013.6760549

38. Zhu, Q.M. and Billings, S.A. (1996). Fast orthogonal identification of nonlinear stochastic models and radial basis function neural networks, *International Journal of Control*, 64: 871-886; https://doi.org/10.1080/00207179608921662

39. Zingano, J.P., Zingano, P.R., Silva, A.M. and Zingano, C.P. (2021). Herd immunity for COVID-19 in homogeneous populations; arXiv:2105.01808; https://arxiv.org/pdf/2105.01808.pdf

40. Zou, H. and Hastie, T. (2005). Regularization and variable selection via the elastic net, *J.R. Stat. Soc. B*, 67: 301-320; https://doi.org/10.1111/j.1467-9868.2005.00503.x

Deep Learning-based Respiratory Anomaly and COVID Diagnosis Using Audio and CT Scan Imagery

Conor Wall[1], Chengyu Liu[2] and Li Zhang[3]

[1] Department of Computer and Information Sciences, Faculty of Engineering and Environment, Northumbria University, Newcastle, NE1 8ST, UK
[2] The School of Instrument Science and Engineering, Southeast University, Nanjing, China
[3] Department of Computer Science, Royal Holloway, University of London, UK

1. Introduction

Deep learning in recent years has become one of the emerging techniques for the development of effective solutions to a range of natural language and computer vision problems [22, 25, 10, 14]. Within the field of medicine, deep learning has also become one of the most imperative areas of research to provide highly cost-effective solutions for medical diagnosis using image, audio, and video inputs. In addition, to tackle the post-pandemic societal requirements, remote and automated diagnostic methods are required to ease the burden of healthcare systems, leading to the need to research in the area to provide effective and accessible solutions.

Therefore, in this research, we first propose a bidirectional long-short-term memory (Bi-LSTM) model for respiratory anomaly detection from audio signals. Secondly, three state-of-the-art pre-trained convolutional neural networks (CNNs), i.e. EfficientNet-B0, EfficientNet-B3, and EfficientNet-B7, are employed for identification of the COVID-19 condition from CT scan imagery. To identify optimal network and learning settings, a grid search optimization mechanism is also proposed to conduct hyper-parameter selection for the CNN and Bi-LSTM networks.

We conduct two sets of experiments for lung condition diagnosis. The first experiment involves utilizing transfer learning on EfficientNet-B0, EfficientNet-B3, and EfficientNet-B7 networks based on the COVID-CT [27] dataset, using lung CT scan images for COCID-19 classification. The second experiment involves designing and training a Bi-LSTM model to classify respiratory diseases, using the ICBHI [14] audio respiratory dataset. The proposed models show impressive performance for diverse respiratory anomaly detection and COVID-19 diagnosis using audio and image inputs, respectively.

2. Related Work

The CNN and LSTM networks are widely adopted deep-learning methods for image and time-series classification problems respectively. The CNN models have been the subject of extensive research regarding image classification and segmentation, and are designed specifically so that they can automatically learn through backpropagation and the spatial hierarchies of features of image and video data [12, 3, 26]. As an example, Wall *et al.* [22] described the use of several pre-trained CNN architectures to classify dermoscopic imagery consisting of two classes, namely, malignant melanomas and benign nevus. Results from this study indicated a high level of performance for all pre-trained models, with the highest level of precision obtained from the ResNeXt50 model, achieving a test accuracy rate of 96.58% for melanoma classification. CNN models have also been used for other medical imaging and healthcare monitoring applications, such as retinal disease diagnosis and motion abnormality monitoring [17, 15]. There are also research studies for the automatic generation of deep networks for diverse image and video classification problems, using Bayesian and evolutionary algorithms [23, 18, 6].

LSTM is a type of recurrent neural network (RNN), which is widely adopted for time series forecasting and audio-classification tasks [5]. This research incorporates the use of a Bi-LSTM model with two LSTM layers. These two layers ensure that the input features can be processed in both forward and backward directions, thus providing a means to better obtain the relations among elements in the input sequence [8]. Many studies adopted LSTM networks for audio classification and time-series forecasting developments. Minh-Tuan and Kim [8 proposed a CNN-Bi-LSTM model for emotion classification from music using the Million Song dataset, which consists of 500 music samples for each of the four emotion classes: anger, sadness, relaxation, and sorrow. Their proposed model achieved an average accuracy rate of 68% across the four classes, in comparison with 63%, 59%, and 50% for the three baseline models, i.e. CNN-LSTM, CNN, and LSTM. Their work indicated that employing a bi-directional LSTM layer may improve classification performance significantly, making it a worthwhile addition to our experimental studies.

Regarding image and audio classification of medical datasets, specifically the two datasets used in this investigation, several studies have been conducted. An RNN model was devised in [12] for both pathological and anomaly-level classifications from respiratory sound recordings using the ICBHI dataset. The pathology-level classification was used to identify two categories of lung conditions, i.e. binary (healthy or unhealthy) and ternary (healthy, chronic, or non-chronic illnesses) cases, while the anomaly-level classification was used to identify four categories of conditions in each respiratory cycle, i.e. normal, wheeze, crackle, and the presence of both crackle and wheeze. Their work used LSTM, GRU, Bi-LSTM, and Bi-GRU networks while specifying seven possible configurations for each network. Several factors, such as window size, window step, windows, framing, and features, differentiated these designs. The findings of their studies revealed several noteworthy observations. The first observation was that all seven settings performed similarly for the four-class anomaly-driven prediction, with ICBHI scores ranging from 0.71 to 0.74, whereas for the pathology-driven predictions, the ICBHI scores ranged from

0.83 to 0.91 under different experimental settings. This demonstrated impressive and robust performance of RNN models, regardless of the setup options used. Over the range of experiments, the LSTM model performed the strongest, while the GRU model was typically the poorest owing to its simplified network architectures.

Concerning the COVID-CT dataset, due to the dataset only being recently made available, there are fewer studies that have utilized the dataset for classification purposes. As an example, [3] demonstrated the use of a Capsule Network (CapsNet) as an alternative to CNN networks. The CapsNet was equipped with the capabilities of recording hierarchical and spatial relationships between image instances. Their system consisted of three stages – the first stage involved lung segmentation using 'U-Net', while the next two steps involved using the aforementioned CapsNet to perform the following tasks, which included extraction of the 'infected' slices from the segmented lung images and subsequent classification of the extracted slices to determine the cases of COVID-19, pneumonia and normality. Finally, a voting mechanism was used to perform a patient-level classification, to determine if the patient has a positive COVID-19 diagnosis or not. Their experimental results indicated a superior performance, and achieved an AUC of 0.9, sensitivity of 0.9455, specificity of 0.8604, and overall accuracy rate of 0.9082.

3. The Proposed Methodologies

In this research, we propose a Bi-LSTM model for respiratory anomaly detection from audio recordings. In addition, three state-of-the-art pre-trained CNNs, i.e. EfficientNet-B0, EfficientNet-B3, and EfficientNet-B7, are also employed for the identification of COVID-19 from CT scan imagery. Optimal settings are also identified, using a grid search mechanism. We introduce each of the proposed systems in detail below.

3.1 Respiratory Anomaly Detection from Audio Recordings Using a Bi-LSTM Network

3.1.1 Preprocessing

We first conduct respiratory anomaly detection from audio signals using a Bi-LSTM network. The ICBHI dataset is used for this task. To determine the methods of preprocessing and extracting effective features from audio inputs, several signal feature extraction algorithms are explored. As evidenced by existing studies [10, 1], we extract Mel-Frequency Cepstral Coefficients (MFCC) features for audio classification using the python library Librosa [7]. The MFCC features are the logarithmic measures of the Mel magnitude spectrum and contain adequate distinguishing characteristics, which make them particularly effective for classifying audio datasets [10].

During preprocessing of audio feature extraction, firstly, the audio files are divided into segments based on the sample rate and duration of each audio clip. The MFCC features are then extracted from each segment and appended to a dictionary with its class label. To be specific, to calculate the frequency spectrum, Fast Fourier Transform (FFT) is performed to each frame. This is accomplished through the use of a method known as the Short-Time Fourier Transform (STFT), which generates the power spectrum. Triangular Mel-scale filters are applied to the power spectrum to extract frequency bands. Equation (1) [28] is then used to compute the Mel frequency, utilizing these frequency bands.

$$\text{Mel}(f) = 1127 * \ln\left(\frac{f}{700}\right) \tag{1}$$

where f denotes the frequency in hertz.

3.1.2   The Bi-LSTM Network

We propose a Bi-LSTM network for the identification of diverse respiratory anomalies from audio recordings using the ICBHI dataset, owing to its superior performance with respect to audio and time-series forecasting. Table 1 illustrates the proposed Bi-LSTM architecture.

Table 1: The proposed Bi-LSTM architecture

Layer	Layer Description	Unit Setting
1	Bi-LSTM	512
2	LSTM	512
3	Dense (ReLU)	256
4	Dropout	0.5
5	Dense	128
Fully Connected	Dense (Softmax)	The number of classes

We employ a Bi-LSTM layer as the initial layer of the proposed model, with the hidden neuron units of 512 in practice being 1024 due to the bidirectional settings doubling the total number of units. The Bi-LSTM layer follows the implementation of a unidirectional LSTM layer, with the hidden unit setting of 512. Next, a dense layer with the activation function 'ReLU' was used as the third layer. A dropout layer is also embedded in the network, which is designed to decrease the amount of overfitting that may occur during the training process [2, 16]. The two final layers are the two dense layers, the first of which has 128 units and the second of which is a fully connected layer with the 'softmax' activation function and with number units equivalent to the number of classes. This proposed Bi-LSTM model is used to identify different lung disease conditions, using audio inputs from the ICBHI dataset.

3.2   COVID Diagnosis Using EfficientNet Models

The second proposed system employs pre-trained CNN models for COVID-19 identification, using the COVID-CT imagery dataset. Several preprocessing steps are performed to ensure enhanced performance. These include performing several augmentation transformations on the training images, to improve network ability to distinguish normal, and abnormal cases, and avoid overfitting. This broadens the scope of training data available to the model and theoretically should improve the training performance. The augmentation transformations include normalizing, random resized cropping, random rotation, random horizontal flip, and centre cropping.

Transfer learning has been conducted on several pre-trained CNN architectures. In other words, these networks are pre-trained on ImageNet and fine-tuned by using the augmented COVID-CT training set. Specifically, we employ three pre-trained models centred around the EfficientNet architecture [19] for COVID diagnosis, i.e. EfficientNet-B0, EfficientNet-B3, and EfficientNet-B7. All three models employ the same EfficientNet backbone. The EfficientNet architecture is based on the principle of a novel scaling approach that uses a simple yet extremely effective compound coefficient to evenly scale all depth, width, and resolution dimensions [19]. This scaling approach has also been tested on other CNN architectures, such as MobileNet, and ResNet to showcase its efficiency. It is used in this research to yield different versions of the networks, i.e. EfficientNet-B0, EfficientNet-B3, and EfficientNet-B7.

Moreover, such a scaling approach with different compound coefficient configurations results in each model having a different number of parameters. EfficientNet-B0, EfficientNet-B3, and EfficientNet-B7 have 5.3M, 12M, and 66M parameters, respectively. The three versions of the EfficientNet model are used to conduct COVID/.-19 diagnosis, using the COVID-CT dataset.

3.3 Model Training

Prior to the training process, the training, validation, and test splits for both datasets need to be determined. Table 2 shows the detailed data splits for both ICBHI and COVID-CT datasets.

Table 2: Training, validation, test splits for each dataset

Datasets	Train (70%)	Validation (10%)	Test (20%)
ICBHI	644	92	184
COVID-CT	244	34	71

To determine the optimal settings of the hyperparameters for the training process, i.e. learning rate, training epoch, and batch size, a grid search was utilized prior to training all the models. The early stopping [13] is also used as a precaution procedure to ensure that optimal performance is achieved during the training process. Early stopping is a validation method for determining whether overfitting starts to occur while the model is still being trained. This is accomplished in our research by including the validation set and the validation loss [13]. It is determined that the 'patience' threshold for the early stopping for training all models is set to 5, in order to prevent overfitting from occurring.

Following the grid search optimization process, the optimal hyperparameter values are identified for each test network, including learning rate, batch size, and training epochs. The optimized hyperparameters for Bi-LSTM and CNNs are illustrated in Tables 3-4. Subsequently, the identified optimal hyperparameters are used to set up all the networks in the training process.

Table 3: Optimal hyperparameters for Bi-LSTM identified using grid search

Model	Hyperparameters	Setting
Bi-LSTM	Epochs	34
	Learning Rate	0.0001
	Batch Size	256

Table 4: Optimal hyperparameters for CNNs identified using grid search

Models	Hyperparameters	Settings
EfficientNet-B0	Epochs	46
	Learning Rate	0.0001
	Batch Size	256
EfficientNet-B3	Epochs	38
	Learning Rate	0.0001
	Batch Size	256
EfficientNet-B7	Epochs	12
	Learning Rate	0.0001
	Batch Size	256

Figures 1-4 illustrate the training and validation losses for all four models, with Figs. 1-3 showing the losses for EfficientNet models from the training of the COVID-CT training set and Fig. 4 showing the losses for Bi-LSTM from the training of the ICBHI training set.

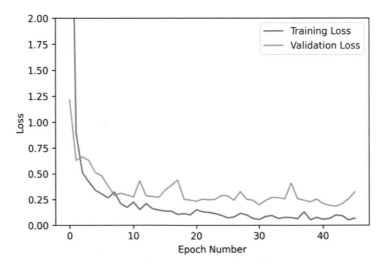

Fig. 1: Training and validation losses of EfficientNet-B0 for the COVID-CT training set

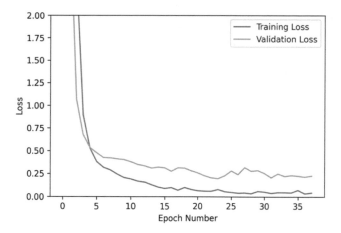

Fig. 2: Training and validation losses for EfficientNet-B3 for the COVID-CT training set

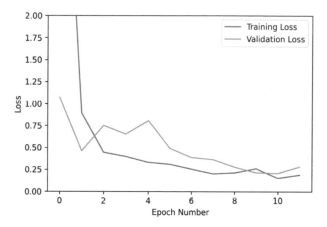

Fig. 3: Training and validation losses for EfficientNet-B7 for the Covid-CT training set

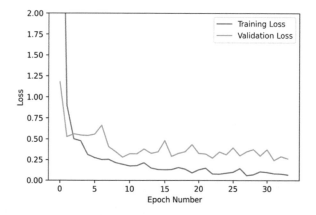

Fig. 4: Training and validation losses for Bi-LSTM for the ICBHI training set

4. Evaluation and Results

As this research is focused on two different types of deep learning methods for classification of medical datasets using audio and image inputs, the reasoning behind the dataset choices is introduced below. The ICBHI audio recording dataset was chosen for this study to aid essential research in providing a low-cost and effective method of early diagnosis, which is critical for reducing fatalities from a variety of respiratory illnesses. The dataset contains 920 audio recordings from 126 subjects whose conditions range from healthy, lower respiratory tract infections (LRTI), upper respiratory tract infections (URTI), chronic obstructive pulmonary disease (COPD), asthma and bronchiectasis. We combine all the disease cases into one class and conduct binary classification, i.e. healthy and unhealthy cases, using Bi-LSTM and this ICBHI dataset.

In addition, the COVID-CT dataset containing lung CT scan images is used in conjunction with three CNN models to classify whether a patient has a positive COVID-19 diagnosis. This automated method of diagnosis could be used as an alternative to the current most widely used PT-PCR test. This is owing to the fact that the PT-PCR test has several issues, such as cost, scalability, the overall reliability and accuracy of the test [21]. Providing a more accurate method of classification would not only be beneficial to early diagnosing of COVID-19, but beneficial to diagnosing other diseases that can be determined using CT scans. The dataset itself contains 349 COVID-19 CT scans from 216 patients and 463 non-COVID CT scans from a number of healthy subjects. These were all verified by a senior radiologist to be suitable for diagnostic use.

Tables 5-6 contain the detailed results for the evaluation of both ICBHI and COVID-CT datasets.

Table 5: The experimental results for the COVID-CT dataset

Model	Sensitivity	Specificity	Score	Accuracy
EfficientNet-B0	0.932	0.913	0.923	0.942
EfficientNet-B3	0.942	0.933	0.938	0.945
EfficientNet-B7	0.945	0.952	0.949	0.956

Table 6: Experimental results for the ICBHI dataset

Model	Sensitivity	Specificity	Score	Accuracy
Bi-LSTM	0.954	0.960	0.957	0.965

Figs. 5-6 show the confusion matrices of the two best performing models for each experiment, i.e. the EfficientNet-B7 for the COVID-CT dataset with respect to the diagnosis of COVID and non-COVID conditions, as well as the Bi-LSTM model for the ICBHI dataset regarding the diagnosis of normal and abnormal respiratory cases.

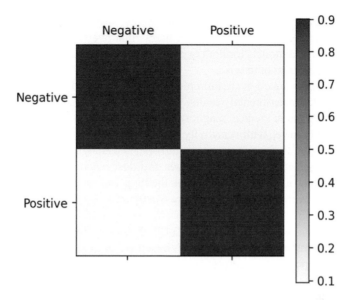

Fig. 5: Confusion matrix for Efficientnet-B7 for the Covid-CT dataset

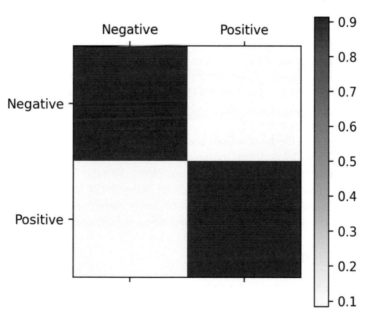

Fig. 6: Confusion matrix for Bi-LSTM for the ICBHI dataset

5. Discussion

Several observations can be made from the results of the two experiments as part of this research. First and foremost, Figs. 1-4 containing the training and validation losses for the training of all four models indicate that little-to-no overfitting occurred during this process. This was imperative to achieve at the start of the experiments and knowing that by the end of the experiments that overfitting was not a factor in the training of all the models for both experiments, it can be said the results are more reliable and robust than otherwise.

Secondly, the adopted evaluation metrics are also introduced below before the discussion on the experimental results. The sensitivity and specificity metrics are commonly employed in medical diagnostics as a benchmark for assessing diagnostic technique performance, with sensitivity referring to the rate of true positives and specificity referring to the rate of true negatives [11]. A high-test accuracy may still include a large percentage of false positives and false negatives, suggesting that the technique of diagnosis is less effective than the high-test accuracy indicates, so using sensitivity and specificity metrics is a more trustworthy approach to assessing model efficiency.

Regarding the results for the COVID-CT dataset, as illustrated in Table 5, the best performing model is EfficientNet-B7, based on the selected evaluation metrics, i.e. sensitivity, specificity, score (the average of sensitivity and specificity), and test accuracy. This model recorded scores of 0.945, 0.952, 0.949, and 0.956 for the respective accuracy metrics mentioned above, surpassing the results for both the EfficientNet-B0 and EfficientNet-B3 models. The likely reason behind this is due to the EfficientNet-B7 model having a comparatively higher number of parameters, possessing more powerful feature learning capabilities, and leading to better overall performance. It can also be seen in Table 4 that this model required the fewest epochs when compared to those of the two other models in the training stage, although given the much higher number of parameters that the model has. Most importantly, the sensitivity and specificity results of 0.945 and 0.952 indicate strongly that the model can confidently classify true positive and true negative cases, resulting in competitive classification performance.

With respect to the results from the ICBHI experiments, as indicated in Table 6, the Bi-LSTM model performs extremely well at classifying normal and abnormal respiratory cases using audio clips of respiratory sounds. With the results from the chosen sensitivity and specificity metrics being 0.954 and 0.960 respectively, this again indicates that the model can confidently classify true positives and true negatives, once again resulting in the model being significantly reliable.

6. Conclusion

In this research, we employ Bi-LSTM and three transfer learning EfficientNet models for the identification of respiratory anomaly and COVID-19 cases, using audio inputs and CT scan imagery respectively. Evaluated use of ICBHI sound recordings and COVID-CT image datasets, the proposed systems illustrate superior performance for breathing abnormality and COVID diagnosis. Besides the effectiveness of the chosen RNN and CNN network architectures for diverse audio and image classification tasks, the feature (such as MFCC) extraction process, augmentation transformations, preventative measures (e.g. early stopping training criteria), as well as optimal hyperparameter identification using grid search, all contribute to the promising and reliable performance of the proposed systems in our experiments. The proposed methods also outperform existing studies, such as [12] for respiratory anomaly detection. Although we employ grid search for hyper-parameter identification, we also aim to adopt evolutionary algorithms [29, 24, 30, 20] for network and learning parameter selection to further enhance performance.

In future directions, we will employ other audio and image medical datasets to further test model efficiency. Hybrid networks [9, 4] in combination with diverse architecture optimization algorithms and quantum computing strategies will also be used to further enhance performance.

References

1. Chen, C. and Li, Q. (2020). A multimodal music emotion classification method based on multifeature combined network classifier, *Mathematical Problems in Engineering*, 2020: 1-11.
2. Davis, S. and Mermelstein, P. (1980). Comparison of parametric representations for monosyllabic word recognition in continuously spoken sentences, *IEEE Transactions on Acoustics, Speech, and Signal Processing*, 28(4): 357-366.
3. Heidarian, S., Afshar, P., Enshaei, N., Naderkhani, F., Rafiee, M.J., Fard, F.B., Samimi, K., Atashzar, S.F., Oikonomou, A., Plataniotis, K.N. and Mohammadi, A. (2021). COVID-fact: A fully-automated capsule network-based framework for identification of COVID-19 cases from chest CT scans, *Frontiers in Artificial Intelligence*, 4.
4. Kinghorn, P., Zhang, L. and Shao, L. (2019). A Hierarchical and Regional Deep Learning Architecture for Image Description Generation, *Pattern Recognition Letters*, 119: 77-85.
5. Kumar, J., Goomer, R. and Singh, A.K. (2018). Long Short-term Memory Recurrent Neural Network (LSTM-RNN) based workload forecasting model for cloud datacenters, *Procedia Computer Science*, 125: 676-682.
6. Lawrence, T., Zhang, L., Lim, C.P. and Phillips, E.J. (2021). Particle swarm optimization for automatically evolving convolutional neural networks for image classification, *IEEE Access*, 9: 14369-14386.
7. McFee, B., Raffel, C., Liang, D., Ellis, D.P., McVicar, M., Battenberg, E. and Nieto, O. (2015). Librosa: Audio and music signal analysis in python, *In: Proceedings of the 14th Python in Science Conference*, vol. 8, pp. 18-25.
8. Minh-Tuan, N. and Kim, Y.-H. (2019). Bidirectional long short-term memory neural networks for linear sum assignment problems, *Applied Sciences*, 9(17): 3470.
9. Mistry, K., Zhang, L., Neoh, S.C., Lim, C.P. and Fielding, B. (2017). A micro-GA embedded PSO feature selection approach to intelligent facial emotion recognition, *IEEE Transactions on Cybernetics*, 47(6): 1496-1509.
10. Nogueira, D.M., Ferreira, C.A., Gomes, E.F. and Jorge, A.M. (2019). Classifying heart sounds using images of motifs, MFCC and temporal features, *J. Med. Syst.*, 43(6): 168.
11. Parikh, R., Mathai, A., Parikh, S., Chandra Sekhar, G. and Thomas, R. (2008). Understanding and using sensitivity, specificity and predictive values, *Indian Ophthalmol.*, 56(1): 45.
12. Perna, D. and Tagarelli, A. (July 2019). Deep auscultation: Predicting respiratory anomalies and diseases via recurrent neural networks, arXiv:1907.05708; accessed: 20 July, 2021, [Online] Available: http://arxiv.org/abs/1907.05708
13. Prechelt, L. (1998). Early stopping-but when? *In: Neural Networks: Tricks of the Trade*, Springer, pp. 55-69.

14. Rocha, B.M., Filos, D., Mendes, L., Vogiatzis, I., Perantoni, E., Kaimakamis, E., Natsiavas, P., Oliveira, A., Jácome, C., Marques, A. and Paiva, R.P. (2018). A respiratory sound database for the development of automated classification. *In:* N. Maglaveras, I. Chouvarda and P. de Carvalho (Eds.), *Precision Medicine Powered by pHealth and Connected Health*, vol. 66, pp. 33-37. Singapore: Springer.
15. Srisukkham, W., Zhang, L., Neoh, S.C., Todryk, S. and Lim, C.P. (2017). Leukaemia diagnosis with bare-bones PSO-based feature optimization, *Applied Soft Computing*, 56: 405-419.
16. Srivastava, N., Hinton, G., Krizhevsky, A., Sutskever, I. and Salakhutdinov, R. (2014). Dropout: A simple way to prevent neural networks from overfitting, *The Journal of Machine Learning Research*, 15(1): 1929-1958.
17. Tan, T.Y., Zhang, L. and Lim, C.P. (2019). Intelligent skin cancer diagnosis using improved particle swarm optimization and deep learning models, *Applied Soft Computing*, 105725.
18. Tan, T.Y., Zhang, L. and Lim, C.P. (2020). Adaptive melanoma diagnosis using evolving clustering, ensemble and deep neural networks. *Knowledge-based Systems*, 187: 1-26. Article ID 104807.
19. Tan, M. and Le, Q. (2019). Efficientnet: Rethinking model scaling for convolutional neural networks. *In: International Conference on Machine Learning*, pp. 6105-6114.
20. Tan, T., Zhang, L., Neoh, S.C. and Lim, C.P. (2018). Intelligent Skin Cancer Detection Using Enhanced Particle Swarm Optimization, *Knowledge-based Systems*, 158: 118-135.
21. Tang, Y.W., Schmitz, J.E., Persing, D.H. and Stratton, C.W. (2020). Laboratory diagnosis of COVID-19: Current issues and challenges, *J. Clin. Microbiol.*, 58(6): e00512-20.
22. Wall, C., Young, F., Zhang, L., Phillips, E.-J., Jiang, R. and Yu, Y. (Oct. 2020). Deep learning-based melanoma diagnosis using dermoscopic images. *In: Developments of Artificial Intelligence Technologies in Computation and Robotics*, pp. 907-914. Cologne, Germany.
23. Xie, H., Zhang, L. and Lim, C.P. (2020). Evolving CNN-LSTM models for time series prediction using enhanced grey wolf optimizer, *IEEE Access*, 8: 161519-161541.
24. Xie, H., Zhang, L. and Lim, C.P. (2021). Feature selection using enhanced particle swarm optimization for classification models, *Sensors*, 21(5): 1-40. Article ID 1816.
25. Yamashita, R., Nishio, M., Do, R.K.G. and Togashi, K. (2018). Convolutional neural networks: An overview and application in radiology, *Insights Imaging*, 9(4): 611-629.
26. Zhang, L., Lim, C.P., Yu, Y. and Jiang, M. (2022). Sound classification using evolving ensemble models and particle swarm optimization, *Applied Soft Computing*, 116: 1-28. Article ID 108322.
27. Yang, X., He, X., Zhao, J., Zhang, Y., Zhang, S. and Xie, P. (2020). Covid-CT-Dataset: A CT Scan Dataset about COVID-19, arXiv:2003.1386; accessed: 20 July, 2021. [Online] Available: http://arxiv.org/abs/2003.13865
28. Zahid, S., Hussain, F., Rashid, M., Yousaf, M.H. and Habib, H.A. (2015). Optimized audio classification and segmentation algorithm by using ensemble methods, *Mathematical Problems in Engineering*, vol. 2015.
29. Zhang, L. and Lim, C.P. (2020). Intelligent optic disc segmentation using improved particle swarm optimization and evolving ensemble models, *Applied Soft Computing*, 92: 106328.
30. Zhang, L., Mistry, K., Neoh, S.C. and Lim, C.P. (2016). Intelligent facial emotion recognition using moth-firefly optimization, *Knowledge-based Systems*, 111: 248-267.

COVID-19 Forecasting in India through Deep Learning Models

Arindam Chaudhuri[1] and Soumya K. Ghosh[2]

[1] Samsung R & D Institute Delhi Noida - 201304 India; NMIMS University
Mumbai - 400056 India
[2] Department of Computer Science Engineering, Indian Institute of Technology Kharagpur,
Kharagpur - 701302 India

1. Introduction

Coronavirus-19 (COVID-19) has been an infectious disease caused by severe acute respiratory syndrome [30, 46, 52] since December 2019. It has been around more than 1.5 years [21] and has become a global pandemic. It was identified in December 2019 in Wuhan, Hubei, China with the first confirmed case being traced in November 2019 [2]. Presently on 19th July, 2021 we have more than 4 million new cases reported across the world with more than 7 thousand deaths [24, 20]. As a result of COVID-19 many countries have closed their borders and imposed partial or full lockdown. This has had a massive impact on the world economy as a whole and will continue in coming years [5, 27, 44]. Agriculture [32, 60], which has been a major source of income for population in rural areas, is worst affected. The continuous lockdown in several countries has created transportation problems for low-income migrant communities [42, 39].

In India, management of COVID-19 [40, 74] has been unique though not very successful till date. The first COVID-19 case in India was reported on 30th January, 2020. India had a lock-down earlier and things were well managed in terms of number of deaths and infections per million population. India currently on 19th July, 2021 has more than 31 million confirmed cases with more than 0.4 million deaths, which is the largest in Asia and second highest in the world, after United States. The fatality rate of COVID-19 in India is among the lowest in world and is steadily declining. India also has one of the fastest recovery rates in the world and figures among the top in world rankings.

Due to complexity of infection spread [4, 71, 57], it has been unreliable to figure out prominent computational and statistical models towards forecasting of COVID-19. This mainly happens because active or novel cases are not taken into consideration without dependencies on population density, logistics and travel and qualitative social aspects [54]. Some of these issues are broken down towards quantitative measurement, which help in forecasting models. Other issues have qualitative nature and lack proper data collection and reporting which makes modeling attempts unreliable. As a result of this, it is required to re-assess the situation with latest data sources and most comprehensive forecasting models [22, 12, 43]. Several other limitations also exist, such as noisy or unreliable data involving active cases [26], mortality rate and asymptotic carriers [70, 6]. There have been situations where models are plagued with a number of limitations and failed in several situations [17]. The infection rate [3] has also been impacted, considering the several country-based mitigation factors in terms of different levels of lockdowns and monitoring. All these issues are present in COVID-19 forecasting models in India. This calls for re-evaluation of the situation in India with latest deep learning models for forecasting COVID-19.

Several versions of deep learning models, such as recurrent neural networks (RNNs) and its variants are well designed towards time series data [25, 67, 33, 58, 19] modeling. RNNs have provided promising results in modeling dynamic systems as compared to feed forward neural networks [50, 51, 29]. Long short-term memory (LSTMs) networks [33] provide better results than RNNs considering long-term dependencies in time series data which span across a few hundreds or thousands of time steps [8]. They have been used towards forecasting COVID-19 [69, 16] with appreciable results in comparison to epidemic models. Convolutional neural networks (CNNs) [68, 66] have also given good results for time series forecasting.

In this research, based on the motivation from [14], we use five versions of LSTM models [13] which are commonly referred as its variants towards forecasting COVID-19 spread for select Indian states. Indian states with COVID-19 hotpots are considered in terms of infection rates and compared with states where infections have been controlled or have gone up to their peak points. Here both univariate and multivariate time series forecasting approaches are considered. The forecasting performance is compared for different states across India through LSTM models and its variants. The results of COVID-19 infections are presented through suitable vizualisations. It has been observed that hierarchical LSTM models show much better results than their non-hierarchical counterparts.

The rest of the paper is structured as follows: Section 2 highlights a review of related work. The computational methodology is presented in Section 3. In Section 4, experiments and results are given. Finally in Section 5, conclusion and direction towards future work are mentioned.

2. Related Work

COVID-19 has raised several concerns from people living in both developed and developing countries. It has changed the way of our lives. It has made a significant impact on the mental health of people across the globe [45, 55, 28]. COVID-19 forced lockdowns and restrictions of movement, giving rise to e-learning [53, 64, 10], telemedicine [41] and created opportunities in applications [74]. The lockdown has shown a positive impact on the environment [72, 48, 38, 56]. It has also been observed that in some countries, comprehensive identification and isolation policies have effectively suppressed COVID-19 spread. In [73] an evaluation of identification and isolation, policies are presented which effectively suppress COVID-19 spread. However, such policies have not been so effective for other countries with similar demographics.

Several statistical and machine learning models have been used towards COVID-19 modeling. In [71] autoregressive integrated moving average (ARIMA) model is used for forecasting COVID-19 in Pakistan. In [57], simple autoregressive neural networks are presented towards forecasting COVID-19 in Egypt which has shown appreciable performance. The Gaussian process regression model has been used in [4] towards forecasting COVID-19 infection in United States. In [16], LSTM neural networks have been used towards COVID-19 time series forecasting in Canada. Hybrid ARIMA and wavelet-based forecasting model [12] has been used for short-term (10 days ahead) forecasts of daily confirmed cases considering Canada, France, India, South Korea and United Kingdom. Autoregressive time series models have been used in [36] as well as in [43], based on two-piece scale mixture normal distributions time series data. In [1], a hybrid of discrete wavelet decomposition and ARIMA models are presented. In [61], weather conditions of different states are incorporated to make more accurate forecasting of COVID-19 cases in different states of India. In [9], LSTM model is developed for 30-day ahead prediction of COVID-19 positive cases in India. In [65], COVID-19 cases are forecasted using prophetic logistic growth model.

In [54], an analysis is performed considering spatiotemporal epidemic variations before utilizing ecological niche models with socioeconomic variables for identifying potential risk zones for megacities. The results have demonstrated that this is capable of being employed as an early forecasting tool towards identification of potential COVID-19 infection risk zones. In [22], machine learning methods, such as Bayesian regression neural network, cubist regression, k-nearest neighbor, quantile random forest and support vector regression with pre-processing based on variational mode decomposition are used for forecasting one, three and six-days-ahead cumulative COVID-19 cases. In [7], a Bayesian framework coupled with LSTM network are used to forecast cases of COVID-19 in developed and developing countries. In [49], cumulative COVID-19 cases are analyzed and forecasted in four selected cities of Brazil, using stacked ensemble forecasting model. In [23], COVID-19 daily cases, deaths caused and recovered cases are forecasted, using LSTM networks for the whole world. In [11], several forecasting models, such as simple average, single exponential smoothing, holt winter method and auto-regressive integrated moving average models are used for time series analysis of COVID-19 pandemic. In [15], a comparison is performed considering ARIMA, LSTM, multi-layer perceptron and convolutional neural network (CNN) models for prediction of COVID-19 cases all over the world. In [37], hybrid machine learning methods of adaptive network-based fuzzy inference systems (ANFIS) and multi-layer perceptron are used for COVID-19 infections and mortality rate in Hungary.

3. Computational Methodology

The computational methodology consists of forecasting COVID-19 in India with some important deep learning models [14], such as LSTM, bidirectional LSTM (Bi-DLSTM), encoder-decoder LSTM (EDLSTM), hierarchical bidirectional LSTM (HBi-DLSTM) and hierarchical encoder-decoder LSTM (HEDLSTM). In order to proceed with forecasting using deep learning models, a time series needs to be reconstructed with multi-step ahead prediction. The reconstruction process creates significant features of original time series [47]. Considering an embedded phase space $X(t) = [y(t), y(t-T), \ldots \ldots \ldots, y(t-(d-1)T)]$, we can generate a time series $y(t)$ with T as time delay, d as embedding dimension having window span $t = 0, 1, 2, 3, 4,$ $\ldots \ldots \ldots, N-dT-1$ and N being length of original time series. The parameters d and T are user defined and experimentally defined. It has been observed that if original attractor is of dimension D, then $d = 2D + 1$.

The hierarchical versions of LSTM have provided better results considering accuracy and precision with respect to training, validation and testing datasets. This happens mainly because of faster convergence of these algorithms to optimal values through leveraging the hierarchical structure of model. The hierarchical structure of model allows better usage of all parameters which increase its strength in comparison to non-hierarchical versions. The hierxarchical model's flexible structure allows addition and removal of layers as per needs and requirements of forecasting problems at hand.

3.1 LSTM Networks

LSTM networks have evolved in order to address limitations of learning long-term dependencies in simple recurrent neural networks (RNN) [63]. LSTM are blessed with solutions relating to problems in vanishing and exploding gradients [34]. They have memory cells and gates with better capabilities in learning long-term dependencies in temporal sequences, as shown in Fig. 1. LSTM units are trained in a supervised fashion on training sequences set using an adaptation of back propagation through time (BPTT) algorithm considering respective gates [63]. LSTM networks calculate hidden state $h(t)$ as follows:

$$i(t) = \sigma(x(t)U^i + h(t-1)W^i) \tag{1}$$

$$f(t) = \sigma(x(t)U^f + h(t-1)W^f) \tag{2}$$

$$o(t) = \sigma(x(t)U^o + h(t-1)W^o) \tag{3}$$

$$C(t) = \tanh(x(t)U^g + h(t-1)W^g) \tag{4}$$

$$C(t) = \sigma(f(t) * C(t-1) + i(t) * C(t)) \tag{5}$$

$$h(t) = \tanh = (C(t) * o(t)) \tag{6}$$

In equations (1) to (6), $i(t)$, $f(t)$ and $o(t)$ refer to input, forget and output gates at time t respectively. $x(t)$ and $h(t)$ refer to number of input features and number of hidden units respectively. W and U are weight matrices adjusted during learning along with b as bias. The initial values are $c(0) = 0$ and $h(0) = 0$. All gates have same dimensions d_h which is the size of hidden state. $C(t)$ is candidate hidden state and $C(t)$ is internal memory.

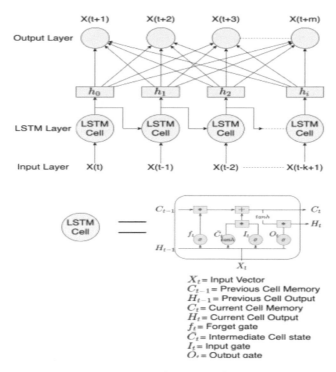

X_t = Input Vector
C_{t-1} = Previous Cell Memory
H_{t-1} = Previous Cell Output
C_t = Current Cell Memory
H_t = Current Cell Output
f_t = Forget gate
$\tilde{C}_t$ = Intermediate Cell state
I_t = Input gate
O_t = Output gate

Fig. 1: LSTM networks

3.2 Bidirectional LSTM Networks

A major shortcoming of conventional LSTM is that they only make use of previous context state for determining future states. Bi-DLSTM [35] process information in both directions with two separate hidden layers which are then propagated forward to same output layer. Bi-DRNN consists of placing two independent LSTM together to allow both backward and forward information about the sequence at every time step. Bi-DLSTM computes forward hidden sequence $h(f)$, backward hidden sequence $h(b)$ and output sequence y by iterating information from backward layer. Then information in other network is propagated from $t = 1$ to $t = T$ in order to update output layer. When both networks are combined, information is propagated in bi-directional manner. Bi-DLSTM [59] were initially being used for word embedding in natural language processing in order to access long-range context or state in both directions. Bi-DLSTM take inputs in two ways, one from past to future and one from future to past. By running information backwards, state information from future is preserved. Hence, with two hidden states combined, at any point in time network, preserves information from both past and future as shown in Fig. 2.

3.3 Encoder-Decoder LSTM Networks

Encoder-decoder LSTM network (EDLSTM) is a sequence-to-sequence model for mapping fixed-length input to fixed-length output [31, 62]. The length of input and

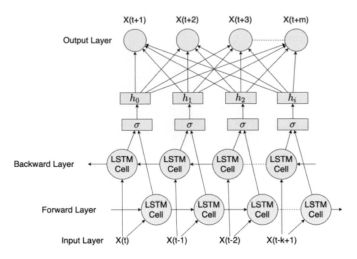

Fig. 2: Bi-directional LSTM networks

output may differ which makes them applicable in automatic language translation tasks. Hence, the input can be sequences of video frames and output is sequence of words. Based on this conditional probability an output sequence is estimated, given an input sequence. In case of multi-step series prediction, both input and outputs are of variable lengths. EDLSTM networks handle variable length input and outputs by first encoding input sequences one at a time, using a latent vector representation and then decoding them. In encoding phase given an input sequence EDLSTM computes a sequence of hidden states. In decoding phase, it defines a distribution over output sequence given input sequence, as shown in Fig. 3.

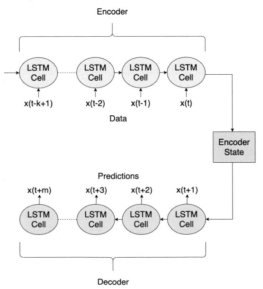

Fig. 3: Encoder-decoder LSTM networks

3.4   Hierarchical Bidirectional LSTM Networks

HBi-DLSTM [14] extends forecasting benefits received from HBi-DLSTM. The method uses feature selection and extraction using unsupervised learning and classification and prediction using supervised learning. The success of this approach is highlighted through high accuracy in prediction results. HBi-DLSTM provides efficient prediction accuracy and computational complexity as data volume grows. The architecture of HBi-DLSTM is presented in Fig. 4. The temporal data sequences are modeled by Bi-DLSTMs which are combined together to form HBi-DLSTM. The model is composed of 7 layers $br_1 - br_2 - br_3 - br_4 - br_5 - fc - sm$. Here br_i; i = 1, 2, 3,4, 5 denote layers with BiDLSTM nodes, fc denotes fully-connected layer and sm denotes softmax layer. Each layer of Bi-DLSTM takes care of prediction tasks which drive towards the whole network's success. To recover any single hierarchy Bi-DLSTM split is run on small subset of data comprising of few values to calculate seed prediction value. The input data subset is produced randomly. This activity starts at layer br_1. Considering initial prediction value, remaining data is placed into seed class where, on an average, it is almost similar. This results in prediction of entire data using only value similarities in smaller subsets. By recursively applying this procedure to each class HBi-DRNN is obtained using small fraction of similarities. The prediction task goes on till br_5. No measurements are observed between classes at previous split in this recursive phase. This results in robust HBi-DRNN that aligns its measurements m to resolve higher resolution in class structure.

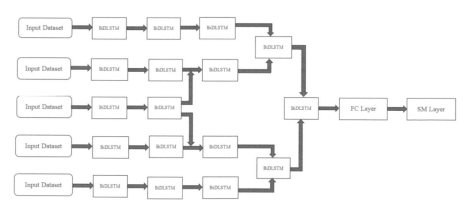

Fig. 4: Architecture of proposed HBi-DLSTM

    HBi-DLSTM is characterized in terms of probability of success in recovering true hierarchy, measurement and runtime complexity. Some restrictions are placed on similarity function such that similarities agree with hierarchy up to some random noise. This condition states that similarity from data point to its class should be in expectation larger than similarity from that data point to other class. This is related to tight classification condition. The within and between-class similarities concentrate away from each other. Here satisfaction results when similarities are constant in expectation perturbed with any subgaussian noise. Considering feature learning stacked Bi-DLSTM extracts temporal features of data sequences. After obtaining features of data sequence classification is performed by fully connected layer *fc* and softmax layer *sm*. The computational architecture effectively addresses vanishing gradient problem. The network neurons are adopted in last recurrent layer br_5. The first four Bi-DLSTM layers use cosh activation function. This provides tradeoff between representation ability and overfitting. The number of weights in network is more than that in cosh neuron. It is easy to overfit network with limited data training sequences. The network hierarchy splits are resolved through fine tuning the algorithm with certain heuristics. All subsampled data points are discarded with lower degree when restricted to sample with removes under-represented classes from sample. In averaging phase, if data points are not similar to any represented class, new class for data point is created.

3.5   Hierarchical Encoder-Decoder LSTM Networks

HEDLSTM [14] extends forecasting benefits received from EDLSTM. The method uses feature selection and extraction through unsupervised learning and classification and prediction through supervised learning. The success of this approach is highlighted through high accuracy in prediction results. HEDLSTM provides efficient prediction accuracy and computational complexity as data volume grows. The architecture of HEDLSTM is presented in Fig. 5. The temporal data sequences are modeled by EDLSTM which are combined together to form HEDLSTM. The model is composed of 7 layers $br_1 - br_2 - br_3 - br_4 - br_5 - fc - sm$. Here br_i; $i = 1$, 2, 3,4, 5 denote layers with EDLSTM nodes, *fc* denotes fully connected layer and *sm* denotes softmax layer. Each layer of HEDLSTM takes care of prediction tasks which drive towards the whole network's success. To recover any single hierarchy, EDLSTM split is run on small subset of data comprising a few values to calculate seed prediction value. The input data subset is produced randomly. This activity starts at layer br_1. Considering initial prediction value, remaining data is placed into seed class, where, on an average, it is most similar. This results in prediction of entire data using only value similarities in smaller subsets. By recursively applying this procedure to each class HEDLSTM is obtained using small fraction of similarities. The prediction task goes on till br_5. No measurements are observed between classes at previous split in this recursive phase. This results in robust HEDLSTM that aligns its measurements *m* to resolve higher resolution in class structure.

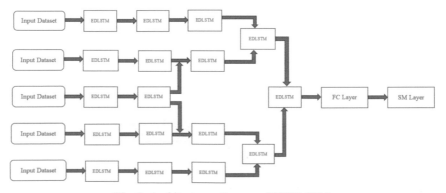

Fig. 5: Architecture of proposed HEDLSTM

HEDLSTM is characterized in terms of probability of success in recovering true hierarchy, measurement and runtime complexity. Some restrictions are placed on similarity function, such that similarities agree with hierarchy up to some random noise. This condition states that similarity from data point to its class should be in expectation larger than similarity from that data point to the other class. This is related to tight classification conditions. The within and between class similarities concentrate away from each other. Here satisfaction results when similarities are constant in expectation perturbed with any sub gaussian noise. Considering feature learning stacked EDLSTM extracts temporal features of data sequences. After obtaining features of data sequence classification is performed by fully connected layer *fc* and softmax layer *sm*. The computational architecture effectively addresses vanishing gradient problems. The network neurons are adopted in last recurrent layer br_5. The first four EDLSTM layers use cosh activation function. This provides tradeoff between representation ability and overfitting. The number of weights in network is more than that in cosh neuron. It is easy to overfit network with limited data training sequences. The network hierarchy splits are resolved through fine tuning the algorithm with certain heuristics. All subsampled data points are discarded with lower degree when restricted to a sample which removes underrepresented classes from the sample. In averaging phase, if data points are not similar to any represented class, new class for data point is created.

4. Experiments and Results

Before we proceed with experimental results on COVID-19 forecasting in different states of India with different variants of LSTM models, we provide a brief overview of the total number of COVID-19 infections for different states in India. Due to certain confidential reasons, experimental datasets are not provided here.

4.1 India COVID-19 Situation Report: 19th June, 2021

The Table 1 provides a rank of top ten Indian states with total cases appearing on 1st of every month. The months considered here are May, 2020, September, 2020, January, 2021 and May, 2021. It is observed that densely populated states, such as Maharashtra [14] has been leading India in the number of total cases since January, 2020. It is noted that Uttar Pradesh, which is another densely populated state, seems to have managed better in months of August, 2020 and September, 2020. Delhi, with relatively smaller population [14] but higher population density, has been one of the leading states with COVID-19 cases.

Figure 6 presents the total number of new weekly cases for different groups of Indian states. All results are shown in a randomly selected time window. It is noticed that the number of cases significantly increased after May, 2020 and certain number of states have most number of new weekly cases, as shown in Fig. 6. It is evident from Fig. 6 that Delhi reached a major peak in weekly new cases in 3rd week of June, 2020. Tamil Nadu began slowing down in increase of new weekly cases around then. However, Maharashtra continued to increase in terms of new cases. Maharashtra began slowing down towards end of July. It is also observed that new cases in West Bengal drastically started increasing from 30th June, 2020 and then somewhat reaching a plateau post 25 August, 2020. Similarly, new cases in Bihar increased drastically from 30 June, 2020 and slowly reached peak by 22 September, 2020 and declined afterwards.

Figure 7 presents daily active cases and cumulative total deaths for key Indian states which remain in top 10 states with most number of cases, as shown in Table 1. All results are shown in a randomly selected time window. It is noticed that active cases began declining in most states post 1 October, 2020 except for Delhi and Karnataka. Delhi has had three major peaks of active cases and third peak began

Table 1: Rank of Indian states in May, 20, September, 20, January, 21 and May, 21

Rank	May, 2020	September, 2020	January, 2021	May, 2021
1	Maharashtra	Maharashtra	Maharashtra	Maharashtra
2	Gujarat	Andhra Pradesh	Kerala	Kerala
3	Delhi	Tamil Nadu	Tamil Nadu	Tamil Nadu
4	Madhya Pradesh	Karnataka	Karnataka	Karnataka
5	Rajasthan	Uttar Pradesh	Telangana	Delhi
6	Tamil Nadu	Delhi	Andhra Pradesh	Gujarat
7	Uttar Pradesh	West Bengal	Delhi	West Bengal
8	Andhra Pradesh	Odisha	Gujarat	Telangana
9	Telangana	Kerala	Madhya Pradesh	Andhra Pradesh
10	West Bengal	Telangana	Uttar Pradesh	Rajasthan

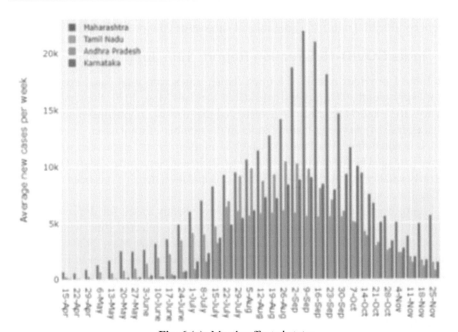

Fig. 6 (a): Mostly affected states

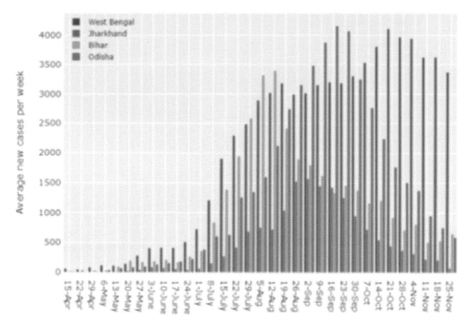

Fig. 6 (b): Eastern states

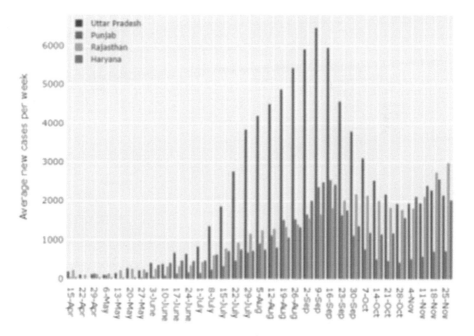

Fig. 6 (c): Northern states

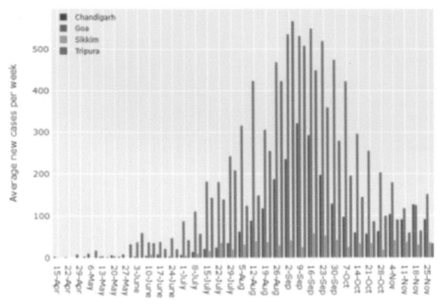

Fig. 6 (d): Smaller states

Fig. 6: Weekly average of new cases for groups of Indian states

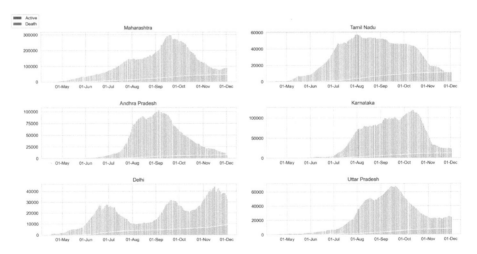

Fig. 7: Daily active cases and cumulative death cases in key Indian states

declining post-mid-November, 2020 while Karnataka reached a plateau and began declining mid-October, 2020 in terms of active cases. In terms of deaths, a sharp increase post-October, 2020 is not seen in most of the states, except for Delhi which can be explained by multiple peaks and a high number of active cases post-October, 2020 when compared to rest of Indian states. It is to be noted that daily deaths are not shown in same graphs since scales between active cases and deaths are quite different.

Figure 8 shows a comparison of India with other countries, such as USA, France, Brazil, Russia and Spain which are considered to have major active cases of COVID-19. All results are shown in a randomly selected time window. It is noticed that India had a steady decline in active cases from October, 2020 without a major rise in cumulative deaths, whereas Spain, Russia, United States and France have steady increase in active cases. Brazil seems to have reached a plateau in active cases from July, 2020 to September, 2020 with major spike in active cases from mid-October, 2020 and with gradual increase in cumulative deaths.

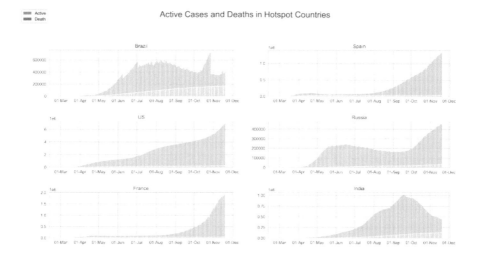

Fig. 8: Daily active cases and cumulative deaths in different countries

The period post-1 December, 2020 showed a decline in COVID-19 cases in most of Indian states. This trend continued till end of February, 2021 after which there has been again an increase in COVID-19 active cases in almost all states of India. This was considered as second wave of COVID-19 with Maharashtra, Delhi, Tamil Nadu, Kerala and Karnataka taking the lead. With constant upward trend till first week of June, 2021 there has been a downward trend in the number of active cases. As on 19 June, 2021 things have been under control.

4.2 Results and Discussions

Based on COVID-19 data available in India till 19 July, 2021 certain significant results of prediction of daily cases in India using different variants of LSTM networks as highlighted in Section 3 are presented in this section. A detailed review of these results is available in [14].

The experimental design considers the prediction task independently by including both univariate and multivariate approaches. A general investigation overview is highlighted through these steps: (a) the evaluation process involves making a dataset split into training and testing subsets. The static split of training samples is adopted by considering datapoints from initial phase till 30 September, 2020. This constitutes the training set. The remaining part of dataset is considered as test dataset. The training and testing sets are developed by randomly shuffling original dataset; (b) an evaluation is also performed towards univariate or multivariate prediction approaches; (c) the results are presented for all Indian states with two leading states having COVID-19 infections, viz Maharashtra and Delhi; (d) an accuracy evaluation is also performed of all forecasting models discussed in Section 3; (e) considering bidirectional and hierarchical versions of LSTM models, an analysis was performed towards two months' outlook where new daily cases predictions are fed into trained models.

The analysis data is adopted from Indian Statistical Institute, Bangalore [14] where Ministry of Health and Family Welfare Government of India had compiled relevant data. The univariate and multivariate time series data are pre-processed into state space vector [14] with time window $D = 7$ and time lag $T = 2$ considering multistep ahead prediction. It is to be noted that each problem with respect to dataset featured *MSA* values as 5 and 13 respectively. The data is averaged with rolling mean of five days. Since a fair number of outliers are present in data, the actual analysis was considered from 16 April, 2020. The dataset is based on new cases per day which is being normalized considering the maximum number of daily new cases. The Adam optimizer is used towards training of all LSTM models with mean squared error (MSE) loss function. Further details about LSTM models' topology are available in [14].

In univariate model case, input comprises one feature window with seven-time steps size backward which is used towards five-time steps ahead prediction. In multivariate model case, input contains five features representing adjacent states with respect to present state into consideration. For example, considering Maharashtra, the states considered are Maharashtra, Gujarat, Madhya Pradesh, Karnataka and Uttar Pradesh and for Delhi, the states considered are Delhi, Rajasthan, Uttar Pradesh, Haryana and Madhya Pradesh. In multivariate model features of all states are considered. The multivariate model considers data with seven-time steps backward window which is used towards five-time steps ahead prediction.

Considering random shuffling training and testing datasets consist of 500 and 200 datapoints respectively. With static split training and testing datasets consist of 480 and 175 datapoints respectively. All datasets are available in [14]. The performance of respective methods is evaluated with respect to scalability and robustness. The root mean squared error (RMSE) [14] is used as performance measure towards prediction accuracy. For each prediction horizon and each problem RMSE is used. The mean errors for respective prediction horizons through time-steps ahead are noted. All experiments have mean and 95% confidence interval considering 30 experiment runs. All LSTM models are being initialized with different initial weights from uniform distribution in range [−0.7, 0.7]. The dropout rate for all LSTM models has been set at 0.5. Tables 2 and 3 present test performance for all LSTM models considering random and static splits considering India, Delhi and Maharashtra.

It was observed that India cases prediction had a unique trend with model improving at increase in prediction horizon in comparison to Maharashtra and Delhi cases. The corresponding cases at random split highlights different trend and better accuracy with lower RMSE for India and Delhi, with the exception of Maharashtra.

Table 2: The results of univariate model on test datasets

Model	India			
	Random Split		Static Split	
	RMSE	Std_Dev	RMSE	Std_Dev
LSTM	22623	809	10536	267
Bi-DLSTM	22609	869	10660	394
EDLSTM	7632	269	12124	1662
HBi-DLSTM	5532	169	10124	1462
HEDLSTM	5332	148	8125	1262
	Delhi			
	Random Split		Static Split	
	RMSE	Std_Dev	RMSE	Std_Dev
LSTM	1402	67	3439	505
Bi-DLSTM	1454	64	3514	874
EDLSTM	1689	76	3487	232
HBi-DLSTM	1489	55	3287	231
HEDLSTM	1289	45	3087	227
	Maharashtra			
	Random Split		Static Split	
	RMSE	Std_Dev	RMSE	Std_Dev
LSTM	5014	241	2696	209
Bi-DLSTM	5204	295	2745	284
EDLSTM	5360	275	2403	265
HBi-DLSTM	5160	196	2103	238
HEDLSTM	5060	176	1903	228

The hierarchical models gave the best performance for cases in India [41] and Maharashtra considering testing dataset. It was observed that hierarchical based LSTM models performed much better than LSTM models for almost all cases. The bi-directional and encoder-decoder-based LSTM models suffered from overtraining. It was also observed that on comparing random with shuffle splits considering univariate and multivariate models for almost all Indian states, static split provides superior performance than random split. In case of Delhi, random split had better results than static split.

The prediction uncertainty was attributed to experiment runs considering different weight initialization in LSTM models. It is to be noted in all cases there was a trend of decline in a number of cases. All LSTM models captured spike and fall in cases appreciably every few days. Also, there was less uncertainty in cases for Delhi in comparison to Maharashtra. In Delhi, there have been four major peaks since inception of COVID-19. The trend for Maharashtra had been identical to trend of India in terms of daily new cases, which sets similar uncertainty trends.

Table 3: The results of multivariate model on test datasets

Model	India			
	Random Split		Static Split	
	RMSE	Std_Dev	RMSE	Std_Dev
LSTM	14945	1403	13100	2276
Bi-DLSTM	16487	767	24396	6169
EDLSTM	21917	3231	14995	4655
HBi-DLSTM	19917	738	12995	4555
HEDLSTM	17917	731	10995	4155
	Delhi			
	Random Split		Static Split	
	RMSE	Std_Dev	RMSE	Std_Dev
LSTM	1319	26	4394	289
Bi-DLSTM	1185	34	5064	234
EDLSTM	1406	224	5238	528
HBi-DLSTM	1206	196	5038	508
HEDLSTM	1006	175	4838	498
	Maharashtra			
	Random Split		Static Split	
	RMSE	Std_Dev	RMSE	Std_Dev
LSTM	4569	272	4953	514
Bi-DLSTM	4928	257	4441	784
EDLSTM	6454	528	5728	579
HBi-DLSTM	6254	508	5528	538
HEDLSTM	6054	498	5028	508

This research tries to incorporate promising forecasting tools using deep learning. It highlights challenges with given limited data and infections spread. The hierarchical versions of LSTM provide best performance for almost all cases in comparison to other LSTM versions. It is found that all Indian states and Maharashtra datasets have similar trends in new cases. The model performance with static split of training and testing data gives better results. However, there remains an ambiguity in selecting univariate and multivariate performances since there remains a dependency on dataset and models. As a result of this, it is difficult to determine adjacent states' effect in multivariate model. With more training data, we will have better forecasting results. As such random split of dataset improves model predictions. Thus, as the model is trained from maximum number of peaks, better prediction results are obtained. The univariate models work well under above stated conditions. Some of the inherent challenges in forecasting for COVID-19 are mainly due to the nature of infections, reporting of cases and lockdown effects. Despite these challenges, all versions of LSTM produced appreciable results for COVID-19.

COVID-19 has had a severe impact on India's gross domestic product (GDP) since its very inception. It has almost broken the backbone of Indian economy. The present Indian government has been a complete failure in controlling spread of COVID-19 in India. In a huge country of India's dimension, a large portion of inter-state migrant workers and a large portion of population live in rural areas with their extended families. All these factors have made it very difficult to contain COVID-19 infections spread. These issues are very hard to be captured by statistical and mathematical models.

5. Conclusion

In this research work, we performed COVID-19 forecasting in India using various versions of LSTM models. With given limited datapoints, forecasting remains a challenging activity with an appreciable amount of biasedness. Several experiments were performed towards creation of training and testing data and models with varying degrees of strengths and limitations. Certain forecasting results present general decline in new cases. This is mainly attributed to number of limitations and assumptions; it does not address robustness in quantification of uncertainty. The future work looks toward further enhancement of respective models such that robust quantification of uncertainty is achieved. As concluding remarks it is worth mentioning that more model features need to be incorporated for reducing model prediction uncertainty.

References

1. Alzahrani, S.I., Aljamaan, I.A. and Al-Fakih, E.A. (2020). Forecasting the spread of the COVID-19 pandemic in Saudi Arabia using ARIMA prediction model under current public health interventions, *Journal of Infection and Public Health*, 13(7): 914-919.
2. Andersen, K.G., Rambaut, A., Lipkin, W.I., Holmes, E.C. and Garry, R.F. (2020). The proximal origin of SARS-COV-2, *Nature Medicine*, 26(4): 450-452.
3. Anderson, R.H., Heesterbeek, H., Klinkenberg, D. and Hollingsworth, T.D. (2020). How will country-based mitigation measures influence the course of the COVID-19 epidemic? *The Lancet*, 395(10228): 931-934.
4. Arias Velasquez, R.M. and Mejia Lara, J.V. (2020). Forecast and evaluation of COVID-19 spreading in USA with reduced-space gaussian process regression, *Chaos, Solitons and Fractals*, 136: 109924.
5. Atkeson, A. (2020). What will be the economic impact of COVID-19 in the US? Rough Estimates of Disease Scenarios, National Bureau of Economic Research, Technical Report.
6. Bai, Y., Yao, L., Wei, Y., Tian, F., Jin, D.-Y., Chen, L. and Wang, M. (2020). Presumed asymptomatic carrier transmission of COVID-19, *Jama*, 323(14): 1406-1407.
7. Battineni, G., Chintalapudi, N. and Amenta, F. (2020). Forecasting of COVD-19 epidemic size in four hitting nations (USA, Brazil, India and Russia) by Fb-prophet machine learning model, *Applied Computing and Informatics*, 6(1): 1-10.
8. Bengio, Y., Simard, P. and Frasconi, P. (1994). Learning long-term dependencies with gradient descent is difficult, *IEEE Transactions on Neural Networks*, 5(2): 157-166.

9. Bhimala, K. R., Patra, G. K., Mopuri, R. and Mutheneni, S. R. (2021). A deep learning approach for prediction of SARS-COV-2 cases using the weather factors in India, *Transboundary and Emerging Diseases*, 33837675: 8250893.

10. Biavardi, N.G. (2020). Being an Italian medical student during the COVID-19 outbreak, *International Journal of Medical Students*, 8(1): 49-50.

11. Bodapati, S., Bandarupally, H. and Trupthi, M. (2020). COVID-19 time series forecasting of daily cases, deaths caused and recovered cases using long short term memory networks, *IEEE 5th International Conference on Computing Communication and Automation*, 525-530.

12. Chakraborty, T. and Ghosh, I. (2020). Real-time forecasts and risk assessment of novel coronavirus COVID-19 cases: A data-driven analysis, *Chaos, Solitons and Fractals*, 135: 109850.

13. Chandra, R., Jain, A. and Chauhan, D.S. (2021). Deep learning via LSTM models for COVID-19 infection forecasting in India, arXiv, arXiv: 2101:11881v1.

14. Chaudhuri, A. (2021). Initial works on COVID-19 in India, Technical Report, Samsung R & D Institute Delhi, India.

15. Chaurasia, V. and Pal, S. (2020). Application of machine learning time series analysis for prediction COVID-19 pandemic, *Research on Biomedical Engineering*, 1-13.

16. Chimmula, V.K.R. and Zhang, L. (2020). Time series forecasting of COVID-19 transmission in Canada using LSTM networks, *Chaos, Solitons and Fractals*, 135: 109864.

17. Chin, V., Samia, N.I., Marchant, R., Rosen, O., Ioannidis, J., Tanner, M.A. and Cripps, S. (2020). A case study in model failure? COVID-19 daily deaths and ICU bed utilization predictions in New York State, arXiv, arXiv, 2006: 15997.

18. Cho, K., Van Merriënboer, B., Bahdanau, D. and Bengio, Y. (2020). On the Properties of Neural Machine Translation: Encoder-Decoder Approaches, arXiv, arXiv, 1409: 1259.

19. Connor, J.T., Martin, R.D. and Atlas, L.E. (1994). Recurrent neural networks and robust time series prediction, *IEEE Transactions on Neural Networks*, 5(2): 240-254.

20. *COVID-19 Dashboard: Center for Systems Science and Engineering (CSSE)*, Johns Hopkins University. https://coronavirus.jhu.edu/map.html

21. Cucinotta, D. and Vanelli, M. (2020). WHO declares COVID-19 a pandemic, *Acta Bio-Medica: Atenei Parmensis*, 91(1): 157–160.

22. Fernandes, N. (2020). Economic effects of Corona virus outbreak (COVID-19) on the world economy, *SSRN Journal*, 3557504.

23. Forecasting Brazilian and American COVID-19 cases based on artificial intelligence coupled with climatic exogenous variables, *Chaos, Solitons and Fractals*, 139: 110027.

24. Da Silva, R.G., Riberio, M.H.D.M., Kleinubing, J.H., Larcher, V.C.M. and Dos Santos Coelho, L. (2020). Forecasting the cumulative cases of COVID-19 in Brazil using machine learning approaches, *Chaos, Solitons and Fractals*, 135: 109853.

25. Dong, E., Du, H. and Gardner, L. (2020). An interactive web-based dashboard to track COVID-19 in real time, *The Lancet Infectious Diseases*, 20(5): 533-534.

26. Elman, J.L. and Zipser, D. (1988). Learning the hidden structure of speech, *The Journal of the Acoustical Society of America*, 83(4): 1615-1626.

27. Fauci, A.S., Lane, H.C. and Redfield, R.R. (2020). COVID-19 – Navigating the Uncharted, *The New England Journal of Medicine*, 382: 1268-1269.

28. Wang, L., Adiga, A., Venkatramanan, S., Chen, J., Lewis, B. and Marathe, M. (2020). Examining deep learning models with multiple data sources for COVID-19 forecasting, arXiv: 2010.14491v2.

29. Gao, J., Zheng, P., Jia, Y., Chen, H., Mao, Y., Chen, S., Wang, Y., Fu, H. and Dai, J. (2020). Mental health problems and social media exposure during COVID-19 outbreak, *PLoS One*, 15(4): e0231924.

30. Giles, C.L., Omlin, C. and Thornber, K.K. (1999). Equivalence in knowledge representation: Automata, recurrent neural networks and dynamical fuzzy systems, *Proceedings of IEEE*, 87(9): 1623-1640.

31. Gorbalenya, E., Baker, S.C., Baric, R.S., De Groot, R.J., Drosten, C., Gulyaeva, A.A., Haagmans, B.L., Lauber, C., Leontovich, A.M., Neuman, B.W., Penzar, D., Stanley Perlman L.L.M.P., Samborskiy, D.V., Sidorov, I.A., Sola, I. and Ziebuhr, J. (2020). The species severe acute respiratory syndrome-related coronavirus: Classifying 2019-NCOV and naming it SARS COV-2, *Nature Microbiology*, 5(4): 536.

32. Graves, A. and Schmidhuber, J. (2005). Framewise phoneme classification with bidirectional LSTM and other neural network architectures, *Neural Networks*, 18(5-6): 602-610.

33. Hart, E., Hayes, D.J., Jacobs, K.L., Schulz, L.L. and Crespi, J.M. (2020). The impact of COVID-19 on Iowa's corn, soybean, ethanol, pork and beef sectors, Center for Agricultural and Rural Development, Iowa State University, CARD Policy Brief.

34. Hochreiter, S. (1998). The vanishing gradient problem during learning recurrent neural nets and problem solutions, *International Journal of Uncertainty, Fuzziness and Knowledge-based Systems,* 6(2): 107-116.

35. Hochreiter, S. and Schmidhuber, J. (1997). Long short-term memory, Neural Computation, 9(8): 1735-1780.

36. Hochreiter, S. and Schmidhuber, J. (1997). Long short-term memory, *Neural Computation*, 9(8): 1735-1780.

37. Huang, Y., Wu, Y. and Zhang, W. (2020). Comprehensive identification and isolation policies have effectively suppressed the spread of COVID-19, *Chaos, Solitons and Fractals*, 139: 110041.

38. Istaiteh, O., Owais, T., Al-Madi, N. and Abu-Soud, S. (2020). Machine learning approaches for COVID-19 forecasting, *International Conference on Intelligent Data Science Technologies and Applications*, 50-57.

39. Kerimray, A., Baimatova, N., Ibragimova, O.P., Bukenov, B., Kenessov, B., Plotitsyn and Karaca, F. (2020). Assessing air quality changes in large cities during COVID-19 lockdowns: The impacts of traffic free urban conditions in Almaty, Kazakhstan, *Science of the Total Environment*, 730: 139179.

40. Kluge, H.H.P., Jakab, Z., Bartovic, J., Anna, V.D, and Severoni, S. (2020). Refugee and migrant health in the COVID-19 response, *The Lancet*, 395(10232): 1237-1239.

41. Lancet, T. (2020). India under COVID-19 lockdown, *The Lancet*, 395(10233): 1315.

42. Leite, H., Hodgkinson, I.R. and Gruber, T. (2020). New development: Healing at a distance – Telemedicine and COVID-19, *Public Money and Management*, 40(6): 483-485.

43. Liem, C., Wang, Y., Weriyanti, C.A. and Latkin, B.J. Hall (2020). The neglected helath of international migrant workers in the COVID-19 epidemic, *The Lancet Psychiatry*, 7(4): 20.

44. Maleki, M., Mahmoudi, M.R., Wraith, D. and Pho, K.-H. (2020). Time series modeling to forecast the confirmed and recovered cases of COVID-19, *Travel Medicine and Infectious Disease*, 37: 101742.

45. Maliszewska, M., Mattoo, A. and Van Der Mensbrugghe, D. (2020). The potential impact of COVID-19 on GDP and trade: A preliminary assessment, *Policy Research Working Paper Number 9211*, World Bank, Washington.

46. Millett, D.A., Jones, A.T., Benkeser, D., Baral, S., Mercer, L., Beyrer, C., Honermann, B., Lankiewicz, E., Mena, L., Crowley, J.S. *et al.* (2020). Assessing differential impacts of COVID-19 on black communities, *Annals of Epidemiology*, 47: 37-44.

47. Monteil, V., Kwon, H., Prado, P., Hagelkruys, A., Wimmer, R.A., Stahl, M., Leopoldi, A., Garreta, E., Del Pozo, C.H., Prosper, F. *et al.* (2020). Inhibition of SARS-COV-2 infections in engineered human tissues using clinical-grade soluble human ace2, Cell, e7, 905-913.

48. Mosavi, A. (2020). COVID-19 pandemic prediction for Hungary: A hybrid machine learning approach, *Mathematics*, 8(6): 890.

49. Muhammad, S., Long, X. and Salman, M. (2020). COVID-19 pandemic and environmental pollution: A blessing in disguise? *Science of the Total Environment*, 728: 138820.

50. Nadler, P., Arcucci, R. and Guo, Y. (2020). A neural SIR model for global forecasting, *Proceedings of Machine Learning for Health NeurIPS Workshop*, PMLR, 136: 254-266.

51. Omlin, C.W. and Giles, C.L. (1992). Training second-order recurrent neural network using hints, *Proceedings of 9th International Conference on Machine Learning*, Morgan Kaufmann, 363-368.

52. Omlin, C.W., Thornber, K.J. and Giles, C.L.(1998). Fuzzy finite state automata can be deterministically encoded into recurrent neural networks, *IEEE Transactions on Fuzzy Systems*, 6(1): 76-89.

53. Organization, W.H. *et al.* (2020). Coronavirus disease 2019 (COVID-19), *Situation Report*, 72.

54. Owusu-Fordjour, C., Koomson, C. and Hanson, D. (2020). The impact of COVID-19 on learning – The perspective of the Ghanaian student, *European Journal of Education Studies*, 7(3): 88-101.

55. Ren, H., Zhao, L., Zhang, A., Song, L., Liao, Y., Lu, W. and Cui, C. (2020). Early forecasting of the potential risk zones of COVID-19 in China's megacities, *Science of the Total Environment*, 729: 138995.

56. Rajkumar, R.P. (2020). COVID-19 and mental health: A review of the existing literature, *Asian Journal of Psychiatry*, 52: 102066.

57. Saadat, S., Ratwani, D. and Hussain, M.A. (2020). Environmental perspective of COVID-19, *Science of the Total Environment*, 728: 138870.

58. Saba, A.I. and Elsheikh, A.H. (2020). Forecasting the prevalence of COVID-19 outbreak in Egypt using nonlinear autoregressive artificial neural networks, *Process Safety and Environmental Protection*, 141: 1-8.

59. Schmidhuber, J. (2015). Deep learning in neural networks: An overview, *Neural Networks*, 61: 85-117.

60. Schuster, M. and Paliwal, K.K. (1997). Bidirectional recurrent neural networks, *IEEE Transactions on Signal Processing*, 45(11): 2673-2681.

61. Siche, R. (2020). What is the impact of COVID-19 disease on agriculture? *Scientia Agropecuaria*, 11(1): 3-6.

62. Singh, S., Parmar, K.S., Kumar, J. and Makkhan, S.J.S. (2020). Development of new hybrid model of discrete wavelet decomposition and autoregressive integrated moving average (ARIMA) models in application to one month forecast the casualty cases of COVID-19, *Chaos, Solitons and Fractals*, 135: 109866.

63. Sutskever, I., Vinyals, O. and Le, Q.V. (2014). Sequence to sequence learning with neural networks, *Advances in Neural Information Processing Systems*, 3104-3112.

64. Takens, F. (1981). Detecting strange attractors in turbulence, Dynamical Systems and Turbulence, Warwick 1980, *Lecture Notes in Mathematics*, 898: 366-381.

65. Ting, D.S.W., Carin, L., Dzau, V. and Wong, T.Y. (2020). Digital technology and COVID-19, *Nature Medicine*, 26(4): 459-461.

66. Tomar, A. and Gupta, N. (2020). Prediction for the spread of COVID-19 in India and effectiveness of preventive measures, *Science of the Total Environment*, 728: 138762.

67. Wang, H.-Z., Li, G.-Q., Wang, G.-B., Peng, J.-C., Jiang, H. and Liu, Y.-T. (2017). Deep learning-based ensemble approach for probabilistic wind power forecasting, *Applied Energy*, 188: 56-70.

68. Werbos, P.J. (1990). Backpropagation through time: What it does and how to do it, *Proceedings of the IEEE*, 78(10): 1550-1560.

69. Xingjian, S., Chen, Z., Wang, H., Yeung, D.-Y., Wong, W.-K. and Woo, W.-C. (2015). Convolutional LSTM network: A machine learning approach for precipitation nowcasting, *Advances in Neural Information Processing Systems*, 802-810.

70. Yang, Z., Zeng, Z., Wang, K., Wong, S.-S., Liang, W., Zanin, M., Liu, P., Cao, X., Gao, Z., Mai, Z. *et al*. (2020). Modified SEIR and AI prediction of the epidemics trend of COVID-19 in China under public health interventions, *Journal of Thoracic Disease*, 12(3): 165.

71. Ye, F., Xu, S., Rong, Z., Xu, R., Liu, X., Deng, P., Liu, H. and Xu, X. (2020). Delivery of infection from asymptomatic carriers of COVID-19 in a familial cluster, *International Journal of Infectious Diseases*, 94: 133-138.

72. Yousaf, M., Zahir, S., Riaz, M., Hussain, S.M. and Shah, K. (2020). Statistical analysis of forecasting COVID-19 for upcoming month in Pakistan, *Chaos, Solitons and Fractals*, 138: 109926.

73. Zambrano-Monserrate, M.A., Ruano, M.A. and Sanchez-Alcalde, L. (2020). Indirect effects of COVID-19 on the environment, *Science of the Total Environment*, 728: 138813.

74. Zhou, C., Su, F., Pei, T., Zhang, A., Du, Y., Luo, B., Cao, Z., Wang, J., Yuan, W., Zhu, Y. *et al*. (2020). COVID-19: Challenges to GIS with Big Data, *Geography and Sustainability*, 1(1): 77-87.

Deep Learning-based Techniques in Medical Imaging for COVID-19 Diagnosis: A Survey

Fozia Mehboob[1], Abdul Rauf[1], Khalid M. Malik[2], Richard Jiang[3], Abdul K.J. Saudagar[4], Abdullah AlTameem[4] and Mohammed AlKhathami[4]

[1] RISE, Research Institute of Sweden, Sweden
[2] Department of Computer Science & Engineering, Oakland University, Rochester, MI, USA
[3] LIRA Center, Lancaster University, Lancaster LA1 4YW, UK
[4] Information Systems, Imam Mohammad Ibn Saud Islamic University (IMSIU)

1. Introduction

COVID-19 is an infectious disease which causes severe respiratory problems. The coronavirus is very contagious and transferred by indirect and direct link with infected people across respiratory dewdrops [4, 36, 39]. Early diagnosis of COVID-19 is fundamental to decrease the spread of coronavirus. Day-by-day increase in cases of COVID-19 worldwide and limitations of present diagnostic techniques impose challenges in detecting and handling the pandemic. Researchers are trying to find the effective diagnosis measures and accelerate vaccine development and treatment.

Since COVID-19 affects the respiratory system, radiology scans of chest are a significant tool for identification of virus. Currently, polymerase chain reaction (PCR) test is considered the standard method for validating COVID-19. Additionally, PCR testing has high false negative rate and turnaround times [28]. Therefore, other testing tools for COVID-19 identification are considered to alleviate the pandemic's influence on many people's lives. CT scan is an appropriate complement to PCR testing and plays a fundamental role in diagnosing COVID-19 viruses with high sensitivity value. Particularly, the emerging variants of COVID-19 are expected to further increase the need for such tools. In medical domain of imaging, deep learning methods have been employed to significantly increase the performance of image exploration [33, 30], particularly, convolutional neural networks are frequently used for medical imaging [1, 7]. In this paper, we analyze the cutting-edge research contributions of deep learning applications for the diagnosis of COVID-19 and highlight the challenges. We critically assess the published papers between the year

2020-2021 on COVID-19 diagnosis using CT scans and X-ray images. Literature collection and preparation protocol is presented in Table 1.

We explored and analyzed datasets, COVID diagnosis, detection, classification, and experimental findings which can be helpful to identify the future research directions in the domain of automatic detection of COVID-19 disease. It is also reflected that there is a crucial need for comprehensive, public, and diverse COVID-19 datasets. Deep learning is considered an emergent field which can play a fundamental role in COVID-19 detection in future. Till now, several researchers have used deep learning methods, transfer learning, etc. for detection of COVID-19 using X-ray or CT scan images and have got promising results. However, researchers are finding improved and advanced architectures for COVID-19 diagnosis. In this survey paper, we have reviewed these new methods or techniques alongside with implementation of these methods for fair comparison. There is no recently published survey paper which focuses on implementation of existing researchers' works for diagnosis of COVID-19. Table 1 describes the literature collection and preparation protocol.

Table 1: Literature collection and preparation protocol

Purpose	To provide a brief overview of existing state-of-the-art methods and identify potential gaps in COVID-19 detection, using CT scan, X-ray images.
Data sources	Google Scholar, Springer, IEEE explorer, Nature journal
Method & Query	Methodological approach was designed, based on data-sources mentioned above and following query strings were used: COVID-19 detection/COVID-19 prediction/AI and COVID-19 detection/deep learning and COVID-19 prediction/ deep learning and COVID-19 diagnosis/deep learning and deep convolutional neural network/deep transfer learning and CT scan images/X-ray images/ deep learning.
Size	A total of 200 papers were retrieved using method and query mentioned above from listed data sources till June, 2021. We have selected only those studies that were relevant and passed the criteria to be 'COVID-19' in the positive set. All other studies were excluded from the final selected papers.
Study types	Journal papers and conference studies were given more importance.

The main contribution of this survey paper are as follows:

- Summarize the work of existing researchers on automatic diagnosis of COVID-19 so that new researchers can identify the knowledge gaps, and can perform comparative analysis of these existing approaches.
- Implement the existing approaches for fair comparison.
- Accumulate different sources of medical images datasets of CT scan or X-ray images.

Fig. 1 illustrates the different methods used for diagnosis for COVID at each stage. The rest of the paper is organized as follows: Section 2 presents a detailed analysis of AI-based approaches for COVID-19 diagnosis. It includes the comprehensive analysis of architectures employed, different deep learning methods and datasets used. Section 3 briefly discusses the performance comparisons of several existing deep learning models with the proposed method. The available datasets used for experimental purposes are also described. In Section 4, we discuss the possible future trends of COVID detection techniques, and finally, we conclude our work in Section 5.

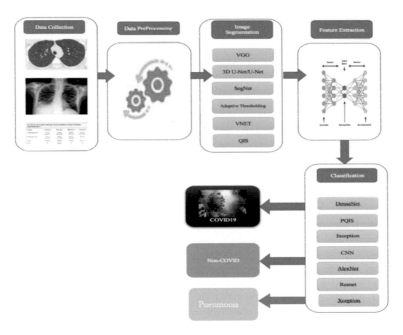

Fig. 1: Pipeline of COVID-19 diagnosis using deep learning techniques

2. AI-based Approaches in Detection of COVID-19 Pandemic

Artificial Intelligence (AI) is all about mechanically replicating the process of the human way of thinking and acting. The idea, like many other big discoveries, came from classical philosophers and is, as such, still a controversial topic in philosophy [21]. The history of Artificial Intelligence is structured into four seasons – spring, fall, summer, and winter [42]. The first real occurrence of Artificial Intelligence goes back to the early 1940s in which the first robot was developed by Gregory Powell and Mike Donavan [42]. The two engineers tried to create a robot that could react on specific orders that were given by humans, thereby following the Three Laws of Robotics that were proposed by Isaac Asimov – to not harm a human being, to obey given orders, and to not put its own existence at risk. At the same time, Alan

Turing, one of the still most influential mathematicians in testing the intelligence of machines, created a machine that was able to decipher the Enigma code that was used by the German military in World War II, which was, until then, an impossible task. In 1950, Alan Turing published his still used and famous Turing Test that considers whether an artificial system can be called intelligent or not; as soon as an artificial system cannot be distinguished from a human, it can be considered to be intelligent.

Table 2: Overview of AI-based approaches

Proposed Model	Accuracy	Dataset	Samples
CNN, MLP, SVM, Random Forest [27]	0.92	-	CT
DenseNet Transfer Learning [17]	-	-	X-Rays
Artificial intelligence-based system [32]	0.9869	Tianchi-Alibaba, LIDC–IDRI,CC- CCII, MosMedData	CT

2.1 Deep Learning-based Approaches using CXR

An Artificial Intelligence-based system was proposed [32] for COVID-19 detection and extensive statistical analysis of CT scans [26, 39] of COVID-19 was performed. Proposed system was evaluated on a large dataset from influenza, COVID-19, non-pneumonia, and pneumonia. Datasets were collected from three different areas of Wuhan, and also used four publicly available databases – Tianchi-Alibaba, LIDC-IDRI, CC-CCII and MosMedData. A diagnosis system based on deep neural network was designed which was trained and tested using datasets from three centers of China. The performance of the proposed system was also compared with experienced radiologists and it revealed that the proposed system outperformed than radiologists in two tasks, with 0.9869, 0.9585 and 0.9727 accuracy for pneumonia or non-pneumonia, influenza or COVID-19 and CAP or COVID-19 [7] tasks. In the reader study, radiologists' average reading time was 6.5 min, although AI system time was 2.73s. that can considerably increase the radiologist's productivity. In addition, lesion area [38] in COVID-19 patients was located, based on CT scan prediction score. Proposed AI system was assessed on the test cohort. Experimental results demonstrated that non-pneumonia attains 0.9752, CAP=0.9804, COVID-19=0.9745 and influenza 0.9885 accuracy. The multi-way accuracy of proposed system was 0.9781 on test cohort whereas 0.9299 and 0.9325 accuracy were achieved on publicly available CC-CCII database and MosMedData database. Major limitation of the proposed system was the generalisability of the AI model to other populations. Few examples were shown in which AI models were cross-referenced with radiologists' findings, though, visualization failed in several examples to provide a clear explanation. Consequently, proposed algorithm was helpful in places having high COVID-19 disease, but is still unlikely to be useful in places having low COVID-19 prevalence. The author proposed an approach based on AI [17] to automatically classify COVID-19 by using CT dataset. Proposed approach employed DenseNet in order to perform classification of COVID-19. Data augmentation and transfer learning were used to train the deep learning pre-trained model with new chest X-ray images. The motivation behind transfer learning was to reuse a developed model for training the novel model with a dissimilar task. Data augmentation technique was performed to increase training dataset size in order to increase the ability and performance of the deep learning model.

A deep convolutional neural network COVID-Net [29] was designed for the detection of COVID-19 from CXR images. A benchmark open-access dataset called COVIDx was generated which comprises of 13,975 CXR images by modifying and combining five chest X-ray data repositories. COVIDx was utilized for training and evaluation of COVID-Net. COVID-Net initial network design makes three predictions, such as no infection (normal), COVID-19 viral infection and non-COVID-19 infection (bacterial, viral, etc.). COVID-Net was pre-trained on the ImageNet dataset and afterward trained on the COVIDx dataset. To evaluate the efficacy of the proposed COVID-Net, we performed both qualitative and quantitative analysis to evaluate the efficacy of the proposed COVID-Net. To inspect the COVID-Net quantitatively, test accuracy, sensitivity, and positive predictive value (PPV) were computed for each infection type and COVID-Net achieved 93.3% test accuracy.

Two deep neural network architectures, namely VGG-19 and ResNet-50, were compared with COVID-Net. It was observed that COVID-Net had lower computational complexity and architectural complexity than the VGG-19 and ResNet-50 architectures. Furthermore, results demonstrated the benefits of high architectural diversity as COVID-Net achieved higher test accuracy and sensitivity than the VGG-19 and ResNet-50 network architectures. Therefore, based on these results, it can be seen that while COVID-Net performs well as a whole in detecting COVID-19 cases from CXR images, there are several areas of improvement that can benefit from collecting additional data, as well as improving the underlying training methodology to generalize better across such scenarios. In order to evaluate COVID-Net qualitatively, an audit was performed which was based on the GSInquire interpretation [29]. The main aim of qualitative analysis was to validate that COVID-Net approach is not making decision based on erroneous information. Based on GSInquire interpretation, the COVID-Net primarily focused on infected lung areas and it was validated that COVID-Net was not relying on inaccurate visual information to make decisions.

A deep learning convolutional neural network, Deep COVID DeteCT (DCD) was developed [21] to automatically predict COVID-19. DCD model was trained and evaluated in 13 international institutions around the world. DCD performance was evaluated with pixel value variations. Participants from countries other than China were investigated using Deep COVID DeteCT method. They trained the model on different sites and tested them using non-China sites. Testing results were compared with the model trained on China sites, illustrating high accuracy except for two COVID–PNA accuracies. Furthermore, they also examined the fine-tuning effects on model performance. Fine tuning was performed on few patients from test site. It was observed that fine tuning increased the performance of model from 73.2% to 80.6%. DCD, a 3D model, tracks and diagnoses the disease course over hospitalization exclusive of complex pre-processing. The DCD model comprises of a 27-layer and I3D-Inception feature extractor. Additionally, Inception model was pre-trained on video dataset Kinetics-600. To predict prognosis, the DCD's convolutional model computed the features from using two consecutive scans and concatenated these features. Additionally, these features were given as input to two fully-connected layers. Prognosis was quantified by the length of hospitalization, which was measured when the scan was taken to discharge time. A longer time of hospitalization was indication of worse prognosis. Time was predicted as binary classification problem instead of regression. Proposed model was generalizable, but creation of prognostic model posed intrinsic challenges, such as complicated clinical variables.

Another latest effort reported on using 10 pre-trained convolutional neural network models for COVID-19 CT scans classification [18]. This study stated that Xception and ResNet-101 delivered best classification accuracy on CT dataset training and testing. Additionally, earlier works on COVID-19 CT scans classification were reported in several studies, such as in [3, 19, 22]. A 3D deep NN known as COVNet was designed for recognition of COVID-19 from chest CT scans [19, 41, 42]. The proposed model was built on pre-trained RestNet50. Both 2D and 3D features were extracted by network from CT scans. Researchers conducted an investigation study [2] on classification of COVID-19, using 16 pre-trained CNNs models. A large dataset of CT scans was collected for the experimental purpose. These pre-trained CNN models are GoogLeNet, SqueezeNet, DenseNet-201, Inception-v3, MobileNet-v2, ResNet-50, ResNet-18, ResNet-101, Inception-ResNet-v2, Xception, ShuffleNet, NasNet-Large, NasNet-Mobile, AlexNet, VGG-19, and VGG-16. These networks were trained on ImageNet database images. Pre-trained CNN models revealed a variety in accuracy and computational complexity on the basis of training and testing. For the data augmentation purpose, translation, scaling, and random reflection were carried out. For the comparison of results, the dataset was arbitrarily split into 20% for testing and 80% for training. The splitting of data was repeated five times for obtaining standard deviation and average for each CNN model. Experimental results revealed that only six epochs are required for training the CNN model, resulting in very high classification performance. Amongst the 16 CNNs models, DenseNet-201 achieves high accuracy, sensitivity, and specificity value, and area under curve. Moreover, transfer learning with whole image slices and without data augmentation delivered better classification accuracy than the using data augmentation. In case of training using data augmentation, DenseNet-201, ResNet-18, ShuffleNet, MobileNet-v2 give the average accuracy above 95%; however, DenseNet-201 attains overall highest accuracy of 96.20%.

GoogLeNet, ResNet-18, ShuffleNet, MobileNet-v2, ResNet-101, ResNet-50, DenseNet-201, and Inception-v3 result in average sensitivity above 95%, whereas ResNet-18 achieves average sensitivity of 98.99%. Results obtained without using augmentation illustrate that ResNet-18, ShuffleNet, MobileNet-v2 and DenseNet-201 result in the highest average F1 score of 0.96. DenseNet-201 was considered as best model for COVID-19 CT data classification without data augmentation.

A semi-supervised neural network model 41] was proposed which comprises PQIS-Net for lung CT images segmentation. The proposed model provides automatic lung CT slices segmentation without integrating pre-trained CNN-based models. In order to perform the classification task in proposed shallow NN-based framework, patch-based classification in segmented slice of PQIS-Net was incorporated. Proposed model was evaluated on publicly available dataset of Brazilian dataset and IEEE CCAP dataset. This dataset comprised five different lung CT images, such as COVID-19, bacterial pneumonia, viral pneumonia, normal lung, and mycoplasma pneumonia.

The IEEE CCAP and Brazilian dataset were divided into 7: 3 ratios for model training, testing, and validation purposes. In addition, fivefold cross-validation was used for the experiments. In order to classify the images, experimentation was designed using ResNet50, proposed model and 3D-UNet model in which the last layer was replaced. Extensive experimentation was performed with the different datasets. Segmentation performance of proposed PQIS Net, 3D-Unet and ResNet50 on these datasets was measured using dice similarity (DS). It was observed that the proposed model performed best in patch-based classification having FC layer. The accuracy achieved by proposed system was similar to ResNet50 whereas precision was similar to 3D-Unet. In addition, the proposed shallow neural network outperformed in KS-test outcome. It was significant that the proposed model using PQIS-Net performed better than 3D-Unet in terms of recall, accuracy, and F1-score on the Brazilian dataset. It was noticed that PQIS-Net using patch-based classification and fully-connected (FC) layers was better than only classification using patch-based and FC layer. Although experimental results revealed that 3D-Unet and ResNet50 slightly outperformed than the proposed model in the segmentation task.

In order to enhance the detection of CT imaging, CovidCTNet method [16] was proposed. The proposed method comprises of several deep learning algorithms which accurately distinguish COVID-19 from pneumonia and other diseases of lungs. CovidCTNet was designed to train on small and heterogeneous sample-size datasets and BCDU-Net as a backbone model. Moreover, SPIE-AAPM-NCI lung-nodule dataset containing 70 CT scans was used. CT images were obtained from various institutions, including five centres in Iran. For identification of COVID-19 in the lungs, pseudo-infection abnormalities were generated in the CT images using BCDU-Net. A validation experiment was conducted to test whether employing Perlin noise and usage of BCDU-Net was essential for pre-processing and tended to model accuracy. The 3D CNN model was implemented with and without usage of Perlin noise and of BCDU-Net. Although model accuracy without using the Perlin noise and BCDU-Net at training stage was very high, but it decreased significantly in validation stage. This endorsed that parameters and features that were chosen by CNN model without using BCDU-Net were insufficient. CovidCTNet correctly identified COVID-19 from other diseases of the lungs. The algorithm output was given as input to CNN algorithm for classification. At validation stage, the area under ROC curve for COVID-19 was 94%, with 93.33% accuracy while CNN classified between two classes, such as COVID-19 and non-COVID-19. Furthermore, CNN achieved 86.66% accuracy when classification was performed between three classes, COVID-19, pneumonia and control. The specificity value of 100% and sensitivity 90.91% were noted for COVID-19. In order to test classification accuracy of the proposed method, a dataset containing 20 cases of control, CAP and COVID-19 were evaluated, using proposed framework and four certified radiologists. The performance accuracy of radiologists was 81%, though CovidCTNet classification model achieved accuracy of 95% while classifying between non-COVID-19 versus COVID-19 classes. In case of detection of three classes, COVID-19, CAP and control, the proposed approach outperformed the radiologists having 85% accuracy as compared to radiologists with 71% accuracy. However, the study was just evaluated on small dataset and mostly they were Iranian patients.

Based on time series-based convolution kernel, lesion detection model from CT images was proposed [9]. The author used a deep learning and image classification-based methods for COVID-19 CT images of patients. Algorithm was compared with Faster RCNN, SSD, and YOLO3 models. This paper studies the CT images from COVID-19 patients taken from different hospitals. The COVID-19 CT lesions detection method was a combination of three parts – object detection, target area selection, and feature extraction using specific convolutional layers. In order to validate the relationship among the complexity of two-way and one-way time series convolution, experimentation was performed which included two models with time-sequence difference. The algorithm used for testing purpose included YOLO3, faster R-CNN, SSD, and proposed space-time convolution model. For experimentation purpose, the test was split into two stages. In addition, each patient's image was gathered at one-time point at first stage, whereas in the second stage, patients with multistage image data were tested. It was revealed from the experiments that YOLO3 achieved high accuracy and sensitivity value. In both structures, the model proposed revealed superior sensitivity and accuracy value than the first stage. Through analysis of experimental data characteristic and model structure it was observed that when same patient had multiple time-image data, the proposed time series convolutional kernel could extract latent image semantic features.

Another study proposed a CNN architecture, known as ResCovNet [31] to detect COVID-19 from chest CT images. Otsu's thresholding was used for pre-processing task. In this study, IEEE Data port, a publicly available CT scan dataset was utilized. The CT volumes were composed of data from multi-center hospital having five categories: viral pneumonia, COVID-19, bacterial pneumonia, normal and mycoplasma pneumonia. At the training stage, data augmentation was performed. In classification task, ResNet152V2 gave the best result and was applied as basic network for the proposed model. Comparing the proposed method with other mechanisms illustrated promising results for diagnosing COVID-19 by distinguishing it from three different types of normal and pneumonia cases. A three-fold cross-validation was employed for performance evaluation. By using the proposed architecture, 88.1% of classification accuracy was achieved. There were few wrong predictions in viral and bacterial pneumonia. These categories were wrongly anticipated as normal cases and COVID-19 respectively.

The author performed a comparative study [34] for identification of a suitable CNN model. VGG19 model was optimized for image modalities in order to show how models are utilized for challenging and highly scarce COVID-19 datasets. The study also demonstrated the challenges in employing current COVID-19 publicly-available datasets for deep learning models and how they impact complex models' trainability. An image pre-processing stage was proposed for creating image dataset in order to develop and test the deep learning models. In selection of suitable deep learning model for multi-model image classification, several CNN models, known as VGG16 and VGG19, Inception V3, Resnet50, Xception, DenseNet, Inception, ResNet and NASNetLarge, were employed. Deep learning [7, 40] techniques have been employed in the medical field for image segmentation that become prevalent diagnostic tools because of key feature demonstration. In the year 2020, various deep learning neural networks were used for COVID-19 detection and contributed promising performance in terms of accuracy [10]. Wang *et al.* [40] introduced a deep learning model with pre-trained U-Net for detection of COVID-19, using lung CT scans. The method reported good sensitivity, accuracy, and specificity value. Yan *et al.* [25] proposed a CNN model presenting Spatial Pyramid Pooling in order to tackle refined infected lesions with wide shape variations, orientation, and overlapping of lung CT scans. Conversely, due to insufficient lung CT images annotation and lack of image adaptability for unanticipated lung image classes, the pre-trained convolutional network model failed to accomplish the required accuracy.

To evade the problem of over-fitting during the process of CNN training, having small datasets, latent representations examining various features of lung CT images was suggested by Kang *et al.* [10]. Additionally, Wang *et al.* [11] introduced a 3D ResNets neural network model. A deep 3D model comprising manifold learning with CT images of chest was suggested for COVID-19 screening [5]. In spite of comparatively less complicated models for COVID-19 CT image segmentation, all these methods relied on exhaustive feature learning through the training process. Authors then designed QIS networks [35] and QIBDS Net [12] for automated segmentation of brain lesion. They also proposed the QIBDS Net optimized form [6], known as Opti-QIBDS Net, that is appropriate for segmentation of brain tumor. In [3], Inception, a pre-trained model, was modified for detection of COVID-19 by using CT scans ROIs which were taken from SARs-COV-2. In another study [22], extraction of respiratory regions was performed from CT scans. ResNet-18 model was employed for extraction of features from images. Lastly, Bayesian function was utilized for image regions classification into COVID-19, irrelevant to infection and influenza, though the dataset used in these studies was not publicly available.

2.2 X-ray-based Studies in Detection of COVID-19 Pandemic

A comprehensive examination was performed in [26] which used 16 CNNs pre-trained models for COVID-19 classification, using CT scans publicly available database. These pre-trained models revealed a variety of accuracy and computational complexity on the basis of training and testing ImageNet database. This investigation findings would enable timely disposition of AI-aided tools for hospitals to fight against the pandemic. X-ray, CT image and ultrasound samples from COVID-19 datasets was used from multiple sources. ImageNet weights was employed by each of the classifiers for transfer learning. In initial results of testing, VGG19 model was chosen for multimodal image classification. From experimentation results, it was found that ultrasound provides best results with 100% sensitivity compared to X-ray with 86% sensitivity value. They demonstrated that VGG19 model could be utilized to develop deep learning based tools for detection of COVID-19. Obtained results from experiments highlight that VGG19 model outperforms in detection of

Table 3: Overview of deep learning-based approaches

Proposed Model	Accuracy	Dataset	Samples
COVID-Net [29]	93.3%	ImageNet dataset, COVIDx dataset	CXR images
I3D-Inception, DeteCT (DCD) [21]	80.6%.	–	CT
DenseNet-201 [42]	96.20%	ImageNet dataset,	CT
PQIS-Net(patch-based classification) [41]	98.4%	Brazilian data set and IEEE CCAP data set.	CT
BCDU-Net [16]	95%	SPIE-AAPM-NCI lung nodule dataset	CT
Space Time1, Space Time2 [9]	95%, 96.7%		CT
ResCovNet [31]	88.1%	IEEE Data port	
Optimized VGG19 model [34]	86%, 100%, 84% for X-Ray, Ultrasound and CT	–	X-Ray, CT image and Ultrasound
Multi-view machine learning technique [10]	95.5%	–	CT
Weakly Supervised UNet [40]	95.9%	–	CT
COVIDSegNet [25]	DICE & Precision 0.726 %, 0.726%	–	CT
3D-ResNets, 3D U-Net [11]	95.7% and 97.3%,	–	
AD3D-MIL [5]	97.9%	–	CT
QIS-Net [35]	0.99%	Nature data sets	MRI
Bi-Opti-QIBDS Net [12]	0.987%	Brain Tumor Dataset	MRI
CAD system based on deep learning [18]	Resnet-101 (99.51%) Xception (99.02%).	16-MDCT scanner (Alexion, Toshiba Medical System, Japan	CT
Inception transfer-learning model [3]	89.5%	–	CT
3-dimensional deep learning model [22]	86.7%	–	CT
Opti-QIBDS Net [6]	0.97%	Brain tumor dataset	MR
COVIDNet CT-2[28]	96.2%/	–	CT
Deep CNN-based system [37]	97.81%	LIDC–IDRI31, Tianchi-Alibaba32, MosMedData33 and CCCCII	CT
SegNet and U-NET [2]	SegNet 0.95%, U-NET shows 0.91 %	http:// medicalsegmentation. com/covid19/.	CT

COVID-19 against normal or pneumonia for all the three lung image modes. The VGG19 model achieved precision of 86%, 100%, and 84% for X-ray, ultrasound and CT scans, respectively. Experimental results also suggested that ultrasound images offered better detection accuracy as compared to CT scans and X-ray. Automatic diagnosis method for COVID-19 using X-ray images was proposed in [13]. A depth-wise CNN-based method was presented for chest X-ray analysis. The proposed method was designed to predict different classes, such as viral, normal, COVID-19, and pneumonia. The dataset was spitted into training and test set with the ratio of 80% and 20%. In addition, effectiveness of the proposed approach was achieved by comparing it with exiting methods, such as DarkCovidNet, EfficientNet and DeTraC-ResNet18. Experimental results revealrf that DarkCovidNet and EfficientNet achieved 93% accuracy; however, 95.12% accuracy was achieved by DeTraC-ResNet18. Moreover, the proposed method improved the accuracy to 95.83%.

3. Experimental Results, Discussion, and Analysis Datasets

The dataset used for experimental purpose is taken from several databases, such as COVIDx dataset [29], COVID+ datasets from China [3], COVID CT scan images datasets from Tongji Hospital, Wuhan, China [42], COVIDx CT-2 benchmark datasets [28] and Italian Society of Medical and Interventional Radiology dataset [2].

The Covid-net approach proposed in [29] is applied on COVIDx dataset to regenerate the results for the comparison purpose. It can be perceived that proposed method COVID-net [29] achieves accuracy, which indicates incredibly few false COVID-19 detections. As we can see in Fig. 3, one non-COVID-19 patient was mistaken as COVID-19 infection and three COVID-19 patients were interpreted as pneumonia infections. High predictive value is vital, given that numerous false positives values would require additional PCR testing. Conversely, additional dataset was required to generalize the proposed COVID-19 detection methodology.

Another study [28] designed a COVID diagnosis model which utilized CT scans for detection of COVID-19. To investigate the proposed COVID-Net CT-2 model's efficacy, quantitative evaluation was performed. Proposed COVID- Net CT2 model code was used to regenerate the results using datasets comprising normal, COVID, pneumonia category images. It becomes in Fig. 2 that algorithm misclassified samples of all categories. It can be seen that 13 patients were misclassified as pneumonia cases whereas 58 normal patients were misclassified as pneumonia cases, as shown in Fig. 4. In addition, 116 pneumonia patients were misclassified as COVID patients. False positive rate was quite high in the case of pneumonia detection; however it was good at COVID-19 detection.

The domain of medical imaging has substantially emerged last year in providing reliable automatic methods for clinical decision making. Researchers also employed 3D models for automated diagnosis of COVID-19. One of the studies [21] presents a 3D-based COVID-19 detection model where CT scans were utilized for prognostic purposes. For diagnosis purposes, we gave the COVID positive CT scans of patients to proposed algorithm. Proposed algorithm was able to differentiate between the normal and COVID-19 CT scans of patients.

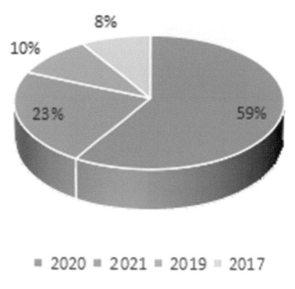

Fig. 2: The Number Approaches Year-wise

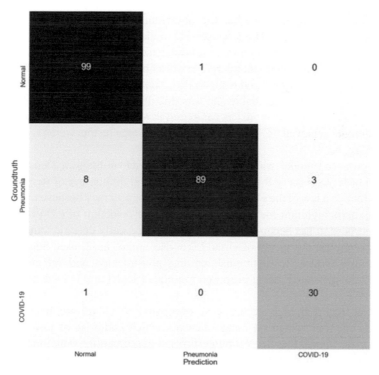

Fig. 3: Confusion matrix of proposed method [2]

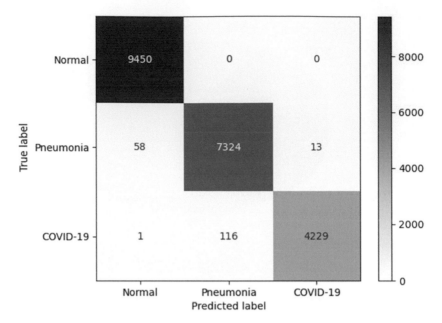

Fig. 4: Confusion matrix of proposed method [34]

In Fig. 5, it can be seen that the algorithm predicts the class along with probabilities of input image. The proposed 3D model was good in terms of accuracy; however, a big sample size was required for the detection purpose. The authors in [42] used 10 pre-tained convolutional neural networks by employing diverse COVID-19 datasets having same training and testing ratio. The dataset used for experimentation purposes was not publicly available and was imbalanced. ResNet-101 model proved to be best as a classification model as compared to other pre-trained network models. Proposed model achieved 95% validation accuracy when results were regenerated as shown in Fig. 6.

The proposed model was applied to test dataset, utilizing a Denset network model. In order to classify the image, six epochs and 30 number of iterations were required, though a few of the classifiers resulted in low accuracy because of imbalance dataset. Experts highlight the significance of earlier detection of COVID-19 virus. For this purpose, it has become imperative to develop early detection and diagnosis methods to prevent the spread of virus. Likewise, author in [2] used SegNet, U-Net deep learning models to distinguish among non-infected and infected COVID cases. Results generated, using proposed methods (SegNt and U-Net) are shown in Figs. 7 and 8.

SegNet achieved high accuracy as compared to U-Net architecture model. SegNet model was trained on image datasets which comprise of just lung areas. The author mentioned that SegNet outperformed in segmenting only the lung region than the infectious region. It was claimed that COVID-19 lesions were considered difficult to discriminate from the chest wall. However, increase in batch size results in degrading network performance.

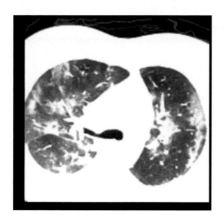

Fig. 5: Diagnosis of COVID using proposed method [21]. Class probabilities: [2.0716771e-09 5.7085012e-06 9.9999428e-01], class prediction: 2

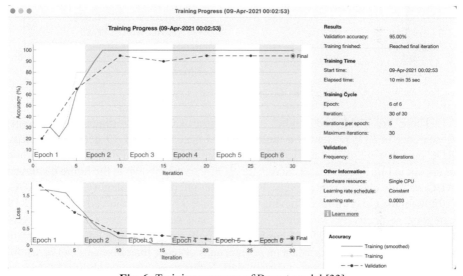

Fig. 6: Training progress of Denset model [33]

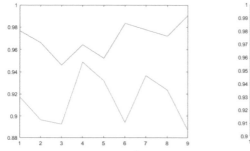
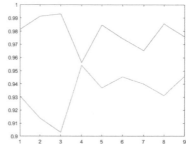

Figs. 7 and 8: Left: Unit2 accuracy result; Right: Segment2 accuracy result

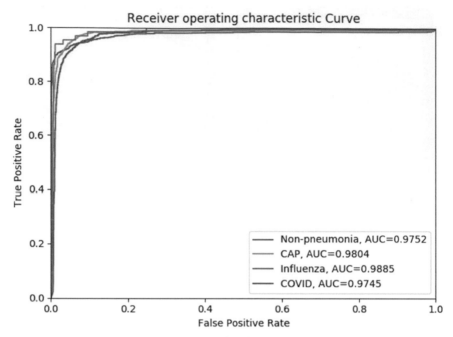

Fig. 9: ROC curve of AI-based system [37]

There are published studies on AI-based COVID-19 diagnostic systems. Researcher in [37] presented a multi-class AI-based COVID-19 detection system. ROC curves were plotted to evaluate the detection accuracy of the proposed method as shown in Fig. 9. On the test dataset, the ROC curve as shown in the figure displayed AUC of four classes respectively. For the non-pneumonia category, 0.9752 AUC, 0.9804 for the CAP, influenza having 0.9885 and 0.9745 was achieved for COVID-19 category, respectively. Moreover, specificity and sensitivity for COVID-19 were 0.9660 and 0.8703, and 0.9781 has been for multi-way AUC. The study had certain shortcomings, such as more data on pneumonias or other lung diseases can be useful for exploring the DL/AI-based system with better diagnostic capability. Moreover, Grad-CAM can extract only the attention region alternatively lesion segmentation, while phenotype feature analysis would better be done on accurate segmentation. The accuracy of all approaches employed in several studies for the diagnosis of COVID are also plotted in the graph as shown in Fig. 10. It was observed that PQIS-Net, U-Net, Xception, Resnet101 achieved accuracy of more than 95%. However, Alexnet achieved the lowest accuracy in automatic diagnosis of COVID.

4. Conclusion

In this study, a comprehensive review of various deep learning methods used for COVID-19 detection is presented. Furthermore, pre-trained CNN architectures and datasets that were employed by different researchers are discussed. For a fair comparison, these models were implemented to and recent challenges related with

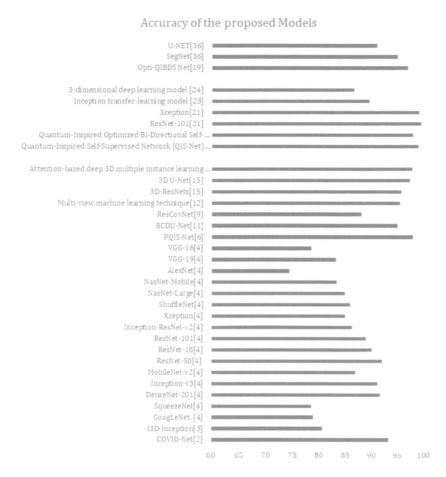

Fig. 10: Accuracy of employed methods

current techniques are highlighted. This survey revealed the substantial potential of deep learning-based methods in automatic detection of COVID-19, using available datasets. A public, diverse and comprehensive image datasets need to be established that are authenticated by experts and annotated with equivalent lesions of lung diseases. As coronavirus is spreading at a rapid rate worldwide, faster and accurate detection has become crucial. In addition, these models are not proved to be reliable as RT-PCR test and researchers are trying to make improvements in these detection techniques. From a survey, it becomes obvious that X-ray images dataset is commonly available than CT scans dataset as CT scans are more time-consuming and costly. So, several researchers employed X-ray images datasets for diagnosing COVID-19. By analyzing numerous research studies in this domain, it was found that there is a lack of sufficient annotated images of coronavirus-affected people because annotated images can have a substantial role in boost up of model's performance. Moreover, it has been revealed by survey, that model performance has been affected by segmentation process of images.

Acknowledgement

The authors extend their appreciation to the deputyship for research and Innovation, Ministry of Education in Saudi Arabia for funding this research work through project number 959.

References

1. Altaf, F., Islam, S.M.S., Akhtar, N. and Janjua, N.K. (2019). Going deep in medical image analysis: Concepts, methods, challenges, and future directions, *IEEE Access*, 7: 99540-99572.
2. Anwar, S.M., Majid, M., Qayyum, A., Awais, M., Alnowami, M. and Khan, M.K. (2018). Medical image analysis using convolutional neural networks: A review, *J. Med. Syst.*, 42(11).
3. Ardakani, A.A. *et al.* (2020). Application of deep learning technique to manage COVID-19 in routine clinical practice using CT images: Results of 10 convolutional neural networks, *Comput. Biol. Med.*, 121: 103795.
4. Chan, J.F.W., Yuan, S., Kok, K.H., To, K.K.W., Chu, H., Yang, J., ... and Yuen, K.Y. (2020). A familial cluster of pneumonia associated with the 2019 novel coronavirus indicating person-to-person transmission: A study of a family cluster, *The Lancet*, 395(10223): 514-523.
5. Chen, N., Zhou, M., Dong, X., Qu, J., Gong, F., Han, Y., Qiu, Y., Wang, J., *et al.* (2020). Epidemiological and clinical characteristics of 99 cases of 2019 novel coronavirus pneumonia in Wuhan, China: A descriptive study, *The Lancet*, 395(10223: 507-513.
6. Chollet, F. (2017). Xception: Deep learning with depthwise separable convolutions. *In: Proceedings of the IEEE Conference on Computer Vision and Pattern Recognition*, pp. 1251-1258.
7. Dastider, A.G., Subah, M.R., Sadik, F., Mahmud, T. and Fattah, S.A. (2020, November). ResCovNet: A Deep Learning-based Architecture for COVID-19 Detection from Chest CT Scan Images. *In: 2020 IEEE Region 10 Conference (TENCON)*, pp. 57-60.
8. Ghinai, I., McPherson, T.D., Hunter, J.C., Kirking, H.L., Christiansen, D., Joshi, K., Rubin, R., Morales-Estrada, S., Black, S.R., Pacilli, M. *et al.* (2020). First known person-to-person transmission of severe acute respiratory syndrome coronavirus 2 (SARS-CoV-2) in the USA, *The Lancet*, 395(10230): 1137-1144.
9. Gunraj, H., Sabri, A., Koff, D. and Wong, A. (2021). COVID-Net CT-2: Enhanced Deep Neural Networks for Detection of COVID-19 from Chest CT Images Through Bigger, More Diverse Learning, arXiv preprint arXiv:2101.07433
10. Han, Z., Wei, B., Hong, Y., Li, T., Cong, J., Zhu, X., Wei, H. and Zhang, W. (2020). Accurate screening of COVID-19 using attention-based deep 3D multiple instance learning, *IEEE Trans. Med. Imag.*, 39(8): 2584-2594. doi: 10.1109/tmi.2020.2996256
11. Horry, M.J., Chakraborty, S., Paul, M., Ulhaq, A., Pradhan, B., Saha, M. and Shukla, N. (2020). COVID-19 detection through transfer learning using multimodal imaging data, *IEEE Access*, 8: 149808-149824.
12. Huang, C., Wang, Y., Li, X., Ren, L., Zhao, J., Hu, Y., ... and Cao, B. (2020). Clinical features of patients infected with 2019 novel coronavirus in Wuhan, China, *The Lancet*, 395(10223): 497-506.
13. ImageNet.http://www.image-net.org

14. Javaheri, T., Homayounfar, M., Amoozgar, Z., Reiazi, R., Homayounieh, F., Abbas, E. and Rawassizadeh, R. (2021). COVID-CTNet: An open-source deep learning approach to diagnose COVID-19 using small cohort of CT images, *NPJ Digital Medicine*, 4(1): 1-10.

15. Jin, C., Chen, W., Cao, Y., Xu, Z., Tan, Z., Zhang, X., ... and Feng, J. (2020). Development and evaluation of an artificial intelligence system for COVID-19 diagnosis, *Nature Communications*, 11(1): 1-14.

16. Jin, C., Chen, W., Cao, Y., Xu, Z., Tan, Z., Zhang, X. and Feng, J. (2020). Development and evaluation of an artificial intelligence system for COVID-19 diagnosis, *Nature Communications*, 11(1): 1-14.

17. Kang, H., Xia, L., Yan, F., Wan, Z., Shi, F., Yuan, H., Jiang, H., Wu, D., Sui, H., Zhang, C. and Shen, D. (2020). Diagnosis of coronavirus disease 2019 (COVID-19) with structured latent multi-view representation learning, *IEEE Trans. Med. Imag.*, 39(8): 2606-2614. doi: 10.1109/TMI.2020.2992546

18. Konar, D., Panigrahi, B.K., Bhattacharyya, S., Dey, N. and Jiang, R. (2021). Auto-diagnosis of COVID-19 using lung CT images with semi-supervised shallow learning network, *IEEE Access*, 9: 28716-28728.

19. Konar, D., Bhattacharyya, S., Gandhi, T.K. and Panigrahi, B.K. (2020). A quantum-inspired self-supervised network model for automatic segmentation of brain MR images, *Appl. Soft Comput.*, 93: 106348. doi: 10.1016/j.asoc.2020.106348

20. Konar, D., Bhattacharyya, S., Dey, S. and Panigrahi, B.K. (2019). Opti-QIBDS Net: A quantum-inspired optimized bi-directional self-supervised neural network architecture for automatic brain MR image segmentation. *In: Proc. TENCON–IEEE Region 10th Conf. (TENCON)*, 11942: 87-95. doi: 0.1109/tencon.2019.8929585

21. Konar, D., Bhattacharyya, S., Dey, S. and Panigrahi, B.K. (2019). Opti-QIBDS net: A quantum-inspired optimized bi-directional self-supervised neural network architecture for automatic brain MR image segmentation. *In: Proc. TENCON-IEEE Region 10th Conf. (TENCON)*, Kochi, India, pp. 761-766. doi: 10.1109/tencon.2019.8929585

22. Lee, E.H., Zheng, J., Colak, E., Mohammadzadeh, M., Houshmand, G., Bevins, N. and Yeom, K.W. (2021). Deep COVID DeteCT: An international experience on COVID-19 lung detection and prognosis using chest CT, *NPJ Digital Medicine*, 4(1): 1-11.

23. Li, L. *et al.* (2020). Artificial Intelligence distinguishes COVID-19 from community acquired pneumonia on chest CT, *Radiology*. https://doi.org/10.1148/radiol.2020200905 (2020)

24. Litjens, G. *et al.* (2017). A survey on deep learning in medical image analysis, *Med. Image Anal.*, 42: 60-88.

25. Liu, J., Zhang, Z., Zu, L., Wang, H. and Zhong, Y. (2020, October). Intelligent detection for CT image of COVID-19 using deep learning. *In: 13th International Congress on Image and Signal Processing, Bio-Medical Engineering and Informatics* (CISP-BMEI), pp. 76-81, IEEE.

26. Liu, J., Liao, X., Qian, S., Yuan, J., Wang, F., Liu, Y., Wang, Z., Wang, F.-S., Liu, L. and Zhang, Z. (2020). Community transmission of severe acute respiratory syndrome coronavirus 2, Henzhen, China, 2020, *Emerging Infectious Diseases*, 26(6): 1320-1323.

27. Long, C., Xu, H., Shen, Q., Zhang, X., Fan, B., Wang, C., Zeng, B., Li, Z., Li, X. and Li, H. (2020). Diagnosis of the Coronavirus disease (COVID-19): rRT- PCR or CT? *European Journal of Radiology*, 126: 108961.

28. Mei, X., Lee, H.C., Diao, K.Y., Huang, M., Lin, B., Liu, C., ... and Yang, Y. (2020). Artificial intelligence-enabled rapid diagnosis of patients with COVID-19, *Nature Medicine*, 26(8): 1224-1228.

29. Ning, W., Lei, S., Yang, J., Cao, Y., Jiang, P., Yang, Q. and Wang, Z. (2020). Open resource of clinical data from patients with pneumonia for the prediction of COVID-19 outcomes via deep learning, *Nature Biomedical Engineering*, 4(12): 1197-1207.

30. Ou, X., Liu, Y., Lei, X., Li, P., Mi, D., Ren, L. and Qian, Z. (2020). Characterization of spike glycoprotein of SARS-CoV-2 on virus entry and its immune cross-reactivity with SARS-CoV, *Nat. Commun.,* 11: 1620. https://go. nature. com/2Z1gLjX

31. Pham, T.D. (2020). A comprehensive study on classification of COVID-19 on computed tomography with pre-trained convolutional neural networks, *Scientific Reports*, 10(1): 1-8.

32. Saood, A. and Hatem, I. (2021). COVID-19 lung CT image segmentation using deep learning methods: U-Net versus SegNet, *BMC Medical Imaging*, 21(1): 1-10.

33. Shin, H.C. *et al.*, (2016). Deep convolutional neural networks for computer-aided detection: CNN architectures, dataset characteristics and transfer learning, *IEEE Trans. Med. Imaging*, 35(5): 1285-1298.

34. Singh, K.K. and Singh, A. (2021). Diagnosis of COVID-19 from chest X-ray images using wavelets-based depth-wise convolution network, *Big Data Mining and Analytics*, 4(2): 84-93.

35. Wang, L., Lin, Z.Q. and Wong, A. (2020). COVID-net: A tailored deep convolutional neural network design for detection of COVID-19 cases from chest X-ray images, *Scientific Reports*, 10(1): 1-12.

36. Wang, J., Bao, Y., Wen, Y., Lu, H., Luo, H., Xiang, Y., Li, X., Liu, C. and Qian, D. (2020). Prior-attention residual learning for more discriminative COVID-19 screening in CT images, *IEEE Trans. Med. Imag.*, 39(8): 2572-2583. doi: 10.1109/tmi.2020.2994908

37. Wang, X., Deng, X., Fu, Q., Zhou, Q., Feng, J., Ma, H., Liu, W. and Zheng, C. (2020). A weakly-supervised framework for COVID-19 classification and lesion localization from chest CT, *IEEE Trans. Med. Imag.*, 39(8): 2615-2625. doi: 10.1109/tmi.2020.2995965

38. Wang, S. *et al.* (2020). A deep learning algorithm using CT images to screen for corona virus disease (COVID-19). medRxiv. https://doi.org/10.1101/2020.02.14.20023028

39. Xu, X. *et al.* (2002). Deep learning system to screen coronavirus disease 2019 pneumonia. arXiv:2002.09334.

40. Yan, Q., Wang, B., Gong, D., Luo, C., Zhao, W., Shen, J., Shi, Q., Jin, S., Zhang, L. and You, Z. (2020). COVID-19 chest CT image segmentation – A deep convolutional neural network solution, arXiv:2004.10987 [Online]. Available: http://arxiv.org/abs/2004.10987

41. Zhao, J., Zhang, Y., He, X. and Xie, P. (2020). *COVID-CT-Dataset: A CT Scan Dataset about COVID-19*. arXiv:2003.13865.

Embedding Explainable Artificial Intelligence in Clinical Decision Support Systems: The Brain Age Prediction Case Study

Angela Lombardi[1], Domenico Diacono[2], Nicola Amoroso[3], Alfonso Monaco[2], Sabina Tangaro[4] and Roberto Bellotti[1]

[1] Dipartimento Interateneo di Fisica, Università degli Studi di Bari and Istituto Nazionale di Fisica Nucleare, Sezione di Bari, Via E. Orabona 4, 70125 Bari (Italy)
[2] Istituto Nazionale di Fisica Nucleare, Sezione di Bari, Via E. Orabona 4, 70125 Bari (Italy)
[3] Dipartimento di Farmacia - Scienze del Farmaco, Università degli Studi di Bari and Istituto Nazionale di Fisica Nucleare, Sezione di Bari, Via E. Orabona 4, 70125 Bari (Italy)
[4] Dipartimento di Scienze del Suolo, della Pianta e degli Alimenti, Università degli Studi di Bari and Istituto Nazionale di Fisica Nucleare, Sezione di Bari, Via E. Orabona 4, 70125 Bari (Italy)

1. Introduction

Clinical decision support systems are artificial systems that exploit multivariate personalized information to predict health status of individuals and provide clinicians and specialists with knowledge that can be used to support clinical operational decisions [31]. The efficiency of these systems is based on the evolution of increasingly advanced machine learning (ML) and deep learning (DL) algorithms, with high levels of accuracy and low error rates [37].

As an example, in the brain imaging research field, highly advanced systems have been developed to predict the brain age of individuals and to use the deviation of the predicted value from the chronological age as a clinical index for several neurodegenerative and psychiatric diseases [5, 4, 26]. In particular, a large number of nonlinear and complex models, such as convolutional neural networks and deep learning architectures have been proposed to exploit raw image as input to automatically extract brain age predictions for different tasks [12, 22, 32]. These methods are being increasingly applied as they can achieve high levels of accuracy with a relatively small number of pre-processing steps and can handle also non-linear relationships between the features in high-dimensional problems.

While the final efficiency of the predictions produced by clinical decision support systems is essential to ensure their applications in the diagnostic domains, it is equally important to ensure other characteristics, such as transparency and interpretability of the decisions [1].

Both ML and DL approaches should provide trustworthy and explainable decisions for human experts in order to enhance the trust of clinical decision support systems and increase their use. Explainable Artificial Intelligence (XAI) has been recently introduced to provide explainable algorithms in different domains [29, 15, 2]. Several works have showed that XAI methods can significantly improve personalized treatments and targeted interventions [11, 21].

Recently, we investigated the reliability and efficiency of a DL framework embedding XAI techniques to predict the age of a healthy cohort of subjects from ABIDE I database [10] by using several morphological features extracted from their MRI scans [24]. In this work, we formalize the main findings of the developed framework and propose a new version that is able to link the explanations of the model to the quality metrics of resonance imaging. The main objective is to show how XAI explanations are affected, not only by the variable of interest to be predicted (i.e. the age of the subjects), but also by other external factors that experts need to take into account in order to improve the overall quality of the system.

2. Materials

2.1 Cohort of Subjects

A cohort of $T = 378$ male typically-developing subjects was selected from the Autism Brain Imaging Data Exchange (ABIDE I). This database consists of T1-weighted MRI publicly available scans and other variables (including phenotype information and quality metrics) collected from 17 international sites from both typically-developing controls and individuals affected by autism. The MRI scans were collected by using different scanners and protocols. More details about images and acquisition protocols from each site are available on the web page of the consortium.[1] The demographic information of the included subjects are listed in Table 1 for each of the 17 sites. All participating sites received local Institutional Review Board approval for acquisition of the contributed data.

Table 1: Demographic information of the subjects per site

Site	Samples	Age Range (Years)
CMU	2	21 – 25
KKI	23	8 – 13
Leuven 1	13	18 – 29
Leuven 2	14	12 – 17

---

[1] http://fcon_1000.projects.nitrc.org/indi/abide

MaxMun	24	7–48
NYU	77	6–32
Olin	13	10–23
Pitt	22	12–33
SBL	14	20–42
SDSU	14	12–17
Trinity	25	12–25
UCLA 1	28	9–18
UCLA 2	11	10–14
UM 1	32	8–19
UM 2	17	13–29
USM	43	8–40
Yale	6	8–17
TOTAL	378	6–48

2.2 Quality Metrics

We used the following quality assessment metrics provided by the publicly available ABIDE pre-processed project [7]:

- Contrast to Noise Ratio (CNR) computed as the mean of the gray matter values minus the mean of the white matter values, divided by the standard deviation of the air values.
- Entropy Focus Criterion (EFC), i.e. the Shannon's entropy used to summarize the principal directions distribution.
- Foreground to Background Energy Ratio (FBER) that represents the mean energy of image values within the head relative to outside the head.
- Smoothness of Voxels (FWHM), i.e., the full-width half maximum of the spatial distribution of the image intensity values in terms of voxels.
- Percent of Artifact Voxels (QI1) is computed as the proportion of voxels with intensity corrupted by artifacts normalized by the number of voxels in the background.
- Signal to Noise Ratio (SNR), i.e. the mean of image values within gray matter divided by the standard deviation of the image values within air (i.e. outside the head) [28].

2.3 Brain Morphological Features

All the T1 raw scans were processed by using the recon-all pipeline from the software FreeSurfer v.5.3.0 [8, 14, 13] on ReCaS datacenter[2] [25]. The pipeline output consists of 68 segmented cortical regions and 40 sub-cortical regions as specified by the Desikan-Killiany atlas [9] and Aseg Atlas [13]. In particular, we exploited the

---

[2] https://www.recas-bari.it/index.php/en/

statistical features related to each morphological descriptor of the regions (i.e. curvature, thickness and white matter volumes for the cortical regions and volumes for the sub-cortical regions) and some global brain metrics (e.g. surface and volume statistics of each hemisphere; total cerebellar gray and white matter volume, brainstem volume, corpus callosum volume, and white matter hypointensities). Specifically, we conducted a whole brain analysis without any a priori assumptions about the most important features for age prediction. Thus, we considered the following metrics as in our previous work [23]:

- global metrics (21 features);
- cortical metrics (544 features);
- sub-cortical metrics (240 features);
- white matter metrics (408 features);

We finally obtained the matrix of the features of dimension $T \times P$ with $T = 387$, and $P = 1213$, where each row represents a single subject described by P morphological features.

3. The XAI Framework

The overall XAI framework is depicted in Fig. 1. Each step is explained in the following sections.

3.1 Cross-validation Scheme

A cross-validation strategy is required to reduce overfitting and improves the generalization of the ML/DL algorithms. According to the simplest scheme, the dataset could be divided into two subsets, whereby one subset is used for training the algorithm and the other one for testing it. However, this approach exacerbates the limitations of the small sample size if the overall number of samples is small since each sample is used once. A cross-validation scheme could overcome this limitation as multiple partitions are generated, allowing each sample to be used multiple times and improving the statistical reliability of the results. Some examples of cross-validation methods include k-fold and leave-one-out [36]. In this work, we selected a leave-one-site cross-validation scheme to test the data from one site while using the data from all the other sites to compose the training set. This strategy has been widely adopted in multi-site analysis as the generalization of the ML/DL models could be tested to new sites, while exploring the correlation between the performance of the models and the variability of the characteristics of the sites [16].

3.2 ML/DL Algorithms

The output for most decision support systems is a classification prediction (e.g. benign vs malignant) or regression (e.g. a numerical score) for each instance described by the attributes or features. The core of these systems is supervised ML/DL algorithms. Here, we developed fully connected deep neural networks models to predict the brain age, Y of a healthy cohort of subjects by using their morphological features, X. We tuned each model with tenfold Grid Search cross-validations on training sets, using

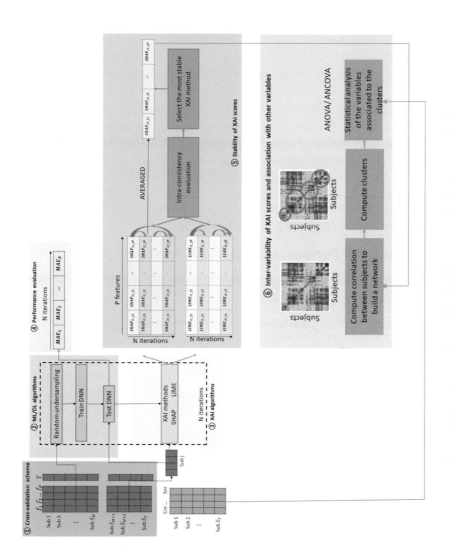

Fig. 1: Schematic overview of the XAI framework for brain age prediction

the left out site as a completely independent test set. A final configuration was reached at the end of the Grid Search with the following features:

- Input layer with 1213 units;
- Four-layers with 512 units per layer;
- ReLU as activation function;
- SGD optimiser with learning rate $= 5e - 5$;
- momentum $= 0.9$;
- Huber as loss function;
- dropout $= 0$;
- output layer with a single unit and no activation function to perform the regression task.

Since the ML/DL algorithms may return unstable performance and features ranking against small changes in the composition of the training set, for each cross-validation round, we randomly under-sampled the training set $N = 100$ times by selecting 80% of the samples to produce small variations of the training set and for each iteration we trained a DNN model to predict the chronological age of the subjects, Y. We tested the DNN models on each sample of the test set, collecting $N = 100$ performance values for each subject. We applied a consensus strategy by averaging $N = 100$ performance values to obtain a single stable value for each subject.

3.3   Evaluation of Performance of ML/DL Models

The performance of the models were evaluated by means of the Mean Absolute Error (MAE):

$$MAE = \frac{1}{t}\sum_{i=1}^{t}|y_i - \hat{y}_i| \qquad (1.1)$$

with t being the sample size for the specific test site, y_i the chronological age and $\hat{y}_i$ the predicted brain age. The correlation coefficient between the chronological age and the predicted age of the subjects was assessed as an index of the performance of the models over the whole dataset:

$$R = \frac{\sum_{i=1}^{T}(y_i - \overline{y})(\hat{y}_i - \overline{\hat{y}})}{\sqrt{\sum_{i=1}^{T}(y_i - \overline{y})^2}\ \sqrt{\sum_{i=1}^{T}(\hat{y}_i - \overline{\hat{y}})^2}} \qquad (1.2)$$

where $\hat{y}_i$ and $\overline{\hat{y}}$ denote the sample mean of the chronological age and the predicted brain age, respectively. The statistical significance of above-chance predictive performance for the overall model was computed by means of a non-parametric permutation test: the age outcomes of the subjects were permuted 1000 times for assessing the performance values (MAE and R) within each permutation round. We finally assigned the final p value for each metric by dividing the number of times for which model performance based on the true age was lower than the performance for the permuted age outcomes by the number of permutations, i.e. 1000 [18].

3.4 XAI Methods

Several methods exist to compute local explanations for each single instance. Here, we used SHAP [27] and LIME [33] to explain the decisions of the DNN models on each test sample as they represent the most popular local explanation algorithms. Such algorithms are model-agnostic as they explain predictions at individual level regardless of the selected models. Both methods learn an interpretable linear model around each test instance and estimate feature importance at local level. For a dataset $D = [(x_1, y_1), (x_2, y_2), ..., (x_T, y_T)]$, where $x_i \in R^P$ is the feature vector for the sample i and y_i the corresponding age, the generic pre-trained model f returns a prediction $f(x_i)$ based on a single input sample s_i. SHAP and LIME aim at finding a linear model g to explain f by using a simplified inputs x' that map the original inputs through a mapping function $x = h_x(x')$ trying to ensure $g(z') \approx f(h_x(z'))$ whenever $z' \approx x'$.

Both methods try to generate an explanation for x that approximates the behavior of the model accurately within the neighborhood of x, while achieving lower complexity [35]. In more simple words, the methods explain the prediction of the instance x by computing the contribution of each feature to the prediction ([38]). The difference between the two methods lies in the set of equations to be set to minimize the object function. More mathematical details about the two methods can be found in the seminal works of [27] and [33].

3.5 Stability of XAI Scores

Both SHAP and LIME are post-hoc local XAI methods as they use a pre-trained ML model to compute approximations of the model's inner decision logic by producing feature importance scores for each independent test sample that represents the contribution of each feature to the final prediction of the ML/DL model [30]. These methods are different from feature selection methods. Indeed the latter methods exploit the overall training set to assess the impact of each feature on a performance metric producing a single feature importance vector as output. On the contrary, local XAI methods produce a feature importance vector for each test sample. A stability analysis of XAI scores is then required to quantify the variation of the score values by slightly varying the training set. In detail, we applied both SHAP and LIME algorithms to extract the age-related feature importance vector for each subject collecting the two matrix S and L of dimension $[N \times P]$, whose generic element s_{nk} (l_{nk}) indicates the SHAP (LIME) value for the k feature within the n iteration. Accordingly, we investigated the reliability of both SHAP and LIME values by computing the intra-consistency coefficient of the scores, i.e. the correlation between each couple of score vectors $s_k = [s_{k1}, s_{k2}, ..., s_{kP}]$ and $s_z = [s_{z1}, s_{z2}, ..., s_{zP}]$, with $k, z = 1, ..., N$ within each subject:

$$IC_{kz} = \frac{\sum_{p=1}^{P}(s_{kp} - \bar{s}_k)(s_{zp} - \bar{s}_z)}{\sqrt{\sum_{p=1}^{P}(s_{kp} - \bar{s}_k)^2}\sqrt{\sum_{p=1}^{P}(s_{zp} - \bar{s}_z)^2}} \tag{1.3}$$

where $\bar{s}_k$ and $\bar{s}_z$ denote the sample means of the two vectors and k and z denote the indices of different model training iterations. We also computed IC_{kz} for each couple of LIME vectors obtaining a distribution of $N\dfrac{(N-1)}{2}$ intra-consistency values for each XAI method. The intra-consistency coefficient varies between 0 (zero) and 1 (one), hence we compared the IC distributions between the two XAI methods by using the Wilcoxon rank-sum test and Cohen's d coefficient in order to choose the most reliable and stable algorithm.

For the best algorithm, we averaged the $N = 100$ realizations of both values to obtain a single representative XAI vector (SHAP $S_t = [s_{t,\,A1}, ..., s_{t,\,AP}]$ or LIME $L_t = [l_{t,\,A1}, ..., l_{t,\,AP}]$) for each subject t, where:

$$S_{t,Ap} = \frac{1}{N}\sum_{n=1}^{N}s_{np} \tag{1.4}$$

is the pth averaged SHAP value for the feature p.

3.6 Inter-similarity of XAI Scores and Association with QA Metrics

The inter-subject similarity was firstly computed as the correlation between the SHAP (LIME) score vectors S_t and S_u (L_t and L_u) for each couple of subjects u and t, with $t,\ u = 1, ..., T$:

$$IS_{ut} = \frac{\sum_{p=1}^{P}(s_{u,Ap} - \overline{S}_u)(s_{t,Ap} - \overline{S}_t)}{\sqrt{\sum_{p=1}^{P}(s_{u,Ap} - \overline{S}_u)^2}\sqrt{\sum_{p=1}^{P}(s_{t,Ap} - \overline{S}_t)^2}} \tag{1.5}$$

where $\overline{S}_u$ and $\overline{S}_t$ denote the sample means of the two vectors. Then, the inter-similarity matrix IS was obtained for the best XAI method, where the entry $(u, t) = IS_{ut}$ indicates the similarity value between the scores of subjects u and t.

We applied the stability-based k-medoid criterion proposed by [17] to find the best partition into clusters of the inter-similarity matrix. This criterion assesses the cluster-wise stability of a dataset by resampling it several times with different methods such as bootstrap or sub-setting and by identifying the most stable clusters across the iterations. More in detail:

- the dataset is bootstrapped and a new dataset is created;
- the k-medoid algorithm is applied to find clusters; this step is performed for different number of clusters k and the Silhouette coefficients are computed;
- the number of clusters k is selected as corresponding to the highest Silhouette coefficient [34];
- for every given cluster in the original clustering, the most similar cluster is searched in the new partition and the Jaccard similarity value is recorded;
- the cluster stability of each cluster is assessed by the average similarity taken over the resampled datasets.

This algorithm returns the most stable k clusters, where k is automatically computed by using the quality of the partitions by means of the Silhouette coefficient.

To analyze the differences between the identified clusters for each of the QA variables, the analysis of covariance (ANCOVA) was conducted for each metric with age as covariate, followed by post hoc Tukey-Kramer tests in case of significant group effects. This statistical technique was chosen as the means of an outcome variable (i.e. each QA metric) between two or more groups (i.e. the identified clusters) can be compared while taking into account the variability of the other variables (i.e. the age of the subjects).

3.7 Results and Discussion

In this work, we reviewed the main points of a XAI framework developed to perform the brain age prediction on healthy subjects by highlighting the expected properties of a clinical decision support system embedding XAI methods.

3.7.1 Performance of ML/DL Models

First of all, the performance of the ML/DL models should match or exceed the state of the art. Fig. 2 shows the histograms of the predictive performance (MAE and R) based on permuted data ($N = 1.000$ permutations; blue) in relation to the predictive performance based on the true non-permuted data (red vertical line). For the whole dataset, the proposed DL models achieved MAE and R values that compare favorably with the literature showing the overall performance $MAE = 2.7$ and $R = 0.86$ [39, 6, 3, 23]. Moreover, both performance metrics were found to be significantly different from the chance level, resulting $p = 0$ from the non-parametric permutation test.

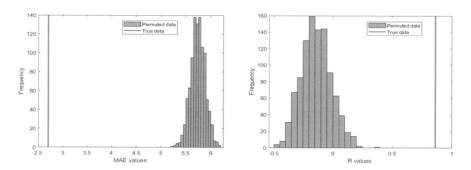

Fig. 2: Histograms of the performance of the DNN models resulting from 1.000 permutations of the age outcomes compared to the actual overall performance (reported with red vertical lines)

3.7.2 Stability of XAI Scores

The selected XAI methods were applied locally to each sample to compute the local explanations, i.e. the feature importance scores which quantify the contributions of each feature to the final prediction of the ML/DL models. Indeed, local XAI algorithms are particularly suitable for clinical applications, as they can be applied to a new patient regardless of pre-trained models on other training sets, with several advantages, such as generalization and universality. However, it is worth ensuring that the XAI scores provided by a local algorithm vary as little as possible by introducing small variations in the training set. The stability of the feature importance is a critical requirement for the reliability of the final ranking of the features in different biological contexts [19, 20]. In this work, we used the intra-consistency criterion to quantify the stability of the XAI methods for each subject with small variations in the training set. We compared the intra-consistency values of both SHAP and LIME scores across the sites as reported in Fig. 3 and Table 2. It is clearly evident that the SHAP intra-consistency scores are significantly higher than the LIME scores for all the sites; hence, we selected the SHAP algorithm in the final version of the framework.

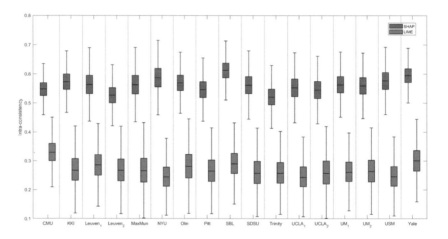

Fig. 3: Boxplots of the IC values for the SHAP and LIME methods for each site

3.7.3   *Inter-variability of XAI Scores and Association with QA Metrics*

We defined an inter-similarity score to compare the SHAP scores across the subjects. Firstly, we averaged the SHAP vectors resulting from N under-sampling rounds of the training set to obtain a final consistent feature importance vector for each subject. Then, we correlated the SHAP vectors between each couple of subjects to assess the similarity of their XAI scores. Finally, the inter-similarity matrix IS_{SHAP} was assessed for the SHAP algorithm and partitioned into clusters to detect groups of subjects with more similar SHAP scores. We found $k = 10$ clusters by using the stability-based k-medoid criterion. We conducted ANCOVA for each QA metric, setting the age variable as a covariate for testing whether the imaging quality metrics would affect the XAI scores. We obtained a significant effect ($p < 0.0001$) for all the QA metrics, meaning that the means of the QA values adjusted for the age are significantly different between the identified clusters. This statistical analysis can be generalized for any other variable that the experts plan to test. It is particularly useful for providing clear indications about all the other variables that have simultaneous effects on the features of interest. In the discussed case, we can infer from Fig. 4 that

Table 2: Comparison between the distributions of the IC coefficients of the SHAP and LIME methods for each site: p-values resulting from the Wilcoxon rank sum test and Cohen's d coefficients

Site	p value	Cohen's d
CMU	10^{-7}	5.47
KKI	10^{-7}	6.31
Leuven 1	10^{-7}	4.91
Leuven 2	10^{-7}	5.46
MaxMun	10^{-7}	3.92
NYU	10^{-7}	5.38
Olin	10^{-7}	5.84
Pitt	10^{-7}	5.42
SBL	10^{-7}	7.24
SDSU	10^{-7}	5.84
Trinity	10^{-7}	5.58
UCLA 1	10^{-7}	6.13
UCLA 2	10^{-7}	5.61
UM 1	10^{-7}	6.15
UM 2	10^{-7}	4.76
USM	10^{-7}	6.92
Yale	10^{-7}	6.67

the 10 identified clusters do not differ in the values of QA metrics, but a subset of them presents significant differences that can be considered to harmonize the effect of the acquisition sites. Future development of the proposed framework will exploit the results of the statistical analysis to embed the XAI algorithms in a feed-forward logic to eliminate the batch effects for the age prediction task.

4. Conclusion

In this work, we proposed a clinical decision support system embedding XAI for age prediction with brain morphology. We evaluated each step of the system by analyzing some critical aspects, concerning both reliability and accuracy of the algorithms. We also emphasized how XAI algorithms can be used to explain not only the contributions of the features on the variable to be predicted, but also the effect of the marginal variables related to the imaging quality of MRI scans. Future developments will build on the last findings to develop systems increasingly oriented to the multivariate quantification of both biological and batch effects on the final decisions made by the algorithms.

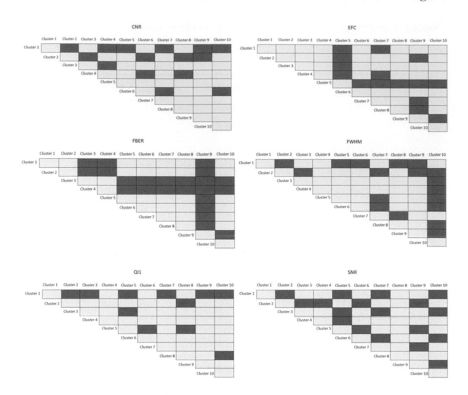

Fig. 4: Matrices of p-values resulting from post hoc Tukey-Kramer test for the QA metrics for the ten clusters of matrix IS_{GHAP}. Significant p-values are marked with darker boxes

Acknowledgements

This work was supported in part by the research project '*Biomarcatori di connettività cerebrale da imaging multimodale per la diagnosi precoce e stadiazione personalizzata di malattie neurodegenerative con metodi avanzati di intelligenza artificiale in ambiente di calcolo distribuito*' (project code 928A7C98) within the Program "Research for Innovation – REFIN' funded by Regione Puglia (Italy) in the framework of the 'POR Puglia FESR FSE 2014-2020 – Asse X - Azione 10.4'. Code development/testing and results were obtained on the IT resources hosted at ReCaS data center. ReCaS is a project financed by Italian MIUR (PONa3_00052, Avviso 254/Ric.

References

1. Antoniadi, A.M., Du, Y., Guendouz, Y., Wei, L., Mazo, C., Becker, B.A. and Mooney, C. (2021). Current challenges and future opportunities for XAI in machine learning-based clinical decision support systems: A systematic review, *Applied Sciences,* 11(11): 5088.

2. Arrieta, A.B., Díaz-Rodríguez, N., Del Ser, J., Bennetot, A., Tabik, S., Barbado, A., García, S., Gil-López, S., Molina, D., Benjamins, R., *et al.* (2020). Explainable artificial intelligence (XAI): Concepts, taxonomies, opportunities and challenges toward responsible AI, *Information Fusion,* 58: 82-115.
3. Ball, G., Beare, R. and Seal, M.L. (2019). Charting shared developmental trajectories of cortical thickness and structural connectivity in childhood and adolescence, *Human Brain Mapping*, 40(16): 4630-4644.
4. Cole, J.H., Marioni, R.E., Harris, S.E. and Deary, I.J. (2019). Brain age and other bodily 'ages': Implications for neuropsychiatry, Molecular Psychiatry, 24(2): 266-281.
5. Cole, J.H., Poudel, R.P.K., Tsagkrasoulis, D., Caan, M.W.A., Steves, C., Spector, T.D. and Montana, G. (2017). Predicting brain age with deep learning from raw imaging data results in a reliable and heritable biomarker, *Neuroimage*, 163c: 115-124. doi: 10.1016/j.neuroimage.2017.07.059
6. Corps, J. and Rekik, I. (2019). Morphological brain age prediction using multi-view brain networks derived from cortical morphology in healthy and disordered participants, *Scientific Reports*, 9(1): 9676.
7. Craddock, C., Benhajali, Y., Chu, C., Chouinard, F., Evans, A., Jakab, A., Khundrakpam, B.S., Lewis, J.D., Li, Q., Milham, M., *et al.* (2013). The neuro bureau preprocessing initiative: Open sharing of preprocessed neuroimaging data and derivatives, *Frontiers in Neuroinformatics,* 7.
8. Dale, A., Fischl, B. and Sereno, M. (1999). Cortical surface-based analysis: I. Segmentation and surface reconstruction, *Neuroimage*, 9(2): 179-194.
9. Desikan, R., Ségonne, F., Fischl, B., Quinn, B., Dickerson, B., Blacker, D., Buckner, R., Dale, A., Maguire, R., Hyman, B., *et al.* (2006). An automated labeling system for subdividing the human cerebral cortex on MRI scans into gyral based regions of interest, *Neuroimage*, 31(3): 968-980.
10. Di Martino, A., Yan, C.G., Li, Q., Denio, E., Castellanos, F.X., Alaerts, K., Anderson, J.S., Assaf, M., Bookheimer, S.Y., Dapretto, M., *et al.* (2014). The autism brain imaging data exchange: Towards a large-scale evaluation of the intrinsic brain architecture in autism, *Molecular Psychiatry*, 19(6): 659-667.
11. Fellous, J.M., Sapiro, G., Rossi, A., Mayberg, H. and Ferrante, M. (2019). Explainable artificial intelligence for neuroscience: Behavioral neurostimulation, *Frontiers in Neuroscience,* 13: 1346.
12. Feng, X., Lipton, Z.C., Yang, J., Small, S.A., Provenzano, F.A., Initiative, A.D.N., Initiative, F.L.D.N., *et al.* (2020). Estimating brain age based on a uniform healthy population with deep learning and structural MRI, *Neurobiology of Aging*, 91: 15-25.
13. Fischl, B., Salat, D., Busa, E., Albert, M., Dieterich, M., Haselgrove, C., Kouwe, A.V.D., Killiany, R., Kennedy, D., Klaveness, S., *et al.* (2002). Whole brain segmentation: automated labeling of neuroanatomical structures in the human brain, *Neuron*, 33(3): 341-355.
14. Fischl, B., Sereno, M. and Dale, A.M. (1999). Cortical surface-based analysis: Ii: inflation, flattening, and a surface-based coordinate system, *Neuroimage*, 9(2): 195-207.
15. Guidotti, R., Monreale, A., Ruggieri, S., Turini, F., Giannotti, F. and Pedreschi, D. (2018). A survey of methods for explaining black box models, *ACM Computing Surveys (CSUR)*, 51(5): 1-42.
16. Heinsfeld, A.S., Franco, A.R., Craddock, R.C., Buchweitz, A. and Meneguzzi, F. (2018). Identification of autism spectrum disorder using deep learning and the abide dataset, *NeuroImage: Clinical,* 17: 16-23.
17. Hennig, C. (2007). Cluster-wise assessment of cluster stability, *Computational Statistics & Data Analysis*, 52(1): 258-271.

18. Hilger, K., Winter, N.R., Leenings, R., Sassenhagen, J., Hahn, T., Basten, U. and Fiebach, C.J. (2020). Predicting intelligence from brain gray matter volume, *Brain Structure and Function*, 225(7): 2111-2129.

19. Kalousis, A., Prados, J. and Hilario, M. (2007). Stability of feature selection algorithms: A study on high-dimensional spaces, *Knowledge and Information Systems*, 12(1): 95-116.

20. Kuncheva, L.I. (2007). A stability index for feature selection. *In: Artificial Intelligence and Applications*, pp. 421-427.

21. Langlotz, C.P., Allen, B., Erickson, B.J., Kalpathy-Cramer, J., Bigelow, K., Cook, T.S., Flanders, A.E., Lungren, M.P., Mendelson, D.S., Rudie, J.D., *et al.* (2019). A roadmap for foundational research on artificial intelligence in medical imaging: From the 2018 nih/rsna/acr/the academy workshop, *Radiology*, 291(3): 781-791.

22. Levakov, G., Rosenthal, G., Shelef, I., Raviv, T.R. and Avidan, G. (2020). From a deep learning model back to the brain – Identifying regional predictors and their relation to aging, *Human Brain Mapping*, 41(12): 3235-3252.

23. Lombardi, A., Amoroso, N., Diacono, D., Monaco, A., Tangaro, S. and Bellotti, R. (2020). Extensive evaluation of morphological statistical harmonization for brain age prediction, *Brain Sciences*, 10(6): 364; doi10.3390/brainsci10060364. URL https://www.mdpi.com/2076- 3425/10/6/364

24. Lombardi, A., Diacono, D., Amoroso, N., Monaco, A., Tavares, J.M.R., Bellotti, R. and Tangaro, S. (2021). Explainable deep learning for personalized age prediction with brain morphology, *Frontiers in Neuroscience*, 15.

25. Lombardi, A., Lella, E., Amoroso, N., Diacono, D., Monaco, A., Bellotti, R. and Tangaro, S. (2019). Multidimensional neuroimaging processing in RECAS datacenter. *In: International Conference on Internet and Distributed Computing Systems*, pp. 468-477, Springer.

26. Lombardi, A., Monaco, A., Donvito, G., Amoroso, N., Bellotti, R. and Tangaro, S. (2020). Brain age prediction with morphological features using deep neural networks: Results from predictive analytic competition 2019, *Frontiers in Psychiatry*, 11: 1613.

27. Lundberg, S.M. and Lee, S.I. (2017). A unified approach to interpreting model predictions. *In:* I. Guyon, U.V. Luxburg, S. Bengio, H. Wallach, R. Fergus, S. Vishwanathan and R. Garnett (Eds.), *Advances in Neural Information Processing Systems*, 30: 4765-4774, Curran Associates, Inc. URL. https://proceedings.neurips.cc/paper/2017/file/8a20a8621978632d76c43dfd28b67767- Paper.pdf

28. Magnotta, V.A., Friedman, L. and Birn, F. (2006). Measurement of signal-to-noise and contrast-to-noise in the fibirn multicenter imaging study, *Journal of Digital Imaging*, 19(2): 140-147.

29. Miller, T. (2019). Explanation in artificial intelligence: Insights from the social sciences, Artificial Intelligence, 267: 1-38.

30. Moradi, M. and Samwald, M. (2021). Post-hoc explanation of black-box classifiers using confident item sets, *Expert Systems with Applications,* 165-113941.

31. Osheroff, J.A., Teich, J.M., Middleton, B., Steen, E.B., Wright, A. and Detmer, D.E. (2007). A roadmap for national action on clinical decision support, *Journal of the American Medical Informatics Association*, 14(2): 141-145.

32. Peng, H., Gong, W., Beckmann, C.F., Vedaldi, A. and Smith, S.M. (2021). Accurate brain age prediction with lightweight deep neural networks, *Medical Image Analysis*, 68: 101871.

33. Ribeiro, M.T., Singh, S. and Guestrin, C. (2016). Why should I trust you? Explaining the predictions of any classifier. *In: Proceedings of the 22nd ACM SIGKDD International Conference on Knowledge Discovery and Data Mining*, KDD, 16: 1135-1144, Association

for Computing Machinery, New York, NY, USA. doi10.1145/2939672.2939778. URLhttps://doi.org/10.1145/2939672.2939778

34. Rousseeuw, P.J. (1987). Silhouettes: A graphical aid to the interpretation and validation of cluster analysis, *Journal of Computational and Applied Mathematics*, 20: 53-65.

35. Slack, D., Hilgard, S., Jia, E., Singh, S. and Lakkaraju, H. (2020). Fooling lime and shap: Adversarial attacks on post hoc explanation methods. *In: Proceedings of the AAAI/ACM Conference on AI, Ethics, and Society*, pp. 180-186.

36. Stone, M. (1974). Cross-validatory choice and assessment of statistical predictions, *Journal of the Royal Statistical Society: Series B (Methodological)*, 36(2): 111-133.

37. Sutton, R.T., Pincock, D., Baumgart, D.C., Sadowski, D.C., Fedorak, R.N. and Kroeker, K.I. (2020). An overview of clinical decision support systems: Benefits, risks, and strategies for success, *NPJ Digital Medicine*, 3(1): 1-10.

38. Wang, M., Zheng, K., Yang, Y. and Wang, X. (2020). An explainable machine learning framework for intrusion detection systems, *IEEE Access*, 8: 73127-73141.

39. Zhao, Y., Klein, A., Castellanos, F.X. and Milham, M.P. (2019). Brain age prediction: Cortical and subcortical shape covariation in the developing human brain, *Neuroimage*, 202: 116149.

Machine Learning-based Biological Ageing Estimation Technologies: A Survey

Zhaonian Zhang[1], Richard Jiang[1], Danny Crookes[2] and Paul Chazot[3]

[1] School of Computing and Communication, Lancaster University, Lancaster, UK
[2] ECIT Institute, Queen's University Belfast, Belfast, UK
[3] Department of Biosciences, Durham University, Durham, UK

1. Introduction

Aging of the population is an important challenge facing the world in the 21st century and has a profound impact on all aspects of society. As we age, the molecules, cells, tissues and organs in the human body change. It is very complicated to analyze the human aging process from a biological point of view [29], and no obvious features have been found so far. Generally speaking, aging is the gradual accumulation of harmful biological changes accompanied by the gradual loss of functions [21]. However, like most species, aging does increase the risk of morbidity and death in humans. Moreover, the process of human aging is different, especially the external manifestations of aging (for example, skin wrinkles, whitening of hair, cataract) are significantly different for each person and for age-related diseases, the age onset of each patient is also different. These show that age is not the only measure of aging, which has prompted scientists to work hard to measure aging from a biological point of view, with the goal of predicting human BA through 'aging biomarkers' and using BA as a standard for measuring aging. Compared with chronological age (CA), it can better predict remaining life and disease risk [4].

With the development of ML, BA's prediction technology has also made great progress. Computers can complete predictions quickly and accurately. Supervised ML covers many different algorithms that learn to recognize patterns and relationships between many input variables (features) in order to estimate one or more output variables (labels) as accurately and robustly as possible [17]. In this article, the input variables are different aging biomarkers, and the output variable is the predicted BA. The models are always training on healthy individuals, and in the analyzing process, we can compare the sample's predicted age and CA. If a sample's predicted age is older than his real age, it is thought to reflect poorer health with the risk of age-related diseases. If a sample's predicted age is younger than his real age, it shows that the subjects are healthy, have good living habits, pay attention to exercise and meditation, and may be highly educated [30, 40] (*see* Fig. 1).

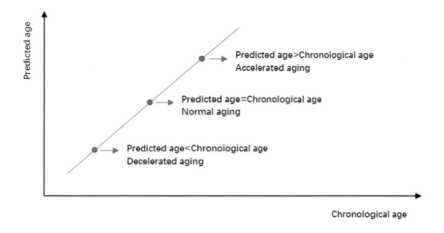

Fig. 1: Compare predicted age and chronological age

This technology can be used to predict the risk of age-related diseases, mortality and monitor biological aging over time. Not only for health detecting, but this technology can also play a very important role in research and development of medicine. At present, many pharmaceutical companies all over the world are committed to research of medicine for the treatment of age-related diseases, but generally speaking, the effect of these medications will not be obvious in the short term; even experienced doctors cannot judge whether the drugs have played a role in the short term, as it may take several years to follow up. This makes it very difficult for pharmaceutical companies to collect medicine data, which restrict the research and development of medicines for age-related diseases. However, this technology finds a new way to solve this problem. The models are always very sensitive; even if there are slight changes in the biomarkers, the models will show them up by the predicted age. This enables pharmaceutical companies to conduct follow-up investigations from the time the patients begin taking drugs, so as to know the effect of drugs in time and get patients' data quickly for further research.

Here, we will show the potential of supervized ML methods to estimate BA, focusing on the latest developments in computing a systemic (blood-based) brain-specific-age and facial-age estimation. Especially, due to the availability of blood test, brain imaging data and clinical events, these models show the best performance in terms of accuracy and prediction of mortality risk, and can be more easily applied to large longitudinal population studies. This article aims to provide references for researchers who are studying age-related diseases, and hopes to help them make contributions in this field.

2. ML for Blood Biomarkers

Blood is an important part of the human body. For a healthy person, the contents of various substances contained in the blood is within a certain range. If it is out of this range, it may mean a dangerous signal. So the blood test can detect a person's health and also the initial characteristics of certain diseases. In addition, as people age, the levels of specific markers in the blood usually fluctuate, such as glucose will increase and hemoglobin will decrease [39]. For these reasons, BA estimation based on the blood biomarker group has always been a research hotspot in aging research [39, 31, 25, 34, 36].

Recently, a method for estimating BA based on deep learning has been proposed. This innovative method is faster and more accurate than traditional methods. It uses healthy individuals as training data, blood circulation biomarkers as input features, and the sample's CA as a label [31, 25, 34, 36]. Deep learning is a specific branch of ML, which is usually used for identification, classification, and to explain the relationship between multiple variables in big data scenarios. Deep neural network is the most commonly used and most representative method in deep learning technology. It recognizes patterns in a large amount of data by imitating the structure and function of the brain. Generally, a deep neural network model has an input layer, one or more hidden 'decision-making' layers and output layers [6] (*see* Fig. 2).

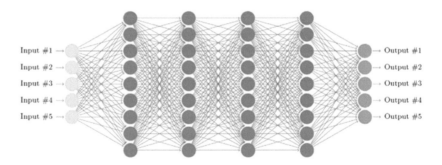

Fig. 2: The structure of deep neural network

In this article, the data of the input layer are the various blood biomarkers of the sample, and the output layer finally returns the CA predicted by the model. These algorithms can capture hidden underlying features and learn complex representations of highly multidimensional data [32]. Deep neural networks have powerful feature extraction capabilities, can capture hidden underlying features, and have strong fault tolerance. They perform better than ML on huge datasets.

2.1 Research Review on Blood Biomarkers

The first study about using blood biomarkers for BA prediction was conducted by Putin and colleagues [36. In this ground-breaking study, their data source was the anonymous blood biochemical records of 62,419 Russia's general population. Specifically, they used the ML method, with 41 standardized blood markers and sex as inputs and BA as outputs. Among all the algorithms, the deep neural network had the best performance in predicting BA. The final results showed that its average absolute error (MAE) was 6.07 years, and the Pearson correlation coefficient (r) between CA and BA was 0.9, coefficient of determination (R^2) was 0.8. The authors continued to experiment with the same data, by changing the hyperparameters of the deep neural network, such as the number of hidden layers, the number of neurons in the hidden layer, the 21 best performing models were obtained. And finally, they built an ensemble model through these models, whose performance has been greatly improved, MAE=5.55, r=0.91, and R^2 =0.83. After that, they tested the model with a core set of the 10 most predictive circulating biomarkers. The final result showed that the accuracy of the model was still very good ($R^2 = 0.63$), which indicated that the model has strong robustness [36].

Mamoshina *et al.* [31] used similar algorithms for experiments, but they selected three datasets for the population of specific countries to train models. Specifically, they are the Canadian dataset containing 20,699 samples, the South Korean dataset containing 65,760 samples, and the East European dataset containing 55,920 samples. The researchers used 19 blood biomarkers and gender as input data for the model, and age as the label. The model was trained in each dataset and tested on independent test sets of all available populations. The results showed that when these models were trained and tested on the same population, they showed great performance with their MAE ranging from 5.59 to 6.36 years, and R^2 ranging from 0.49 to 0.69. However, when these models were trained in one population and tested in another population, the performance of the models decreased significantly, with their MAE ranging from 7.1 to 9.77 years, and R^2 ranging from 0.24 to 0.34. Similarly, when the neural network used all the data combined from the three datasets for training, including the population label as additional feature, the performance of the model was greatly improved. Researchers used the single population to test this model (each of the three datasets was tested), found its MAE ranging from 5.60 to 6.22 years, and its R^2 ranging from 0.49 to 0.70. When the model was tested on the combined dataset, MAE = 5.94 and R^2 = 0.65 (*see* Table 1).

Table 1: Biological age estimates based on blood biomarkers

Input features	Algorithm	Population	Num (training:test)	MAE	Pearson r	R^2
41 blood biomarkers	DNN	Russian	62,419 (90:10)	6.07	0.9	0.8
19 blood biomarkers	DNN	Canadian	20,699 (80:20)	6.36	0.7	0.52
		South Korean	65,760 (80:20)	5.59	0.7	0.49
		Eastern European	55,920 (80:20)	6.25	0.84	0.69
		All	142,379 (80:20)	5.94	0.8	0.65

The above results indicated that blood biomarkers have population specificity [31], which may be due to the different environments of populations, or genetic differences caused by different ancestors. Through feature significance analysis, the five features that had the greatest impact on the results were gender, albumin, glucose, hemoglobin and urea levels. In addition, the content of blood biomarkers was related to a person's age and whether they have age-related diseases [7], and the prediction results for female age group were more accurate than that for men [31].

It is worth noting that a recent study by Mamoshina *et al.* [33] proved a positive correlation between smoking and BA. For women, the BA estimate for smokers was twice that of non-smokers, while for men it was 1.5 times. The difference was significant under age 40 and continued until age 55; subsequently, it gradually disappeared, possibly because of the increasing ability of smokers to resist the dangers of smoking after age 55 [33].

3. ML for Understanding Brain Age

Brain Age (BrA) is also a very important kind of BA. The aging brain functions decline and neurodegenerative diseases bring increasingly serious economic, old-age, medical, and other social problems to our society. Therefore, it is an important task for researchers to accurately and quickly predict the BrA of subjects. Although brain aging is a natural process, there are significant individual differences in changes in brain volume, cortical thickness and white matter microstructure during this process. In addition, the deviation degree between the individual brain aging trajectory and the average trajectory of healthy brain aging can reflect the individual's future risk of neurodegenerative diseases [7, 15]. Therefore, building models based on the characteristic patterns of brain aging contained in neuroimaging data and detecting the aging trajectories of individual brains can provide a new perspective for studying individual differences in brain aging.

In previous studies, researchers have built various models with different types of data. The common types of data include sMRI (structural magnetic resonance imaging), fMRI (functional magnetic resonance imaging), and DTI (diffusion tensor imaging) (*see* Fig. 3). Here, we will focus on the most recent and accurate developments about the research based on sMRI [7, 8, 9, 10, 11, 12, 13].

There are two main algorithms – one is Gaussian Process Regressions and the other is Convolutional Neural Networks (CNN). GPR obtains the NxN similarity matrix by normalizing gray matter (GM) and white matter (WM) images, and then predicts BA through regression tasks, and makes a large number of features conform to the multivariate Gaussian distribution. This algorithm can reflect the local pattern of covariance between individual data points. As for CNN, it is a class of deep neural network that has proved very powerful in image classification (*see* Fig. 4). CNN has many advantages. It can effectively learn the corresponding features from a large number of samples, avoiding the complicated feature extraction process. It uses a simple non-linear model to extract more abstract features from the original image, and only a small amount of human involvement is required in the whole process. CNN has two characteristics – local perception and parameter sharing. Local perception makes each neuron of the CNN to avoid the need to perceive all the pixels

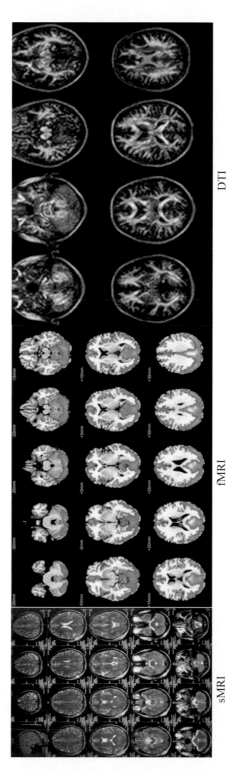

Fig. 3: Examples of sMRI, fMRI and DTI

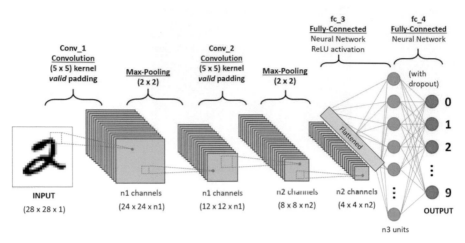

Fig. 4: Structure of a CNN

in the image, but only perceive the local pixels of the image, and then combine these in a higher layer. The partial information is merged to obtain all the characterization information of the image. The neural units of different layers are locally connected, that is, the neural units of each layer is only connected to some of the neural units of the previous layer. Each neural unit only responds to the area in the receptive field and does not care about the area outside the receptive field at all. Such a local connection mode ensures that the learned convolution kernel has the strongest response to the input spatial local mode. The weight-sharing network structure makes it more similar to a biological neural network, which reduces the complexity of the network model and reduces the number of weights. This kind of network structure is highly invariant to translation, scaling, tilt, or other forms of deformation. Moreover, CNN uses the original image as input, which can effectively learn the corresponding features from a large number of samples, avoiding the complicated feature extraction process. Here, the input data is brain sMRI, and output is the predicted BA.

4. Research Review on BrA

A recent study showed that BrA was relatively accurate in predicting BA after experiments performed by using a large healthy control dataset with a sample size of 2001. When using the GPR algorithm, if the data normalized GM images, the test results showed MAE = 4.66 and R^2 = 0.89. The results of using the CNN algorithm were slightly better, with MAE of 4.16 and R^2 of 0.92. When the data uses normalized WM images, the results of the two algorithms were also similar. The GPR algorithm had an MAE of 5.88 and R^2 of 0.84, compared with CNN's MAE of 5.14 and R^2 of 0.88. But when raw (non-parcelated) data was used, the gap between the two algorithms was large. GPR had an MAE of 11.81 and R^2 of 0.32, while CNN still maintained good performance with MAE of 4.65 and R^2 of 0.88. Interestingly, when using the combined data of GM and WM, both algorithms achieved the best performance. Whether it was GPR or CNN, the results said that the value of Intraclass

Correlations Coefficients was very high (>0.9), which showed high levels of within-scanner and between-scanner reliability in BrA estimation (*see* Table 2).

Table 2: Biological Age Estimates Based on Brain sMRI

Input features	Algorithm	Num (training:test)	MAE	Pearson r	R^2
Structural MRI (normalized GM images)	GPR	2,001 (90:10)	4.66	0.95	0.89
Structural MRI (normalized WM images)			5.88	0.92	0.84
Structural MRI (raw data)			11.81	0.57	0.32
Structural MRI (normalized GM+WM images)			4.41	0.96	0.91
Structural MRI (normalized GM images)	CNN		4.16	0.96	0.92
Structural MRI (normalized WM images)			5.14	0.94	0.88
Structural MRI (raw data)			4.65	0.94	0.88
Structural MRI (normalized GM+WM images)			4.34	0.96	0.91

According to a study by Imperial College London, the greater the difference between predicted BrA and CA in older people, the higher the risk of mental or physical problems and early death. Researchers found that the brains of patients suffering from Alzheimer's disease, traumatic brain injury, and psychosis usually accelerate aging [10, 18, 20]. In 2016, Steffener, Luders and others proved that education, meditation, and physical exercise can make the brain young and energetic [30, 40]. In addition, Franke K. and her colleagues [19] studied the relationship between diet and BrA. By comparing healthy groups, they observed that the brains of baboons who experienced malnutrition tended to age prematurely. More recently, Hatton and colleagues [22] reported an association between negative fateful life events and advanced brain aging after controlling for physical, mental, and lifestyle factors.

5. ML for Facial Image Processing

Facial features are the parts of the face that contain the most information. Facial features generally include eyes, nose, mouth, facial wrinkles, as well as more complex attributes, such as gender and emotion. Among them, gender, race, age and other characteristics can be extracted by low-resolution photos [23], but details, such as moles and facial scars need high spatial frequency images to be identified [24].

Facial age is also a kind of BA, the estimation of which is one of the most important projects in the future of computer science applications. However, estimation systems face many challenges. For example, first, the data for each age is not evenly distributed [3, 26]. Secondly, the difference between images is huge, the resolution of the pictures, the light, the posture of people, etc. are all different [26, 38, 27, 35]. Thirdly, aging of the human face is affected by many external factors, such as living environment, diseases and other factors [1]. Finally, the face is also directly related to factors, such as gender and race [16]. Therefore, after continuous attempts by researchers, several methods to improve the accuracy of facial age prediction models have been summarized. The first thing is that it is better to train a model with a constrained dataset than an unconstrained dataset. Second, in addition to facial images, it will be better to add gender and race as the input data. The third is that in the choice of CNN, the more the layers, the higher the accuracy of the model. In addition, the age range of the data should be wide, so that the sample distribution of the training set is more balanced. Finally, before training the model, the pictures can be preprocessed, such as by aligning faces.

The age estimation system sequence is shown in Fig. 5. In general, the two most important parts of the whole process are feature extraction and age estimation. The face detection step before this is to ensure that the input data are face objects. The subsequent face alignment step can improve the prediction accuracy of the model. Fig. 6 shows the meaning of face detection and face alignment.

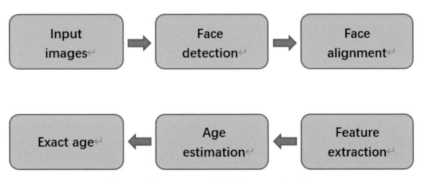

Fig. 5: The process of facial age estimation

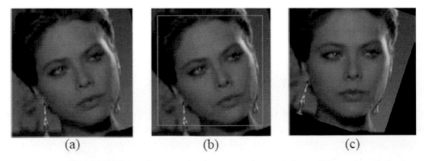

Fig. 6: (a) input image, (b) face detection, (c) face alignment

After this, the computer will extract the features of the input picture, and finally give the predicted age classification or calculation result. Feature extraction generally has two methods, namely, manual algorithm and deep learning. Manual algorithm can predict age even if the training data is small, but its accuracy is not high, and it is necessary to manually define the features to be extracted in advance. Deep learning needs to be trained on a large amount of data, but it has high accuracy and can automatically extract features.

As for age estimation, there are four basic methods – classification, regression, ranking, and hybrid. Classification is suitable for data with few age categories, but not for data with accurate and continuous ages. And for each age group, it needs a large number of samples. The regression is more suitable for cases with target exact age. However, if the age distribution of samples is unbalanced, it will be more sensitive and less robust than other methods. Ranking is very suitable for unbalanced data distribution, but its accuracy is the worst among all the methods. Hybrid is the combination of classification and regression, so its performance is the best, but correspondingly, it has higher complexity.

6. Research Review on Facial Age Estimation

The first study on facial age estimation was completed by Kwon and Lobo in 1999. They used the algorithm of Snakelets + classification to classify facial images into age groups (baby, young, old) on a small dataset which they collected by themselves. The accuracy of the final model was 100% [14]. In 2015, Ali *et al.* used Canny Edge Detection + Classification to divide the age of the sample into three groups (10-30, 30-50 and 50+) by three regions of the face image (cheeks, eye corners and forehead). The accuracy of the final model ranged from 62.86 to 72.48 [2]. Wan *et al.* used two datasets for the experiment, and added sex and race as additional conditions to the CNN model to estimate the exact age. The final MAE was 2.93 for the constrained dataset and 5.22 for the unconstrained dataset [42]. In 2019, Xie *et al.* used two pre-trained CNN models (Alexnet, VGG-16) to create a framework that uses age-related facial attributes to estimate the age of humans. The results said that on the constrained dataset, MAE was 2.69; on the unconstrained dataset, MAE was 5.74 [43]. In a recent study by Liu *et al.*, two datasets were used to test the accuracy of the multi-task learning + hybrid to predict the true age. The results showed that MAE ranged from 2.30 to 3.02 on the constrained dataset, and 5.67 on the unconstrained dataset [28] (*see* Table 3).

7. Conclusion

This article reviews the development of projects using ML for age prediction. From the perspective of medical and clinical values, experiments based on blood biomarkers and brain sMRI are more valuable, but facial-based age prediction can also provide help for advanced technologies, such as face recognition. Now, with the exploration of deep learning, more and more algorithms have been developed, which provide powerful tools for research in various fields. But these algorithms

Table 3: Facial Age Estimation Research

Research	Algorithm	Dataset	Target	Accuracy (%)	MAE
Kwon and Lobo, 1999	Snakelets + classification	47 images	Age group	100	–
Ali *et al.*, 2015	Canny edge detection+ classification	885 images	Age group	62.86-72.48	–
Wan *et al.*, 2018	CNNs+regression	Morph II CACD	Real age	–	2.93 5.22
Xie *et al.*, 2019	Alexnet, VGG-16 +classification	Morph II AgeDB	Real age	–	2.69 5.74
Liu *et al.*, 2020	Multi-task learning+hybrid	Morph II	Real age	–	2.30– 3.02
		UTKFace			5.67

Note:
Morph II is a constrained dataset containing 55134 images, the age range is from 16 to 77, 46,645 males and 8,489 females [37].
CASD is an unconstrained dataset containing 163446 images, the age range is from 16 to 62. It also has additional information such as the name of the celebrity, date of birth, and estimated year of which the photo was taken [5].
AgeDB is an unconstrained dataset containing 16488 images, the age range is from 1 to 101, 9788 males and 6700 females [43].
UTKFace's number of total images>20000, it is an unconstrained dataset containing, the age range is from 0 to 116. It also has additional information such as gender, ethnicity, and datetime of each image [41].

actually still have limitations. First, the process of ML giving results is invisible, and it is difficult to explain with mathematical or biological knowledge. Secondly, adjusting the parameters of the model is a complex process, which generally needs experienced people to do. Finally, the input data is likely to have errors, possibly due to the instruments or other factors.

In future, as it becomes easier to obtain data, researchers can shift their focus to BA prediction for specific organs. For example, predicting the age of the lungs by vital capacity, or predicting the age of the heart by the frequency of heartbeats, this kind of research has important value for analyzing diseases of specific organs.

Overall, we want to help people who are currently studying BA predictions through a summary of existing technologies, and hope that in future, we can take part in the big data revolution together.

References

1. Albert, A.M., Ricanck, K. and Patterson, E. (2007). A review of the literature on the aging adult skull and face: Implications for forensic science research and applications, *Forensic Sci. Int.*, 172(1): 1-9.

2. Ali, S.M., Darbar, Z.A. and Junejo, K.N. (2015). Age estimation from facial images using biometric ratios and wrinkle analysis, *5th Natl. Symp. Inf. Technol. Towar. New Smart World (NSITNSW)*, pp. 1-5, doi: 10.1109/NSITNSW.2015.7176403.

3. Al-Shannaq, A.S. and Elrefaei, L.A. (2019). Comprehensive analysis of the literature for age estimation from facial images, *IEEE Access*, 7: 93229-93249.

4. Baker, G.T. and Sprott, R.L. (1988). Biomarkers of aging, *Exp. Gerontol.*, 23: 223-239.

5. Chen, B.C., Chen, C.S. and Hsu, W.H. (2015). Face recognition and retrieval using cross-age reference coding with cross-age celebrity dataset, *IEEE Trans. Multimed.*, 17(6): 804-815.

6. Ching, T., Himmelstein, D.S., Beaulieu-Jones, B.K., Kalinin, A.A., Do, B.T., Way, G.P., *et al.* (2018). Opportunities and obstacles for deep learning in biology and medicine, *J. R. Soc. Interface*, 15: 20170387. doi: 10.1098/rsif.2017.0387

7. Cole, J.H., Ritchie, S.J., Bastin, M.E., Valdés Hernández, M.C., Muñoz Maniega, S., Royle, N., *et al.* (2018). Brain age predicts mortality, *Mol. Psychiatr.*, 23: 1385-1392. doi: 10.1038/mp.2017.62

8. Cole, J.H., Marioni, R.E., Harris, S.E. and Deary, I.J. (2018). Brain age and other bodily 'ages': Implications for neuropsychiatry, *Mol. Psychiatr.*, 24: 266-281. doi: 10.1038/s41380-018-0098-1

9. Cole, J.H., Poudel, R.P.K., Tsagkrasoulis, D., Caan, M.W.A., Steves, C., Spector, T.D., *et al.* (2017). Predicting brain age with deep learning from raw imaging data results in a reliable and heritable biomarker, *Neuroimage*, 163: 115-124. doi: 10.1016/j.neuroimage.2017.07.059

10. Cole, J.H., Leech, R. and Sharp, D.J. (2015). Prediction of brain age suggests accelerated atrophy after traumatic brain injury, *Ann. Neurol.*, 77: 571-581. doi: 10.1002/ana.24367

11. Cole, J.H., Annus, T., Wilson, L.R., Remtulla, R., Hong, Y.T., Fryer, T.D., *et al.* (2017). Brain-predicted age in down syndrome is associated with beta amyloid deposition and cognitive decline, *Neurobiol. Aging.*, 56: 41-49. doi: 10.1016/j.neurobiolaging.2017.04.006

12. Cole, J.H., Jolly, A., De Simoni, S., Bourke, N., Patel, M.C., Scott, G., *et al.* (2018). Spatial patterns of progressive brain volume loss after moderate severe traumatic brain injury, *Brain*, 141: 822-836. doi: 10.1093/brain/awx354

13. Cole, .J.H., Underwood, J., Caan, M.W.A., De Francesco, D., Van Zoest, R.A., Leech, R., *et al.* (2017). Increased brain-predicted aging in treated HIV disease, *Neurology*, 88: 1349-1357. doi: 10.1212/WNL.0000000000003790

14. Kwon, Young Ho and Niels da Vitoria Lobo (1999). Age classification from facial images, *Comput. Vis. Image Underst.*, 74: 1-21.

15. Di Giuseppe, R., Arcari, A., Serafini, M., Di Castelnuovo, A., Zito, F., De Curtis, A., *et al.* (2012). Total dietary antioxidant capacity and lung function in an Italian population: A favorable role in premenopausal/never smoker women, *Eur. J. Clin. Nutr.*, 66: 61-68. doi: 10.1038/ejcn.2011.148

16. Duan, M., Li, K. and Li, K. (2018). An ensemble CNN2ELM for age estimation, *IEEE Trans. Inf. Forensics Secur.*, 13(3): 758-772.

17. Fabris, F., de Magalhães, J.P. and Freitas, A.A. (2017). A review of supervised machine learning applied to ageing research, *Biogerontology*, 18: 171-188. doi: 10.1007/s10522-017-9683-y

18. Franke, K. and Gaser, C. (2012). Longitudinal changes in individual Brain Age in healthy aging, mild cognitive impairment, and Alzheimer's Disease, *GeroPsych. J. Gerontopsychology Geriatr. Psychiatry*, 25: 235-245.

19. Franke, K., Clarke, G.D., Dahnke, R., Gaser, C., Kuo, A.H., Li, C., *et al.* (2017). Premature brain aging in baboons resulting from moderate fetal undernutrition, *Front Aging Neurosci.*, 9: 92. doi: 10.3389/fnagi.2017.00092

20. Gaser, C., Franke, K., Kloppel, S., Koutsouleris, N. and Sauer, H. (2013). BrainAGE in mild cognitive impaired patients: Predicting the conversion to Alzheimer's disease, *PLoS One*, 8.
21. Harman, D. (2001). Aging: Overview, *Ann. NY Acad. Sci.*, 928: 1-21.
22. Hatton, S.N., Franz, C.E., Elman, J.A., Panizzon, M.S., Hagler, D.J., Fennema-Notestine, C., *et al.* (2018). Negative fateful life events in midlife and advanced predicted brain aging, *Neurobiol. Aging*, 67: 1-9. doi: 10.1016/j.neurobiolaging.2018.03.004
23. Klare, B.F., Li, Z. and Jain, A.K. (2011). Matching forensic sketches to mug shot photos, *IEEE Trans. Pattern Anal. Mach. Intell.*, 33(3): 639-646.
24. Klare, B. and Jain, A.K. (2010). On a taxonomy of facial features. *In: Proc. 4th IEEE Int. Conf. Biometrics, Theory Appl. Syst. (BTAS)*, pp. 1-8.
25. Klemera, P. and Doubal, S. (2005). A new approach to the concept and computation of biological age, *Mech. Ageing Dev.*, 127: 240-248. doi: 10.1016/j.mad.2005.10.004
26. Li, K., Xing, J., Hu, W. and Maybank, S.J. (2017). D2C: Deep cumulatively and comparatively learning for human age estimation, *Pattern Recognit.*, 66: 95-105.
27. Liu, H., Lu, J., Feng, J. and Zhou, J. (2017). Group-aware deep feature learning for facial age estimation, *Pattern Recognit.*, 66: 82-94.
28. Liu, N., Zhang, F. and Duan, F. (2020). Facial age estimation using a multi-task network combining classification and regression, *IEEE Access*, 8: 92441-92451.
29. Lopez-Otin, C., Blasco, M.A., Partridge, L, Serrano, M. and Kroemer, G. (2013). The hallmarks of aging, *Cell*, 153: 1194-1217.
30. Luders, E., Cherbuin, N. and Gaser, C. (2016). Estimating brain age using high-resolution pattern recognition: Younger brains in long-term meditation practitioners, *NeuroImage*, 134: 508-513. http://dx.doi.org/10.1016/j.neuroimage.2016.04.007
31. Mamoshina, P., Kochetov, K., Putin, E., Cortese, F., Aliper, A., Lee, W.S., *et al.* (2018). Population specific biomarkers of human aging: A big data study using South Korean, Canadian and Eastern European patient populations, *J. Gerontol. Ser. A*, 00: 1-9. doi: 10.1093/gerona/gly005
32. Mamoshina, P., Vieira, A., Putin, E. and Zhavoronkov, A. (2016). Applications of deep learning in biomedicine, *Mol. Pharm.*, 13: 1445-1454. doi: 10.1021/acs.molpharmaceut.5b00982
33. Mamoshina, P., Koche, K., Cortese, F. and Kova, A. (2019). Blood biochemistry analysis to detect smoking status and quantify accelerated aging in smokers, *Sci. Rep.*, 15: 142. doi: 10.1038/s41598-018-35704-w
34. Murabito, J.M., Zhao, Q., Larson, M.G., Rong, J., Lin, H., Benjamin, E.J., *et al.* (2018). Measures of biologic age in a community sample predict mortality and age-related disease: The Framingham offspring study, *J. Gerontol. Ser. A Biol. Sci. Med. Sci.*, 73: 757-762. doi: 10.1093/gerona/glx144
35. Osman, O.F.E. and M.H. Yap (2019). Computational intelligence in automatic face age estimation: A survey, *IEEE Trans. Emerg. Top. Comput. Intell.*, 3(3): 271-285.
36. Putin, E., Mamoshina, P., Aliper, A., Korzinkin, M. and Moskalev, A. (2016). Deep biomarkers of human aging: Application of deep neural networks to biomarker development, *Aging* (Albany, NY), 8: 1021-1033. doi: 10.18632/aging.100968
37. Ricanek, K. and Tesafaye, T. (2006). MORPH: A longitudinal image database of normal adult age-progression, *FGR 2006 Proc. 7th Int. Conf. Autom. Face Gesture Recognit.*, 2006: 341-345.
38. Rodríguez, P., Cucurull, G., Gonfaus, J.M., Roca, F.X. and Gonzàlez, J. (2017). Age and gender recognition in the wild with deep attention, *Pattern Recognit.*, 72: 563-571.
39. Sebastiani, P., Thyagarajan, B., Sun, F., Schupf, N., Newman, A.B., Montano, M., *et al.* (2107). Biomarker signatures of aging, *Aging Cell*, 16: 329-338. doi: 10.1111/acel.12557

40. Steffener, J., Habeck, C., O'Shea, D., Razlighi, Q., Bherer, L. and Stern, Y. (2016). Differences between chronological and brain age are related to education and self-reported physical activity, *Neurobiol. Aging*, 40: 138-144. http://dx.doi.org/10.1016/j.neurobiolaging.2016.01.014
41. UTKFace dataset (2020). [Online]. Available: https://susanqq.github.io/UTKFace/. Accessed: 2020.
42. Wan, J., Tan, Z., Lei, Z., Guo, G. and Li, S.Z. (2018). Auxiliary demographic information assisted age estimation with cascaded structure, *IEEE Trans. Cybern.*, 48(9): 2531-2541.
43. Xie, J.C. and Pun, C.M. (2019). Chronological age estimation under the guidance of age-related facial attributes, *IEEE Trans. Inf. Forensics Secur.*, 14(9): 2500-2511.

Review on Social Behavior Analysis of Laboratory Animals: From Methodologies to Applications

Ziping Jiang¹, Paul L. Chazot² and Richard Jiang¹

¹ School of Computing and Communications, Lancaster University, InfoLab21,
   South Drive Lancaster University, Bailrigg, Lancaster LA1 4WA, United Kingdom
² Department of Biosciences, Durham University, Durham, UK

1. Introduction

Behavior describes the way animals interact with each other as well as the environment. The study of animal behavior is a long-standing topic that can provide insights into various objectives. In phylogenetics, the behavior pattern is a critical feature for comparing and grouping different species, especially laboratory animals and microorganisms. In ethology, animal behaviors are recorded to distinguish fixed action patterns and evolve behaviors to investigate the effects of environments on organisms. Additionally, behavior analysis is also essential in genetic research. With the progress of biotechnology, state-of-the-art genetic engineering technologies can transfer genes within and across species boundaries to produce improved organisms.

The behavioral differences between experimental groups and comparison group are a determining factor for analyzing the underlying genotype of a modified organism. The most commonly used laboratory animals are fruit fly and house mice, also known as *Drosophila melanogaster* and *Mus musculus*. Fruit fly can exhibit a wide range of complex social behaviors through only neurons, making them an ideal target for test genetic research. House mouse, on the other hand, has a similar genotype to that of humans; therefore, it is widely used for research in psychology, medicine, and other scientific disciplines. However, the labeling behavior of laboratory animals is a labor-intensive job. First, as the behaviors are performed randomly, most datasets contain dozens of hours of videos to provide a statistically solid result. Second, most of the videos are required to be annotated frame-by-frame due to the fast speed and short duration of laboratory animal behaviors. At last, this job requires professional knowledge that only biologists with related work experience can fulfill. To alleviate their burden, computer vision approaches are then introduced for automatic annotating.

Automating the analysis of animal behaviors, similar to that of human behavior, includes tracking and detecting. At different stages of the progress of artificial intelligence research, the proposed approaches are based on different techniques. In this work, we aim to provide a thorough investigation of related work, furnishing biologists with a scratch of efficient animal behavior annotation methods. Apart from that, we also discuss the strengths and weaknesses of those algorithms to provide some insights for those who are engaged in this field.

This review is organized as follows. In Section 2, we list the remarkable dataset along with the original objective when proposed. From Sections 3 to 5, we provide a general introduction on methodologies of related topics, including feature extraction, statistical learning, and deep learning. In Sections 6 and 7, we review the application of these methods in tracking and detecting the behavior of laboratory animals. At last, Section 8 talks about the miscellaneous works that are relevant to the topic.

2. Datasets

The Caltech Resident-Intruder Mouse dataset (CRIM13) [4] is introduced for the study of mice social behavior involved in aggression and courtship. The dataset consists of 237 videos, recorded from side view and top, synchronously. The scene of each video is similar. At the beginning of each video, a male resident mouse stays in the cage, and at some point, an intruder is introduced into the experiment to investigate the social interaction between the mice. The actions contained in the

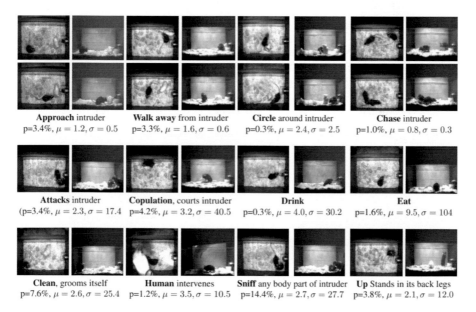

Fig. 1: Sample image from CRM13. Behavior categories: frame examples, descriptions, frequency of occurrence, (*p*), and duration mean and variance expressed in seconds (μ, σ). All behaviors refer to the cage resident, main mousey need to use the optional argument [t] with the side caption command.

video are categorized into 12 specified actions and one category to label the frames where no behavior of interest occurs. Mazur-Milecka and Ruminski [33] introduced a small dataset using for analysis of rat aggressive behavior. Instead of using the visual camera, they recorded the rat movement with a thermal camera. Since the cage is made from plexiglass without nesting, water nor food placed within, the heat of rats can be observed clearly.

The objectives of most proposed fruit fly datasets are to identify the phenotype differences between genetically edited objects. The Rubin GAL4 collection [20] is one of the largest datasets that provide a variety of phenotypes caused by different transgenic lines. In particular, the dataset has 14,524 16-minute videos of 2,144 different lines. However, the minor behavior differences can hardly be detected by current techniques, therefore the Rbuin GAL4 collection is only adopted in tracking tasks. On the contrary, Caltech fly-vs-fly interactions [11] are proposed for detecting aggressive behaviors of genetically edited flies. It contains 22 hours of fruit fly social behaviors, including courtship, threat, tussle, etc. The dataset is split into three subsets, according to the gender of the test subjects. Along with the dataset, the authors also provide a feature detector system that is able to extract the feature representations. FlyBowl [40] records videos of groups of flies freely behaving in

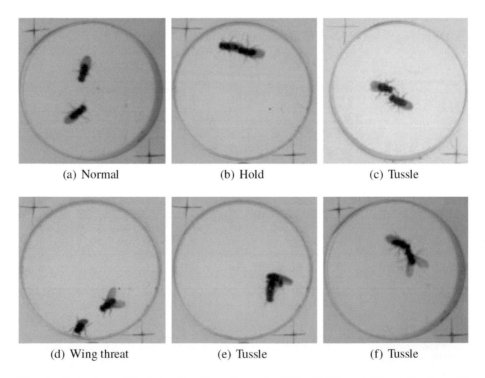

(a) Normal   (b) Hold   (c) Tussle

(d) Wing threat   (e) Tussle   (f) Tussle

Fig. 2: Examples of fly behaviors from fly-vs-fly [11]. (a) Normal flies without social behavior; (b) hold behavior between flies; (c) tussle behavior between flies; (d) wing threat behavior of a fly; (e) confusable tussle behavior (against hold); (f) confusable tussle behavior (against wing threat)

an open-field walking arena. The system is designed with automated tracking and behavior analysis algorithms for observing the social behaviors of multiple flies in a complex dynamic environment.

3. Methodologies

Research in tracking and labeling laboratory animal behavior has a long history. As an empirical topic, the methodologies of proposed models vary at different periods. In this section, we provide a brief discussion of related works, including (1) computer vision-based feature extraction methods that gathering useful information from raw data, (2) statistical learning methods that classify actions according to obtained feature, and (3) deep learning methods that provide an end-to-end solution to the tasks using a deep model with high computational power.

3.1 Feature Extraction

Intuitively, video data record the information about the changes of objects as well as the environment. To provide a convincing prediction of the recorded action in a sequence of frames, spatial features as well as temporal features should be taken into account.

Optical flow is one of the earliest approaches [8, 1]. It measures the apparent motion of objects in a sequence of frames. However, the validity of such measurement depends on the recording conditions. In fact, it can be influenced by variations in the background as well as the burst movements of multiple objects, especially when they are in occlusion.

Spatio-temporal interest point detection [26, 46]; on the other hand, focus on the points with high neighborhood variations in video, regardless of the size, frequency, and velocity of moving patterns. For those some of the data where useful information is also contained in the background, a descriptor is introduced for illustrating the surrounding of the interest points [27, 25].

3.2 Statistic Learning and Tracking Algorithms

Statistic learning dominates the research of Artificial Intelligence for decades. One of the most appealing properties of statistical learning methods is their explain ability and transparency. In particular, given a trained model $\mathcal{H}(\theta)$ and a feature vector $x = (x_1, x_2, \ldots, x_n)$, one can find out how the predicted result $\mathcal{H}(x, \theta)$ is influenced by each of the features. Moreover, since the training and prediction process can be observed, the model is open for amendment once the result is not satisfying.

The limitations of statistical learning methods are also straightforward. First, their capability is limited, since most of the real-world problems are too complex to be described by a model. Second, to obtain a satisfying result, the inputs of a model should be cautiously designed, indicating the pre-processing of data that requires great effort. At last, as a natural consequence of the high requirement of pre-processing, it lacks transferability.

Support Vector Machines (SVMs) is one of the most influential classification algorithms in the 1990s, which was widely used in visual pattern recognition. Denote

$(x_1, y_1), (x_2, y_2), \ldots, (x_n, y_n)$ as a sample set where $x_i \in R^m$ is the feature vector while $y_i \in \{-1, 1\}$ is the class label. The intuitive idea of SVMs is to separate the dataset according to their ground truth label, using a hyperplane. Mathematically, $wx + b = 0$. However, for real-world problems with high complexity, a linear classifier is not sufficient to provide an accurate prediction. A kernel function is then introduced to provide non-linearity by transferring the features. The SVMs are first applied in recognizing human action [39], where the features are extracted using local measurement of spatio-temporal interest points.

Hidden Markov Model (HMM) assumes that the current state of a random variable depends on its previous state. However, the state cannot be observed directly but can influence the behavior of an observable process. The hidden state space consists of one of N possible values, modeled by a distribution. This implies that the state of the object at time t is selected from one of the hidden states, and a transition matrix is introduced to describe the probability of state at time $t + 1$ given the current state. For each of the N possible states, there is an emission function that illustrates the observed value. During training, the parameters of emission function, as well as the transition matrix, are determined according to the observed features.

3.3 Deep Learning Methods

A neural network with a deep structure, which is known as deep model, has a long history and was popular in the 1980s and 1990s [48]. However, due to the limitation of datasets and computation power, its fashion fell out in the 2000s. In recent years, with the emergence of large annotated datasets, such ILSVR [38], PASCAL VOC [9], and the development of high-performance computing techniques, deep models were proved effective by many proposed models, such as VGG16 [42], ResNet [16], etc.

Object detection aims to select the objects from a given image. For the large-scale image object detection task, R-CNN [13] introduced an inspiring two-stage architecture by combining a proposal detector and region-wise classifier. SPP-Net [15] and Fast R-CNN [12] were then introduced with the idea of region-wise feature extraction which significantly accelerated the overall detector. Girshick *et al*. [37] proposed a Regional Proposal Network, which is almost cost-free by sharing conv features with a detection network, for object-bound prediction. A multistage detector Cascade R-CNN [5] was then proposed which improved the accuracy of detection by setting increasing IoU thresholds for a sequence of detectors.

Action recognition has similar objective with object detection but is performed video-wise. Academic research in action recognition has made great progress in recent years [28]. Karpathy *et al*. [24] studied the performance of CNN and found that CNN architecture is capable of action recognition in large-scale video. Ng *et al*. [47] adopted a CNN for feature extraction, followed by an LSTM for action classification. Simonyan *et al*. [41] proposed a two-stream ConvNet, consisting of separate networks for the frame and optical flow, incorporating spatial and temporal networks. Ji *et al*. [21] proposed CNN-based human detector and head tracker. Tran *et al*. [44] proposed a simple but effective 3D CNN architecture for video classification.

4. Tracking Models

The starting point of behavior analysis is tracking the target objects. Prior to the age of deep learning, statistical learning methods dominated the empirical research of Artificial Intelligence.

Smart vivarium [2, 3] is one of the representative works that applies tracking algorithms into the semi-natural enclosure. Different from an open environment, occlusion between objects happens more frequently in an enclosure. To address this issue, the algorithm combines Bayesian Multiple Blob tracker [19] and contour tracker [30]. To increase the robustness of the model, bootstrap filtering is performed using blob observations, and course observations for each mouse independently. At last, the output of the algorithm is computed and sampled independently for each mouse and weighs their importance. The advantage of adopting particle filtering is that the important region for each mouse is calculated independently, preventing loss of track of an occluded object.

One of the most inspiring contributions of the smart vivarium is that it fills the deficiency that prior tracking algorithms have difficulty in detecting objects in occlusion with a lower computational burden. However, it also inherits some of the weaknesses of the original works. For instance, the bounding region does not always match the posture of objects since the templates are fixed. Second, it is sometimes stuck in local optima due to the roughness of the likelihood function of the contour tracker. At last, there is a chance that the identity labels swap when there are multiple occlusions, which is a usual failure mode, that is observed in various algorithms.

The EthoVision [43] a video tracking system that builds on straightforward computer vision methods. The system offers three detection methods: (1) gray scaling separates the image into background and foreground according to the pixel brightness; (2) subtraction method first records a reference image for the background, and then subtracts the value of the reference image from the live image to acquire the object location; (3) the third method defines the color of an object by its hue and saturation and tracking the object by matching the pattern with the given image. However, their method assumes that the background of the video is stationary, and there are limited objects that hardly overlap. BioTracker [35] is an open-source computer vision framework. With a complete video I/O, graphics overlay and mouse and keyboard interfaces, the framework provides a user-friendly environment to those who have limited computer vision knowledge.

Janelia Automatic Animal Behavior Annotator (JAABA) is [23] an intuitive, interactive system for annotating laboratory animals, including mouse, fruit fly and larva. Given a dataset, the interface of JAABA allows users to label a selected animal in a selected frame of behavior. The system then computes per-frame features from the trajectory data as well as windowed features that provide temporal context for the current frame. These features are passed to the JAABA machinery, which learns a new behavior pattern for labeling the future data.

4.1 Behavior Detection

Prior to the progress of deep learning methods, research in action detection can be viewed as a two-step process: spatial-temporal feature extraction and action detection based on the extracted features.

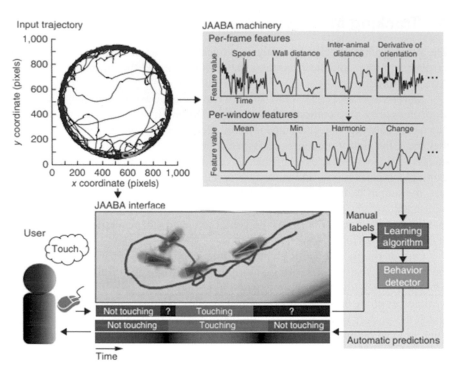

Fig. 3: JAABA overview. Top left: is the input trajectory: the *x,y* position of a fly over 1000s of video is plotted in black. *Bottom*: The interface of JAABA where users interact with the interface to annotate the behavior. *Gray shading*: The underlying JAABA machinery where the per-frame features and per-window features are computed and used as inputs for training the detector

Dollar *et al.* [7] is one of the earliest works that illustrate the effectiveness of using spatial-temporal features in mouse behavior detection. They suggest that despite the overall appearance of two instances of the same behavior might be different, the motion of most of the interest points should be similar. Based on that observation, they extract spatial-temporal cuboids that are centered at the interest points and create a collection of cuboids phenotypes using clustering algorithms. Their experiments show that the algorithm has better performance in identifying emotions than detecting mouse behavior, even though the behaviors are contained in the dataset are simple. The following works address the inadequacy of the model by improving the feature extractor, as well as the classification methods.

Hong *et al.* [17] provide a detailed example of applying Artificial Intelligence methods to analyze genetic influences on the social behavior of mice. In particular, they integrate a visual camera with a depth camera to obtain pose information. The information is then used to extract features from body size to relative angles. Although the classier they applied for behavior detection are standard methods with limited modification, they provide a thorough investigation of how genetic differences influence the behavior pattern of the mouse.

Dankert *et al.* [6] propose a method based on machine vision methods to measure the aggression and courtship of fruit flies. Based on the monitored visual data, they introduce a method to compute the location, orientation, and wing posture of each fly. Once the features are computed, they estimate the social behavior using statistical methods. Based on their work, Eyjolfsdottir *et al.* [11] introduced additional features that could better describe the status of interacting flies. Apart from that, they suggested that classifying the actions frame-by-frame can result in a noisy result, and introduced a sliding window that takes the features in adjacent frames into account to increase the robustness of their method. Before the training of the model, the features are then transferred into different features to provide a thorough description of the current status of the objects, including temporal region features, harmonic features, boundary features, etc. The prediction is performed by a hidden semi Markov model. As the deep learning methods shows its potential in various fields, recent attempts have been made to introduce deep neural network into biology research. The advantage is that deep learning models are able to integrate feature extraction and classification into an end-to-end pipeline, providing better transferability. However, it is obtained at the expense of interpretability. Jiang *et al.* [22] are one of the earliest methods that introduce a 2D-3D hybrid framework for detecting social behavior of fruit flies, where the 2D convolutional network is used for extracting the features frame-wise, while the 3D convolutional blocks fuse the spatial-temporal features. However, it is observed that due to the high similarity between different actions, the model performance is not satisfying.

Another example is provided by Eyjolfsdottir *et al.* [10], in which the authors introduce the recurrent neural network to process the sequential motion data. The network has a discriminative part for action classification and a generative part for motion predicting. The recurrent cells of the generative part are laterally connected, which enables the network to learn a high-level representations of behavioral phenomena.

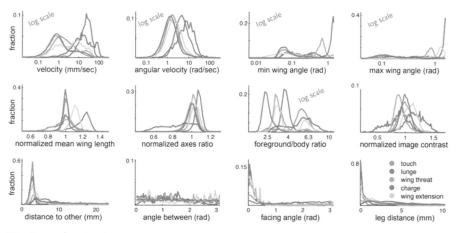

Fig. 4: Frame-wise feature distribution for the actions of the boy meets boy sub-dataset from fly-vs-fly [11]. The features are used to train a hidden semi-Markov model

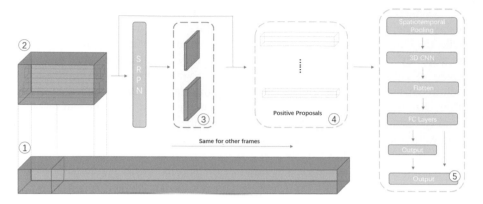

Fig. 5: Model structure from [22]. (1) Feature map of a batch. (2) Bout features for classifying behaviors in frame $k + 1$, containing features of frame 1 to $2k + 1$. Likewise, the last bout in the batch is used for classifying behaviors in frame $L - k$, containing features frame $L - 2k$ to frame L. Thus, there are $L - 2k$ bouts in a batch of input. (3) Output of SRPN consist of a 'score map' and a 'coordinates map', stands for the probability of 'behavior score' and adjusted bounding boxes coordinates of the corresponding anchor box. (4) Proposed bounding boxes are then sorted according to their 'behavior score', M_{clx} bounding boxes with highest score are selected and then fed into classifier. (5) The classifier starts with spatio-temporal pooling layer to pool features in bounding boxes into a fixed size, followed by 3D convolutional layers and fc layers.

5. Other Applications

5.1 Sensor-based Tracking and Detecting

The development of sensor technologies enables a variety of types of data other than visual data. In this section, we provide a discussion on the novel usage of advanced monitor technologies.

Mazur-Milecka and Ruminski [34] propose an efficient deep learning-based method for tracking laboratory mice. The model uses U-net and V-net architectures to convert heat maps to the gray-scale images as well as the segmentation map. However, their model is only valid when the enclosure is clear, otherwise, the heatmap can be affected by other objects.

Weissbrod *et al.* [45] introduce a video radio-frequency-identified (RFID) tagging system and tracking system in a semi-natural environment for tracking laboratory animals and identifying their behavior. RFID technology is able to identify the location and movements of tagged targets and transmit the information to antenna receivers. The collected positional and movement information is then used for social behavior analysis.

However, there are several limitations of such an approach. First, setting up an RFID system requires hardware support, which cannot always be matched for every laboratory. Second, the system is useful in tracking objects, while with only position information, it is hard to provide a straightforward demonstration of detected behaviors.

5.2 Pose Estimation

DeepLabCut [32] is one of the first works that applies deep learning methods into the pose estimation. The model is developed from the DeepCut algorithms [36, 18] and shows that a limited number of training data is sufficient to train the model with human-level accuracy.

LiftPose3D [14] is a deep network-based method that reconstructs 3D poses from a single 2D camera view. The model is developed from a network that was initially designed to estimate human posture [31]. Given a 3D poses library, a neural network is trained to match its key points at different 2D angles. Once the matching between key points from the 2D frame and the 3D model is established, the model is able to reconstruct the posture of the target from its 2D image at any angle.

OptiFlex [29] is the first video-based architecture for estimating animal pose. The model consists of a flexible base model that accounts for variability in animal body shape, and an optical flow model that obtains temporal context information from nearby frames. Similar to prior works, the base model is applied to provide a baseline estimation of the target pose, while the novel optical flow module takes a sequence of heatmap frames to enhance the robustness of prediction.

References

1. Ali, S. and Shah, M. (2008). Human action recognition in videos using kinematic features and multiple instance learning, *IEEE Transactions on Pattern Analysis and Machine Intelligence*, 32(2): 288-303.
2. Belongie, S., Branson, K., Dollár, P. and Rabaud, V. (2005). Monitoring animal behavior in the smart vivarium. *In: Measuring Behavior*, pp. 70-72, Wageningen, The Netherlands.
3. Branson, K. and Belongie, S.(2005). Tracking multiple mouse contours (without too many samples). *In: 2005 IEEE Computer Society Conference on Computer Vision and Pattern Recognition (CVPR'05)*, vol. 1, pp. 1039-1046, IEEE.
4. Burgos-Artizzu, X.P., Dollár, P., Lin, D., Anderson, D.J. and Perona, P. (2012). Social behavior recognition in continuous video. *In: 2012 IEEE Conference on Computer Vision and Pattern Recognition*, pp. 1322-1329, IEEE.
5. Cai, Z. and Vasconcelos, N. (2017). *Cascadercnn: Delving into High Quality Object Detection*, arXiv preprint arXiv:1712.00726
6. Dankert, H., Wang, L., Hoopfer, E.D., Anderson, D.J. and Perona, P. (2009). Automated monitoring and analysis of social behavior in Drosophila, *Nature Methods,* 6(4): 297-303.
7. Dollár, P., Rabaud, V., Cottrell, G. and Belongie, S. (2005). Behavior recognition via sparse spatio-temporal features. *In: 2005 IEEE International Workshop on Visual Surveillance and Performance Evaluation of Tracking and Surveillance*, pp. 65-72, IEEE.
8. Efros, A.A., Berg, A.C., Mori, G. and Malik, J. (2003). Recognizing action at a distance. *In: Computer Vision, IEEE International Conference*, vol. 3, pp. 726-726, IEEE Computer Society.
9. Everingham, M., Van Gool, L., Williams, C.K., Winn, J. and Zisserman, A. (2010). The Pascal visual object classes (voc) challenge, *International Journal of Computer Vision*, 88(2): 303-338.

10. Eyjolfsdottir, E., Branson, K., Yue, Y. and Perona, P. (2016). Learning recurrent representations for hierarchical behavior modeling, arXiv preprint arXiv, 1611:00094.

11. Eyjolfsdottir, E., Branson, S., Burgos-Artizzu, X.P., Hoopfer, E.D., Schor, J., Anderson, D.J. and Perona, P. (2014). Detecting social actions of fruit flies. *In: European Conference on Computer Vision*, pp. 772-787, Springer.

12. Girshick, R. (2015). Faster r-CNN. *In: Proceedings of the IEEE International Conference on Computer Vision*, pp. 1440-1448.

13. Girshick, R., Donahue, J., Darrell, T. and Malik, J. (2014). Rich feature hierarchies for accurate object detection and semantic segmentation. *In: Proceedings of the IEEE Conference on Computer Vision and Pattern Recognition*, pp. 580-587.

14. Gosztolai, A., Günel, S., Ríos, V.L., Abrate, M.P., Morales, D., Rhodin, H., Fua, P. and Ramdya, P. (2021). Liftpose 3d, a deep learning-based approach for transforming 2d to 3d pose in laboratory animals, *BioRxiv*, pp. 2020-09.

15. He, K., Zhang, X., Ren, S. and Sun, J. (2014). Spatial pyramid pooling in deep convolutional networks for visual recognition. *In: European Conference on Computer Vision*, pp. 346-361, Springer.

16. He, K., Zhang, X., Ren, S. and Sun, J. (2016). Deep residual learning for image recognition. *In: Proceedings of the IEEE Conference on Computer Vision and Pattern Recognition*, pp. 770-778.

17. Hong, W., Kennedy, A., Burgos-Artizzu, X.P., Zelikowsky, M., Navonne, S.G., Perona, P. and Anderson, D.J. (2015). Automated measurement of mouse social behaviors using depth sensing, video tracking, and machine learning, *Proceedings of the National Academy of Sciences*, 112(38): E5351-E5360.

18. Insafutdinov, E., Pishchulin, L., Andres, B., Andriluka, M. and Schiele, B. (2016). Deepercut: A deeper, stronger, and faster multi-person pose estimation model. *In: European Conference on Computer Vision*, pp. 34-50, Springer.

19. Isard, M. and MacCormick, J. (2001). Bramble: A bayesian multiple-blob tracker. *In: Proceedings Eighth IEEE International Conference on Computer Vision. ICCV,* vol. 2, pp. 34-41, IEEE.

20. Jenett, A., Rubin, G.M., Ngo, T.T., Shepherd, D., Murphy, C., Dionne, H., Pfeiffer, B.D., Cavallaro, A., Hall, D., Jeter, J., *et al.* (2012). A gall-driver line resource for Drosophila neurobiology, *Cell Reports*, 2(4): 991-1001.

21. Ji, S., Xu, W., Yang, M. and Yu, K. (2013). 3d convolutional neural networks for human action recognition, *IEEE Transactions on Pattern Analysis and Machine Intelligence* 35(1): 221-231.

22. Jiang, Z., Chazot, P.L., Celebi, M.E., Crookes, D. and Jiang, R. (2019). Social behavioral phenotyping of Drosophila with a 2d–3d hybrid CNN framework, *IEEE Access*, 7: 67972-67982.

23. Kabra, M., Robie, A.A., Rivera-Alba, M., Branson, S. and Branson, K. (2013). Jaaba: Interactive machine learning for automatic annotation of animal behavior, *Nature Methods*, 10(1): 64-67.

24. Karpathy, A., Toderici, G., Shetty, S., Leung, T., Sukthankar, R. and Fei-Fei, L. (2014). Large-scale video classification with convolutional neural networks. *In: Proceedings of the IEEE Conference on Computer Vision and Pattern Recognition*, pp. 1725-1732.

25. Klaser, A., Marszałek, M. and Schmid, C. (2008). A spatio-temporal descriptor based on 3d-gradients. *In: BMVC 2008 - 19th British Machine Vision Conference*, pp. 275-1, British Machine Vision Association.

26. Laptev, I. (2005). On space-time interest points, *International Journal of Computer Vision*, 64(2): 107-123.

27. Laptev, I., Marszalek, M., Schmid, C. and Rozenfeld, B. (2008). Learning realistic human actions from movies. *In: 2008 IEEE Conference on Computer Vision and Pattern Recognition*, pp. 1-8, IEEE.

28. LeCun, Y., Bengio, Y. and Hinton, G. (2015). Deep learning, *Nature*, 521(7553): 436.
29. Liu, X., Yu, S.Y. and Flierman, N., Loyola, S., Kamermans, M., Hoogland, T.M. and De Zeeuw, C.I. (2020). Optiflex: Video-based animal pose estimation using deep learning enhanced by optical flow, BioRxiv. DOI: 10.1101/2020.04.04.025494.
30. MacCormick, J. (2012). Stochastic Algorithms for Visual Tracking: Probabilistic Modeling and Stochastic Algorithms for Visual Localization and Tracking, Springer Science & Business Media.
31. Martinez, J., Hossain, R., Romero, J. and Little, J.J. (2017). A simple yet effective baseline for 3d human pose estimation. *In: Proceedings of the IEEE International Conference on Computer Vision*, pp. 2640-2649.
32. Mathis, A., Mamidanna, P., Cury, K.M., Abe, T., Murthy, V.N., Mathis, M.W. and Bethge, M. (2018). Deeplabcut: Markerless pose estimation of user-defined body parts with deep learning, *Nature Neuroscience*, 21(9): 1281-1289.
33. Mazur-Milecka, M. and Rumiński, J. (2017). Automatic analysis of the aggressive behavior of laboratory animals using thermal video processing. *In: 2017 - 39th Annual International Conference of the IEEE Engineering in Medicine and Biology Society (EMBC)*, pp. 3827-3830, IEEE.
34. Mazur-Milecka, M. and Ruminski, J. (2021). Deep learning based thermal image segmentation for laboratory animals tracking, *Quantitative InfraRed Thermography Journal*, 18(3): 159-176.
35. Mönck, H.J., Jörg, A., von Falkenhausen, T., Tanke, J., Wild, B., Dormagen, D., Piotrowski, J., Winklmayr, C., Bierbach, D. and Landgraf, T. (2018). Biotracker: An open-source computer vision framework for visual animal tracking, arXiv preprint arXiv:1803.07985
36. Pishchulin, L., Insafutdinov, E., Tang, S., Andres, B., Andriluka, M., Gehler, P.V. and Schiele, B. (2016). Deepcut: Joint subset partition and labeling for multi person pose estimation. *In: Proceedings of the IEEE Conference on Computer Vision and Pattern Recognition*, pp. 4929-4937.
37. Ren, S., He, K., Girshick, R. and Sun, J. (2015). Faster r-CNN: Towards real-time object detection with region proposal networks. *In: Advances in Neural Information Processing Systems*, pp. 91-99.
38. Russakovsky, O., Deng, J., Su, H., Krause, J., Satheesh, S., Ma, S., Huang, Z., Karpathy, A., Khosla, A., Bernstein, M., *et al.* (2015). Imagenet large-scale visual recognition challenge, *International Journal of Computer Vision*, 115(3): 211-252.
39. Schuldt, C., Laptev, I. and Caputo, B. (2004). Recognizing human actions: A local svm approach. *In: Proceedings of the 17th International Conference on Pattern Recognition, ICPR*, vol. 3, pp. 32-36, IEEE.
40. Simon, J.C. and Dickinson, M.H. (2010). A new chamber for studying the behavior of Drosophila, *Plos One*, 5(1): e8793.
41. Simonyan, K. and Zisserman, A. (2014). Two-stream convolutional networks for action recognition in videos. *In: Advances in Neural Information Processing Systems*, pp. 568-576.
42. Simonyan, K. and Zisserman, A. (2014). Very deep convolutional networks for large-scale image recognition, arXiv preprint arXiv, 1409.1556
43. Spink, A., Tegelenbosch, R., Buma, M. and Noldus, L. (2001). The ethovision video tracking system – A tool for behavioral phenotyping of transgenic mice, *Physiology & Behavior*, 73(5): 731-744.
44. Tran, D., Bourdev, L., Fergus, R., Torresani, L. and Paluri, M. (2015). Learning spatiotemporal features with 3d convolutional networks. *In: Proceedings of the IEEE International Conference on Computer Vision*, pp. 4489-4497.

45. Weissbrod, A., Shapiro, A., Vasserman, G., Edry, L., Dayan, M., Yitzhaky, A., Hertzberg, L., Feinerman, O. and Kimchi, T. (2013). Automated long-term tracking and social behavioral phenotyping of animal colonies within a semi-natural environment, *Nature Communications*, 4(1): 1-10.
46. Willems, G., Tuytelaars, T. and Van Gool, L. (2008). An efficient dense and scale-invariant spatio-temporal interest point detector. *In: European Conference on Computer Vision*, pp. 650-663, Springer.
47. Yue-Hei Ng, J., Hausknecht, M., Vijayanarasimhan, S., Vinyals, O., Monga, R. and Toderici, G. (2015). Beyond short snippets: Deep networks for video classification. *In: Proceedings of the IEEE Conference on Computer Vision and Pattern Recognition*, pp. 4694-4702.
48. Zhao, Z.Q., Zheng, P., Xu, S.T. and Wu, X. (2018). Object detection with deep learning: A review, arXiv preprint arXiv:1807.05511

Acute Lymphoblastic Leukemia Diagnosis Using Genetic Algorithm and Enhanced Clustering-based Feature Selection

Siew Chin Neoh[1], Srisukkham Worawut[2], Li Zhang[3] and Md. Mostafa Kamal Sarker[4]

[1] School of Computer Science, Nottingham University Malaysia Campus, Malaysia
[2] Faculty of Science, Chiang Mai University, Thailand
[3] Department of Computer Science, Royal Holloway, University of London, UK
[4] Precision Medicine Centre of Excellence, The Patrick G Johnston Centre for Cancer Research, Queen's University Belfast, Belfast BT9 7AE, UK
E-mail:

1. Introduction

Acute lymphoblastic leukemia (ALL) is an aggressive type of cancer with respect to white blood cells and it can be fatal within days without treatment [1, 5, 12]. It can affect patients of all ages, including infants and children. Early diagnosis may significantly increase survival rates. Owing to the subtle variations between lymphocytic and lymphoblastic cells, it is a challenging task to attain precise diagnosis of normal and ALL cases. Since it is crucial to generate an effective lesion representation to ensure the success of ALL classification, we explore microscopic feature extraction and evolutionary algorithm-based feature selection in this research.

Specifically, we propose evolutionary algorithm-based clustering and classification methods for feature selection pertaining to ALL diagnosis. Our system consists of three key stages, i.e. feature extraction, feature selection, and classification. Firstly, we employ feature descriptors, such as local binary patterns (LBP) for feature extraction from the white blood cell sub-images. Secondly, the genetic algorithm (GA) algorithm is used in conjunction with classification and clustering methods to identify the most significant features while removing irrelevant ones to inform subsequent ALL diagnosis. In this research, we propose three evolutionary feature selection methods, i.e. filter-based enhanced Fuzzy C-Means (EFCM) clustering + GA, (2) filter-based linear discriminant analysis (LDA) + GA, and (3) embedded-based Support Vector Machine (SVM) + GA. In particular, for the EFCM clustering, we take both inter- and intra-cluster variances into account to overcome the limitations of the original FCM, where only the intra-cluster differences are considered. After the identification of the most significant features using the above feature selection methods, a number of classification methods, such as Naive Bayes (NB), K-Nearest Neighbors (KNN), Multilayer Perceptron (MLP) and SVM, are used for the classification of healthy and ALL cases.

A comprehensive evaluation is conducted for ALL identification, using the ALL-IDB2 dataset. The empirical results indicate the efficiency of the proposed EFCM + GA feature selection approach in comparison with LDA + GA, SVM + GA, and other feature projection methods.

The paper is organized as follows. Section 2 discusses related studies on a variety of feature selection methods and deep neural networks for ALL diagnosis. We introduce the proposed three-feature selection methods, i.e. EFCM + GA, LDA + GA, and SVM + GA in Section 3. Evaluation details are presented in Section 4. Section 5 concludes this research and identifies future directions.

2. Related Work

Blood cancer detection has gained increasing research attention. A variety of machine learning and deep learning methods have been developed for leukemia diagnosis. As an example, Neoh *et al.* [12] proposed a stimulating discriminant measures (SDM) method for the segmentation of nucleus and cytoplasm of lymphocytic and lymphoblastic cells. Their model adopted a new loss function where both the inter-cluster and intra-cluster differences were taken into account for the nucleus-cytoplasm segmentation. Discriminative features were subsequently extracted from the segmented nucleus and cytoplasm in the sub-image for ALL diagnosis. On evaluation by using a number of ensemble classification models (e.g. Dempster-Shafer) with bootstrapping and tenfold cross-validation, their work achieved superior performance for healthy and blast cell identification. Tan *et al.* [15] developed two novel PSO methods, i.e. PSO with adaptive coefficients (ACPSO) and PSO with random coefficients (RCPSO), for undertaking feature selection and hyper-parameter fine-tuning in deep networks with respect to skin lesion and blood cancer detection. Their enhanced PSO models employed circle, sine, and helix functions for adaptive and random coefficient generation. Feature descriptors, such as Histogram of Oriented Gradients (HOG), Local Binary Patterns (LBP), and Gray Level Run Length Matrix (GLRLM) were used to extract raw features from the images. ACPSO and RCPSO were subsequently used to identify the most discriminative features from the extracted raw feature sets. In parallel, the proposed PSO models were also used to fine-tune the hyper-parameters of deep convolutional neural networks (CNNs). On evaluating by using a number of skin lesion and blood cancer data sets, their work showed impressive performances than a number of existing methods. Xie *et al.* [21] proposed two modified Firefly Algorithm (FA) algorithms, i.e. inward intensified exploration FA (IIEFA) and compound intensified exploration FA (CIEFA), in conjunction with K-Means (KM) clustering for the identification of lymphocytes and lymphoblasts for ALL diagnosis. The two FA variants employed a multi-dimensional search mechanism and a dispersing strategy to increase search diversity and re-allocate fireflies with high similarities to remote regions to extend the search territory. The models were then used to further enhance the clustering centroids of the KM clustering. On evaluation by using ALL-IDB2, their model outperformed the original KM clustering and a number of classical and advanced FA methods, significantly, for ALL identification and other classification tasks.

Tan *et al*. [16] proposed an adaptive learning PSO (ALPSO) for skin lesion and nucleus-background segmentation of skin lesion and blood cell sub-images. Their work employed a series of search operations, including differential evolution (DE), PSO, simulated annealing (SA), and Levy flight, to adjust the population and search diversity of the original PSO method. Their model was subsequently used to optimize the hyper-parameters of CNNs and the clustering centroids of FCM for the lesion and nucleus-background segmentation. In particular, in comparison with other deep networks and clustering methods in combination with advanced PSO variants, their model achieved impressive performance for nucleus segmentation to inform ALL diagnosis. Srisukkham *et al*. [14] proposed two bare-bones PSO (BBPSO) algorithms for feature selection pertaining to ALL diagnosis using microscopic images. Their BBPSO models enabled the swarm to follow optimal signals, i.e. the averaged solution of global and personal best solutions, while simultaneously moving away from worst regions. Their BBPSO models were subsequently used to perform feature optimization by removing redundant and irrelevant characteristics while identifying the most significant ones to enhance the performance. Their work obtained enhanced performance in comparison with a number of classical and advanced BBPSO and PSO variants for ALL diagnosis. Two PSO-based feature selection methods for classification models, i.e. PSOVA1 and PSOVA2, were also proposed by Xie *et al*. [22] for ALL diagnosis. PSOVA1 adopted random and remote personal best solutions of other particles to diversify the local and global elite solutions to guide the search process. PSOVA2 employed adaptive exemplar breeding mechanisms, nonlinear search coefficients, exponential and scattering schemes for swarm leader, and worst solution enhancement, to overcome stagnation. Their work outperformed 15 baseline classical and advanced search methods for discriminative feature selection pertaining to ALL diagnosis. There are also a number of other studies, such as FA-based feature selection [24] and FA-based ensemble model construction [25] for ALL classification using ALL-IDB2 and other blood cancer data sets. A comprehensive review of automated ALL diagnosis is provided in Ghaderzadeh *et al*. [5].

3. The Proposed GA-based Feature Selection Methods for ALL Diagnosis

In this research, we propose automated ALL identification to promote early and instant diagnosis. Our system consists of three key stages, i.e. feature extraction, feature selection, and classification. In particular, three GA-based feature selection methods, i.e. (1) filter-based EFCM + GA, (2) filter-based LDA + GA, and (3) embedded-based SVM + GA, are proposed to identify the most significant features while removing redundant ones. Several popular machine learning models (e.g. MLP, SVM, NB, and KNN) are subsequently used for ALL classification.

In this research, we employ a total of 180 sub-images with 60 negative (healthy) and 120 positive (unhealthy) cases extracted from the ALL-IDB2 data set to test model efficiency. In our previous studies, Srisukkham *et al.* [14 extracted a total of 80 features consisting of 16 shape, 54 texture, and 10 color features from the segmented nucleus and cytoplasm in each sub-image.

The key features include the areas of cytoplasm and nucleus, nucleus to cytoplasm ratio, length to diameter ratio, filled area, perimeter, solidity, eccentricity, form factor, compactness, roundness of nucleus region, correlation, contrast, difference variance, entropy, cluster prominence, cluster shade, dissimilarity, skewness, kurtosis, mean and standard deviations of the a* and b* components of CIELAB color space, etc.

Although 80 features are extracted from the segmented images, there are features which are significantly more important than others and there is always a possibility of including redundant or unimportant information which may influence the classification performance. As a result, it is necessary to conduct feature selection in order to identify the most discriminant features for ALL detection to improve classification accuracy.

As mentioned earlier, this research develops three GA-based feature optimization algorithms: (1) filter-based EFCM + GA, (2) filter-based LDA + GA, and (3) embedded-based SVM + GA to select the most discriminating features for robust and reliable ALL detection. The first two feature selection processes are combined with clustering methods, i.e. EFCM and LDA, with inter- and intra-cluster variance measurements for normal and blast cell identification. The third feature selection process is used in conjunction with a classification method, i.e. SVM, for distinguishing benign and malignant cases. The GA is used for feature optimization owing to its great efficiency in solving diverse optimization problems in comparison with other classical search methods [11, 13, 26, 17, 18].

Specifically, we propose a new minimization function for the EFCM clustering method pertaining to the clustering performance measurement. In comparison with the original FCM where the objective function only takes intra-cluster variations into account, this proposed new objective function adopts both inter- and intra-cluster variations for clustering performance measurement. Moreover, the intra-clustering variance is calculated as the largest distance between the data sample and its corresponding clustering centroid, while the inter-clustering variance is defined as the smallest distance between the instance of one cluster and the centroid of another cluster.

Algorithm 1 illustrates the detailed data flow using EFCM + GA for feature selection. A similar process is also adopted by LDA + GA. Algorithm 2 shows the feature selection process using SVM + GA. We introduce these algorithms in details given below.

Algorithm 1 illustrates the pseudo-code of the feature selection process using EFCM + GA. At the initial stage, a population, P, that comprises of binary-valued chromosomes, $I_1, I_2,..., I_k$, each with the length of 80 (i.e. the 80 raw features), is generated. Each gene of the chromosome represents a feature where the gene value of '1' indicates the feature that is selected while '0' indicates the feature not selected. The algorithm starts by evaluating the selected features from training images that consist of both normal and abnormal lymphocytes. Having lymphocytes and lymphoblasts as two different clusters, the newly proposed object function is employed to calculate the between-cluster and within-cluster variances based on each selected feature set. Subsequently, a fitness value, F = SW_{EFCM}/SB_{EFCM} is assigned to each set of features represented by I_1, I_2,..., I_k. Then, the chromosomes are ranked according to their fitness. Stochastic universal sampling is used to select chromosomes for crossover and mutation with the probability of 0.7 and 0.3 respectively. The generated offspring are then evaluated and recombined with parent chromosomes based on a generation gap of 0.9. The process of evolution is then continued until the maximum generation is reached. The same algorithm is also used for the development of filter-based selection using LDA + GA. The only difference is that between-cluster and within-cluster variances of LDA, SB_{LDA} and SW_{LDA}, are used instead of those for EFCM.

Algorithm 1: GA for Filter-based Feature Selection using EFCM

Input:

$P(I_1, I_2,...I_k)$;

80 extracted features;

Images for training (normal and abnormal)

Output:

The best proposed features for recognizing normal and abnormal lymphocytic cells

Begin

    Separate training images into two clusters: normal and abnormal;

    for each $I_i \in P$

    Evaluate within-cluster and between-cluster variations for **EFCM**;

    Assign fitness based on the aforementioned new object function;

    end for

    while maximum iteration not reached

    Rank P;

    SE : = select (P); //Select chromosomes based on stochastic universal sampling

    OF : = crossover (SE); //Crossover operation to generate offspring

    OF : = mutation (OF); //Mutation operation to generate offspring

    Evaluate each $I_i \in OF$; //Determine fitness for each new I_i in OF based on **EFCM**

    P: = merge (P, OF);

    end while

    return new updated P;

End

Algorithm 2 depicts the pseudo-code for embedded-based feature selection where the GA is used to interact with the SVM classifier to fine-tune the features for recognizing normal and abnormal lymphocytic cells. As the feature selection process is only conducted in the training stage, 90 randomly selected images among the total of 180 images are used as the training set. In order to ensure more convincing training accuracy, tenfold cross-validation is employed by randomly assigning 90 training images into ten folds where a fold index is assigned to each image. In v-fold cross-validation, the overall dataset is firstly divided into v groups with equal number of samples in each group, then we use v-1 groups of the data for training and the remaining group for testing. This process is repeated v times so that each group can be tested in turn. In this research, the RBF kernel is employed for the SVM classifier. The optimized setting of RBF kernel parameter and soft margin is obtained via trial-and-error. Finally, the mean accuracy of all folds, avg_i, is used to identify the fitness, F_{SVM}, for each chromosome, I_i, as shown in Eq (1).

$$F_{SVM}(I_i) = \begin{cases} \dfrac{1}{avg_i * 100}, & \text{if } avg_i > 0 \\ \dfrac{1}{\beta}, & \text{otherwise} \end{cases} \qquad (1)$$

where $avg_i * 100$ avg_i indicates the percentage of accuracy is based on the mean accuracy of all folds, avg_i, whereas β is a value very close to zero but not zero in order to avoid infinity value for $F_{SVM}(I_i)$ when the accuracy obtained from SVM is zero.

After obtaining the fitness of each chromosome, the GA operations of selection, crossover, and mutation are conducted to produce offspring. These offspring are again evaluated using the similar procedure (indicated as sub-algorithm 2.1) mentioned above. The process of the GA reproduction continues until the maximum generation is achieved. The best result obtained is the proposed feature subset for the recognition of normal and blast cases.

Algorithm 2: GA for SVM-based embedded feature selection

Input:
$P(I_1, I_2, ...I_k)$;
80 extracted features;
Images for training (normal and abnormal)
Output:
The best proposed features for recognizing normal and abnormal lymphocytic cells
Begin
    Randomly separate training images into 10 folds;
    for each $I_i \in P$
    training sample= features combination, I_i, from all indexed training images;
        for f =1:10 // for each fold
    training fold sample= training sample with index $^1 f$; //90% of the training set
        testing fold sample= training sample with index $= f$; //10% of the test set
        Struct = svmtrain(training fold sample, training fold label, RBF kernel,
RBF scaling factor, soft margin);
        Outlabel = svmclassify(Struct, testing fold sample);
        Evaluate accuracy of the fth fold by checking the matching of
Outlabel to the annotation of the testing fold.
      end
    average_accuracy, avg_i= mean of all folds' accuracy;
    Assign fitness, $F_{SVM}(I_i)$.
    end for

sub-algorithm 2.1

while maximum iteration not reached
Rank *P*;
SE : = select (*P*); //Select chromosomes based on stochastic universal sampling
OF : = crossover (*SE*); //Crossover operation to generate offspring
OF : = mutation (*OF*); //Mutation operation to generate offspring
Evaluate each $I_i \in OF$ using **sub-algorithm 2.1**; //Determine fitness for each new I_i in *OF*
$\dot{P}$: = merge (*P, OF*);
end while
return new updated *P*;
End

4. Evaluation

To test model efficiency, we employ a set of 180 sub-images consisting of 60 healthy and 120 blast sub-images extracted from ALL-IDB2 data set [9] for ALL identification. We conduct both filter and embedded feature selection models in combination with four standard classifiers, i.e. NB, KNN, MLP, and SVM for the recognition of normal and abnormal lymphocytic cells. In order to provide more comprehensive comparisons, two types of validation methods are used to compare the results: (1) 50:50 hold-out validation, and (2) 10-fold cross-validation. In the hold-out validation, 90 images are used for training and another 90 non-overlapped unseen images are used for testing. As for tenfold cross-validation, the 90 non-overlapped test images are divided into 10 equal folds for testing.

Table 1 depicts the classification accuracy rates for each feature selection method in combination with different classifiers. As observed in the table, when 80 raw features are used without any feature selection, the recognition accuracy varies among different classifiers. The raw features obtain encouraging results when the SVM classifier is used; however, the results decrease dramatically when other classifiers (especially, NB) are used. In comparison with other feature selection techniques where only partial features are selected, the recognition results are comparable and sometimes higher than those obtained using all the 80 raw features (especially for NB, KNN, and MLP in tenfold cross-validation). Furthermore, these feature selection techniques could significantly improve computational efficiency since fewer features are required for classification.

In addition to the 80 raw features, the 80 projected features are also produced by the NWFE method [8] since projected features may contain more discriminating power. However, the experiments indicate that the use of projection in this experiment does not increase the performance of the original 80 raw features much as feature projection might change some information of the original features as compared to feature selection which is purely based on the original information. Therefore, we employ the proposed three feature selection techniques, i.e. EFCM + GA, LDA + GA, and SVM + GA, to investigate the influence of different feature selection algorithms on the classification performance.

From Table 1, it is clearly observed that the SVM classifier provides the best recognition performance among all the classifiers. In 50:50 hold-out validation, the best result, 95.56% accuracy, is obtained from filter-based feature selection using EFCM + GA in combination with the SVM classifier. As for the tenfold cross-validation, SVM + GA-based embedded feature selection together with the SVM classifier gives the highest accuracy of 98.89%. Fig. 1 illustrates the ROC

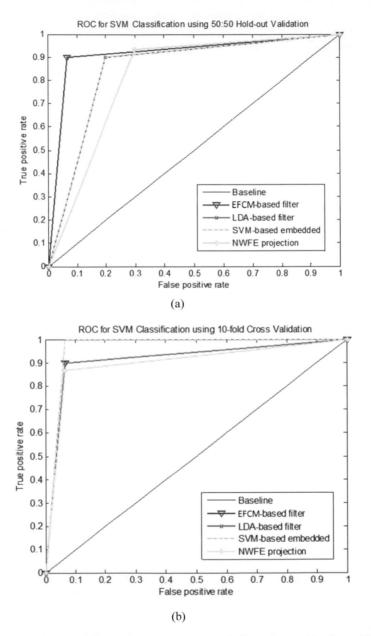

(a)

(b)

Fig. 1: ROC curves of different feature selection techniques in combination with SVM classification where (a) is for 50:50 hold-out validation and (b) is for tenfold cross-validation

Table 1: Comparison of classification accuracy rates using different feature selection approaches and classifiers

Feature Selection Approaches	No. of Features	Validation Methods	Classifiers			
			NB (%)	KNN (%)	MLP (%)	SVM (%)
EFCM+GA-based feature selection	45	50:50 hold-out	73.33	75.56	89.89	**95.56**
		10-fold	78.89	85.56	89.89	94.44
LDA+GA-based feature selection	45	50:50 hold-out	67.78	61.11	92.99	93.33
		10-fold	78.89	71.11	92.99	94.44
SVM +GA based embedded selection	49	50:50 hold-out	67.78	63.33	86.44	93.33
		10-fold	74.44	76.67	86.44	**98.89**
Full feature set	80	50:50 hold-out	67.78	81.11	85.33	93.33
		10-fold	77.78	82.22	85.33	96.67
NWFE feature Projection [8]	80	50:50 hold-out	67.78	81.11	85.78	93.33
		10-fold	77.78	82.22	85.78	92.22

curves of different feature selection techniques for SVM classification. In terms of feature selection, the features proposed by EFCM + GA-based filter selection are more generic and are able to deliver recognition performance more consistently throughout the test classifiers. For instance, both EFCM and LDA select 45 features for classification. However, EFCM outperforms LDA for all classifiers except for MLP. Even though the features selected by LDA achieve good results when MLP is applied, the features are not able to achieve consistent promising recognition performance when other classifiers are used. As for the SVM + GA-based embedded approach, due to classifier-dependent feature selection, the features proposed are able to show the highest recognition accuracy with the SVM classifier during the tenfold cross-validation. However, these features are not able to produce better performance than those produced by EFCM-based selection when different classifiers are used.

Overall, in combination with the SVM classifier, SVM + GA-based embedded feature selection contributes the highest classification performance but with a more classifier-dependent feature subset. Conversely, the EFCM-based filtering contributes a relatively more generic feature subset that is generally promising across most classifiers with the best performance achieved for the 50:50 hold-out validation when the SVM classifier is applied. Both SVM + GA-based embedded model and EFCM + GA-based filter model are able to generate encouraging accuracy rates for distinguishing normal and abnormal lymphocytic cells with efficient computational costs in comparison to those of the processes using the raw features. In addition, the experiments of this research have also proven EFCM's discriminant capability in clustering-based feature selection techniques.

Table 2 illustrates the comparison with recent state-of-the-art existing studies for ALL diagnosis using the ALL-IDB2 dataset. In comparison with a variety of clustering, classification, and deep learning methods, our proposed feature selection methods illustrate superior performance and can be used as alternative competitive solutions for ALL classification.

Table 2: Comparison with existing studies for the ALL-IDB2 dataset

Studies	Methodologies	Results
Tan et al. [15]	RCPSO and ACPSO-based hyper-parameter fine-tuning in CNNs	0.9113 (RCPSO) 0.8799 (ACPSO)
Xie et al. [21]	IIEFA and CIEFA enhanced KM clustering methods	0.7893 (IIEFA) 0.804 (CIEFA)
Xie et al. [22]	PSOVA1 (with modified local/global optimal signals) and PSOVA2 (with adaptive exemplar breeding) based feature selection	0.9241 (PSOVA2) 0.9185 (PSOVA1)
Srisukkham et al. [14]	BBPSO-based feature selection	0.9494
This research	EFCM + GA based feature selection	**0.9556** (hold-out)
This research	SVM + GA based feature selection	**0.9889** (ten-fold)

5. Conclusion

In this research, we have proposed a new clustering-based feature selection method, i.e. EFCM + GA, for ALL diagnosis, where the inter- and intra-cluster variations are taken into account for clustering performance measurement. Two other feature selection methods, such as LDA + GA and SVM + GA are also proposed for performance comparison. Multiple classification models, such as NB, KNN, MLP, and SVM, are used for ALL identification with the selected significant features as inputs. On evaluation using ALL-IDB2 dataset, the EFCM + GA method illustrates promising performance for discriminative feature selection in comparison with other feature selection models as well as existing feature project methods.

In future work, we aim to employ other hybrid search mechanisms [19. 27, 23, 28, 10, 2, 29, 3, 30, 4] in conjunction with EFCM for discriminative feature selection. We will also evaluate the proposed EFCM + GA model using nucleus-cytoplasm segmentation tasks to further test model efficiency. The EFCM clustering method will also be used in conjunction with ensemble classification models and zero-shot learning to inform new unseen blood cancer cases [6, 20, 7].

References

1. Bodzas, A., Kodytek, P. and Zidek, J. (2020). Automated detection of acute lymphoblastic leukemia from microscopic images based on human visual perception, *Frontiers in Bioengineering and Biotechnology*, 8: 1005.
2. Farid, D., Zhang, L., Hossain, A.M., Rahman, C.M., Strachan, R., Sexton, G. and Dahal, K. (2013). An adaptive ensemble classifier for mining concept-drifting data streams, *Expert Systems with Applications*, 40(15): 5895-5906.
3. Fielding, B. and Zhang, L. (2020). Evolving deep dense block architecture ensembles for image classification, *Electronics*, 9(11). 1-31. Article ID 1880.
4. Fielding, B. and Zhang, L. (2018). Evolving image classification architectures with enhanced particle swarm optimization, *IEEE Access*, 6: 68560-68575. ISSN 2169-3536.

5. Ghaderzadeh, M., Asadi, F., Hosseini, A., Bashash, D., Abolghasemi, H. and Roshanpour, A. (2021). Machine learning in detection and classification of leukemia using smear blood images: A systematic review, *Scientific Programming*, 2021: 1-14. Article ID 9933481.

6. Kinghorn, P., Zhang, L. and Shao, L. (2018). A region-based image caption generator with refined descriptions, *Neurocomputing*, 272: 416-424.

7. Kodirov, E., Xiang, T. and Gong, S. (2017). Semantic auto-encoder for zero-shot learning. *In: Proceedings of the IEEE Conference on Computer Vision and Pattern Recognition*, pp. 3174-3183.

8. Kuo, B.C. and Landgrebe, D.A. (2004). Non-parametric weighted feature extraction from classification, *IEEE T. Grosci Remote*, 42: 1096.

9. Labati, R.D., Piuri, V. and Scotti, F. (2011). ALL-IDB: The acute lymphoblastic leukemia image database for image processing, *Paper Presented at the 18th International Conference on Image Processing*, Brussels, Belgium, *IEEE*.

10. Liu, H. and Zhang, L. (2019). Advancing Ensemble Learning Performance through data transformation and classifiers fusion in granular computing context, *Expert Systems with Applications*, 131: 20-29.

11. Mistry, K., Zhang, L., Neoh, S.C., Lim, C.P. and Fielding, B. (2017). A micro-GA embedded PSO feature selection approach to intelligent facial emotion recognition, *IEEE Transactions on Cybernetics*, 47(6): 1496-1509.

12. Neoh, S.C., Zhang, L., Mistry, K., Hossain, M.A., Lim, C.P., Aslam, N. and Kinghorn, P. (2015). Intelligent facial emotion recognition using a layered encoding cascade optimization model', *Applied Soft Computing*, 34: 72-93.

13. Neoh, S.C., Srisukkham, W., Zhang, L., Todryk, S., Greystoke, B., Lim, C.P., Hossain, A. and Aslam, N. (2015). An intelligent decision support system for leukaemia diagnosis using microscopic blood images, *Scientific Reports*, 5(14938), Nature Publishing Group, ISSN 2045-2322.

14. Srisukkham, W., Zhang, L., Neoh, S.C., Todryk, S. and Lim, C.P. (2017). Intelligent leukaemia diagnosis with bare-bones PSO-based feature optimization, *Applied Soft Computing*, 56: 405-419, ISSN 1568-4946.

15. Tan, C.J., Neoh, S.C., Lim, C.P., Hanoun, S., Wong, W.P., Loo, C.K., Zhang, L. and S. Nahavandi (2019). Application of an evolutionary algorithm-based ensemble model to job-shop scheduling, *Journal of Intelligent Manufacturing*, 30: 879-890.

16. Tan, T., Zhang, L., Lim, C.P., Fielding, B., Yu, Y. and Anderson, E. (2019). Evolving ensemble models for image segmentation using enhanced particle swarm optimization', *IEEE Access*, 7: 34004-34019.

17. Tan, T.Y., Zhang, L. and Lim, C.P. (2019). Intelligent skin cancer diagnosis using improved particle swarm optimization and deep learning models, *Applied Soft Computing*, 105725.

18. Tan, T.Y., Zhang, L. and Lim, C.P. (2020). Adaptive melanoma diagnosis using evolving clustering, ensemble and deep neural networks, *Knowledge-based Systems*, 187: 1-26. Article No. 104807.

19. Tan, T.Y., Zhang, L., Neoh, S.C. and Lim, C.P. (2018). Intelligent skin cancer detection using enhanced particle swarm optimization, *Knowledge-based Systems*, 158: 118-135.

20. Wang, W., Zheng, V.W., Yu, H. and Miao, C. (2019). A survey of zero-shot learning: Settings, methods, and applications, *ACM Transactions on Intelligent Systems and Technology (TIST)*, 10(2): 1-37.

21. Xie, H., Zhang, L., Lim, C.P., Yu, Y., Liu, C., Liu, H. and Walters, J. (2019). Improving K-means clustering with enhanced Firefly Algorithms. *Applied Soft Computing*, 84: 105763.

22. Xie, H., Zhang, L. and Lim, C.P. (2020). Evolving CNN-LSTM models for time series prediction using enhanced grey wolf optimizer, *IEEE Access*, 8: 161519-161541.
23. Xie, H., Zhang, L. and Lim, C.P. (2021). Feature selection using enhanced particle swarm optimization for classification models, *Sensors*, 21(5): 1-40. Article No. 1816.
24. Zhang, L., Lim, C.P. and Han, J. (2019). Complex deep learning and evolutionary computing models in computer vision', *Complexity*, 2019: 1-3. Article ID 1671340.
25. Zhang, L. and Lim, C.P. (2020). Intelligent optic disc segmentation using improved particle swarm optimization and evolving ensemble models, *Applied Soft Computing*, 92: 106328.
26. Zhang, L., Lim, C.P. and Yu, Y. (2021). Intelligent human action recognition using an ensemble model of evolving deep networks with swarm-based optimization, *Knowledge-based Systems*, 220: 106918.
27. Zhang, L., Mistry, K., Lim, C.P. and Neoh, S.C. (2018). Feature selection using firefly optimization for classification and regression models, *Decision Support Systems*, 106(2018): 64-85.
28. Zhang, L., Mistry, K., Neoh, S.C. and Lim, C.P. (2016). Intelligent facial emotion recognition using moth-firefly optimization, *Knowledge-Based Systems*, 111: 248-267.
29. Zhang, L., Srisukkham, W., Neoh, S.C., Lim, C.P. and Pandit, D. (2018). Classifier ensemble reduction using a modified firefly algorithm: An empirical evaluation, *Expert Systems with Applications*, 93: 395-422.
30. Zhang, Y., Zhang, L., Neoh, S.C., Mistry, K. and Hossain, A. (2015). Intelligent affect regression for bodily expressions using hybrid particle swarm optimization and adaptive ensembles, *Expert Systems with Applications*, 42(22): 8678-8697.

Artificial Intelligence-enabled Automated Medical Prediction and Diagnosis in Trauma Patients

Lianyong Li[1#], Changqing Zhong[1#], Gang Wang[1], Wei Wu[1], Yuzhu Guo[2], Zheng Zhang[1], Bo Yang[1], Xiaotong Lou[1], Ke Li[1]* and Heming Yang[1]*

[1] PLA Strategic Support Force Medical Center, Beijing, China
[2] Department of Automation Sciences and Electrical Engineering, Beihang University, Beijing, China

1. Introduction

Trauma, characterized by potentially life-long disability, high mortality rate, and subsequent significant psychological trauma, has taken a social and economic toll far greater than that of any other clinical disease, becoming one of the most important public health challenges. Medical personnel keep innovating rescue equipment and improving treatment technology in the treatment of trauma; thus the success rate of treatment has improved to some extent. However, the current mode of trauma treatment is mainly dependent on manual rescue in the phase of either pre-hospital, medical evacuation or emergency room. In recent years, Artificial Intelligence, machine learning, and big data analysis technologies have attracted people's attention, and their applications in the medical field are gradually increasing. The earliest application of Artificial Intelligence in trauma patients was pre-hospital medical dispatch and triage, which played a role in saving medical resources, shortening the time of pre-hospital treatment, and making the connection within the hospital more smooth [32, 37, 41]. In addition, Artificial Intelligence for the prediction and diagnosis of trauma-related disorders is gradually emerging. We systematically searched literature and thus discuss the application and progress of AI in trauma-related disorders in our review (Fig. 1). Different from previous reviews, we classified trauma-related disorders according to the injury site, and summarized the application and development of Artificial Intelligence in the prediction and diagnosis of trauma medicine over the past 10 years [28].

* Corresponding Authors: yhming306@163.com; leeker1974@126.com
\# Authors equally contributed to this work.

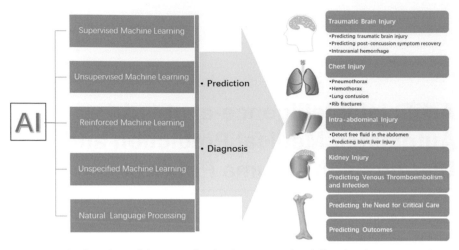

Fig. 1: The flowchart of the AI application in trauma-related disorders in present review

2. Prediction and Diagnosis Trauma-related Disorders

2.1 Traumatic Brain Injury

Traumatic brain injury (TBI) is the main cause of death and disability. Many survivors are left with severe disabilities, which also cause a huge socioeconomic burden. At present, research and application of Artificial Intelligence in traumatic brain injury mainly focus on the prediction of traumatic brain injury and intracranial hemorrhage.

2.1.1 *Predicting Traumatic Brain Injury*

Although traumatic brain injury (TBI) is the most urgent injury, not all injuries require hospitalization. Establishing evaluation tools to identify accurately and quickly, the patients who need further monitoring or hospitalization, can save medical resources, avoid misdiagnosis, and reduce unnecessary medical expenditure. Hale *et al.* [16] prospectively collected a multi-center pediatric emergency head-trauma dataset, including head CT imaging data of 12,902 patients under the age of 18. Information interpreted by radiologists was extracted and prediction of traumatic brain injury that required monitoring or hospitalization was made by Artificial Intelligence. The sensitivity was 99.73% and the specificity was 60.47%. The author believes that the use of artificial neural network methods can provide more reliable and accurate prediction of traumatic brain injury, and can also update the dataset in real time.

2.1.2 *Predicting Post-concussion Symptom Recovery*

Concussion is a mild traumatic brain injury (TBI), which is common clinically. Although it is often mild, there is a risk of serious short-term and long-term sequelae. Therefore, it is also important to judge the risk timely and predict the recovery from a concussion accurately. Fleck *et al.* [12] applied a new Artificial Intelligence

– genetic fuzzy tree (GFTs, Genetic Fuzzy Trees) to analyze the diffusion tensor imaging data of 43 adolescent patients with mild traumatic brain injury or traumatic orthopedic injury. It is used to predict the likelihood of recovery of symptoms after a concussion within a week after injury. Through total of diffusion tensor imaging scans and 225 training sessions, the results showed that the sensitivity and specificity of concussion recovery prediction was 59% and 65%, respectively. This provided preliminary evidence of Artificial Intelligence for predicting traumatic symptoms recovery effectively.

Patients with severe traumatic brain injury are prone to disturbances of consciousness. These patients often need mechanical ventilation to protect the airway and assist in breathing. However, the ventilator can cause lung injury, ventilator-related pneumonia and other related complications, with some patients failing to wean off the ventilator completely. It is also an important task to accurately predict when to wean off the ventilator or determine whether the patients with craniocerebral trauma need to extend mechanical ventilation. Abujaber *et al.* [1] used Logistic Regression (LR), Artificial Neural Network (ANN), Support Vector Machine (SVM), Random Forest (RF), and C.5 Decision Tree (C.5 DT) and other methods to analyze the TBI patients who received mechanical ventilation as well as the severity of the injury to the head area. The prediction model was determined. The results showed that the machine learning method was superior to the traditional multivariate analysis method in predicting mechanical ventilation of traumatic brain injury. The machine learning model showed a higher prediction success rate and discriminative ability and more stable performance.

2.1.3 Intracranial Hemorrhage

Half of the mortality in patients with intracranial hemorrhage (ICH) occurs within the first 24 hours. Enlarged hematoma can lead to worsening of neurological deficits, and irreversible damage may occur as early as the first few hours after the onset of ICH; so, accurate diagnosis is essential for these patients. These patients can be managed according to specific types, locations, and sizes based on neuroimaging of CT examination. Artificial Intelligence can screen acute lesions in CT images, which may help radiologists cope with the rapidly increasing number of cases [6]. One approach is to use these screening tools to prioritize acute ICH cases on the work-list. Therefore, it is important to determine the potential impact of such a system, not only by considering the accuracy of ICH testing, but also by considering the type of bleeding, the location of the lesion, and the severity of bleeding in order to best allocate resources.

Aidoc is one of the first Artificial Intelligence devices approved by the FDA. Its machine learning (ML) algorithm based on convolutional neural networks can detect high-density anomalies on head CTs. The algorithm was tested on 7,112 non-contrast head CTs obtained from two large-scale urban academic and trauma centers, between 2016 and 2017, and its diagnostic specificity was 99%, sensitivity was 95%, and overall accuracy was 98% [25].

Daniel T. Ginat *et al.* [13] analyzed the value of this deep learning software in trauma/emergency, outpatient, and inpatient medical environments for unenhanced brain CT (NCCT) in acute intracranial hemorrhage and its priority in the work-list. It was found that the accuracy of intracranial hemorrhage detection in trauma emergency cases was significantly higher than that in hospitalized cases (96.5% vs 89.4%). Therefore, compared with intracranial hemorrhage caused by other factors, Artificial Intelligence may have certain advantages in the diagnosis of intracranial hemorrhage caused by traumatic brain injury.

Melissa A. Davis *et al.* [7] found that in emergency cases in the first-level trauma center of the medical system, the application of artificial Intelligence made the report turnaround time (RTAT) decrease from 66.9 minutes (SD 77.1 minutes, CI 64.9-68.8 minutes) to 59.1 minutes (SD 74.1 minutes, CI 57.2-61.0 minutes) ($P < 0.001$). However, the study compared the length of stay (LoS) of ICH-positive cases before and after the implementation of the deep learning algorithm from 527 minutes to 491 minutes, but it has not yet reached statistical significance ($P = 0.46$).

2.2 Chest Injury

Common chest traumas include pneumothorax, hemothorax, lung contusion, and rib fractures. CT is the preferred method of examination for chest trauma. Trauma patients may have multiple and varying degrees of injury, which makes the diagnosis challenging. The limited working time of emergency diagnosis can easily lead to misdiagnosis and affect the treatment of patients. The application of deep learning in pulmonary nodular lesions has shown its convenience and efficiency. Deep learning on chest CT-assisted diagnosis in emergency trauma has also gradually attracted attention.

Lyu *et al.* [22] used deep learning in chest CT-assisted diagnosis system to evaluate 403 patients with emergency chest trauma. The sensitivity and specificity of the assisted diagnosis system for detecting pneumothorax, pleural effusion/blood, and rib fracture were 96.6% and 97.6%, 80.0% and 99.7%, 99.2% and 83.9% respectively. The sensitivity of detecting lung contusion was 97.7%. The assisted diagnosis system and manual imaging diagnosis have high consistency in the diagnosis of chest injury. Missed diagnosis of two cases of pneumothorax, three cases of pleural effusion/blood, nine cases of rib fractures, and six cases of other fractures in traditional approach were all detected by the deep learning-assisted diagnosis system. The author concluded that the deep learning chest CT diagnosis system can effectively assist in the detection of chest CT injuries in emergency trauma patients, and it is expected to optimize the diagnosis and treatment process of emergency trauma patients. Weikert *et al.* [40] also found that CT scan based on deep learning has a sensitivity of 87.4% and a specificity of 91.5% in detecting rib fractures.

In the case of massive bleeding caused by chest trauma, rapid intervention to control the bleeding is the key to patients' survival. Patients may receive resuscitation thoracotomy (RT) or resuscitation intravascular balloon occlusion (REBOA) as a treatment for rapid control of bleeding.

Zeineddin [14] performed machine learning on the vital signs, continuously measuring in the pre-hospital environment and accurately predicting the needs of RT and REBOA. He thus provided key information for rapid life-saving decisions.

Although CT is the preferred method of examination for chest trauma, however, not all patients with chest trauma need CT examinations. Excessive CT examinations not only waste medical resources, but also it can cause certain radiation damage to patients. Especially when there are large-scale casualties in emergency situations, such as earthquakes, floods, volcanoes and other natural disasters, if all chest injuries undergo CT examinations, it will cause waiting and render check-up crowded. To solve one of the problems, Kondori [33] applied machine learning models to analyze chest trauma patients. This could help emergency doctors to determine whether patients need CT scans to avoid unnecessary CT scans.

2.3 Intra-abdominal Injury

Most of the emergency abdominal injuries are blunt abdominal trauma (BAT), which has a high incidence of mortality. Most BAT cases are related to motor-vehicle collision (MVC) or car and pedestrian accidents. Among BAT patients in the emergency department, the incidence of intra-abdominal injury is approximately 13%. When BAT occurs, the spleen and liver are the most commonly damaged solid organs. Injuries to the pancreas, bowel and mesentery, bladder, diaphragm, and retroperitoneal structures (kidneys, abdominal aorta) are less common, but must also be considered.

Focused assessment with sonography (FAST) in a traumatic environment has become the standard of care for rapid detection of free fluid in the abdomen in the emergency department and trauma intensive care unit. FAST examination is highly specific for free abdominal fluid in blunt and penetrating trauma. Free fluid in the upper right quadrant has been shown to be a strong independent predictor of the need for therapeutic laparotomy in trauma. However, under conditions of emergency, medical transportation, large-scale casualties, disaster response, and battlefield medicine, FAST faces many obstacles and most of all it is lack of training. Sjogren *et al.* [35] used image segmentation and machine learning to automatically detect free fluid in the abdomen after FAST examination, and found that the detection sensitivity and specificity were 100% (69.2% to 100%) and 90.0% (55.5% to 99.8%), respectively. The author believed that computerized detection of free fluid on abdominal ultrasound images may be sensitive and specific enough to help clinicians interpret FAST examinations.

In addition, Pennell *et al.* [27] applied machine learning to the dataset of the patients of blunt abdominal injury. The model, created by analyzing 19 clinical variables including vomiting, dyspnea, abdominal pain, Glasgow Coma Scale score, etc., could predict whether a patient had intra-abdominal injury. Dreizin *et al.* [8] conducted deep learning on CT and angiography of patients with blunt liver injury in two hospitals, and predicted the degree of major arterial damage by determining the percentage of the liver with lacerations, and provided a basis for predicting the need for emergency hemostasis intervention.

2.4 Fracture

Musculoskeletal trauma accounts for a large part of the emergency patients, and its medical costs account for more than 50% of the total cost of non-fatal injuries. Missed or delayed diagnosis in trauma imaging can lead to increased mortality and morbidity. The peak of misdiagnosis of fractures occurs between 8 p.m. and 2 a.m. This may be caused by the fact that radiologists cannot issue reports immediately, and non-radiology clinicians misjudge the imaging data. Recent studies have also shown that the application of machine learning (ML) and Artificial Intelligence (AI) has improved the diagnostic accuracy of radiologists and reduced the rate of misdiagnosis.

Artificial Intelligence was mostly researched and applied in limb and joint trauma. Duron *et al.* [10] found that the use of Artificial Intelligence (AI) auxiliary tools for high-resolution X-rays improved the sensitivity (increased by 8.7%, $P = .006$) and specificity (4.1% increase, $P = .03$) for emergency doctors and radiologists in the diagnosis of fractures. Additional reading time was not required. England *et al.* [11] used Deep Convolutional Neural Network (DCNN) to detect elbow joint effusion in trauma patients. Through training on a collection of 657 X-ray images, the final model test had a sensitivity of 0.909 (95% CI, 0.788-1.000), a specificity of 0.906 (95% CI, 0.844-0.958), and accuracy of 0.907 (95% CI, 0.843-0.951) for diagnosing elbow joint effusion. In diagnosing traumatic elbow joint effusion, it achieves the same performance as non-radiological emergency medicine clinicians. In the diagnosis of ankle fractures, a neural network-based model can help to classify ankle fractures in detail, and it is believed that the method can be extended to other body parts [26]. The image reading system using Artificial Intelligence has the sensitivity, specificity, and accuracy of 97%, 95.7% and 96.08% in the diagnosis of hip fractures, respectively. This system improves the diagnostic performance of junior doctors to the level of senior doctors [4]. In terms of pelvic fractures, research also found that Artificial Intelligence-based diagnostic systems showed performance comparable to radiologists and orthopedists in detecting pelvic fractures, and can grade the severity of pelvic fractures [5, 9]. However, some scholars also found that a model based on deep learning of radiographic images of proximal femoral fractures had a sensitivity of 61% for femoral neck fractures and 67% for trochanter fractures, the specificity was 67% and 69%, respectively. For rigorous medicine, this result is not satisfactory, and it is too early for widely clinical application [15].

In addition to predicting limb and joint fractures, scholars have also applied machine learning to the prediction of lower limb amputation after trauma-related arterial injury, and the prediction of the outcome of limb revascularization. The prediction models had achieved excellent predictive performance in the verified population. It is believed that the results of limb revascularization (AUROC 0.97) are accurately predicted during the initial wound assessment, and the patients with traumatic arterial injury at high risk of amputation are accurately identified (the accuracy is 0.88). This information can supplement clinical judgments, help make reasonable treatment decisions, and accurately predict efficacy [3, 28].

2.5 Kidney Injury

Patients with severe trauma or burns are at risk of acute kidney injury (AKI). Early recognition of AKI helps guide the selection and dosage adjustment of fluid resuscitation and nephrotoxic drugs in these populations. Unfortunately, traditional biomarkers of renal function, such as creatinine and urine output (UOP), have proven to be suboptimal in predicting AKI. New AKI biomarkers have been proposed, but their application is still limited. Studies have shown that when machine learning was used in combination with neutrophil gelatinase-associated lipocalin (NGAL),

NT-proBNP or creatinine, the ability to predict acute kidney injury was further enhanced [29, 30].

2.6 Predicting Venous Thromboembolism and Infection

Venous thromboembolism (VTE) and infections are common complications of hospitalized trauma patients, having an adverse effect on the patient's prognosis. Predicting the risk of thrombus and infection can serve the purpose of guiding prevention. However, tools for predicting the risk of thrombus and infection in trauma populations are often based on literature review and clinical experience. He *et al.* [17] retrospectively analyzed emergency trauma patients admitted to the trauma center of a tertiary hospital, extracted patient electronic medical record data, combined a variety of feature screening methods and random forest (RF) algorithm to construct a venous thromboembolism prediction model, which showed good results of predictive effect. In the study of infection prediction after severe trauma, the machine learning algorithm was applied to a multi-biomarker prediction model established on the data of 128 adult blunt-trauma patients, and it could accurately predict the occurrence of multiple infection events after blunt trauma [39].

2.7 Predicting the Need for Critical Care

For trauma patients, accurately identifying and judging the type and severity of the injury is the key to effective evaluation. Although vital signs play an important role in monitoring trauma patients and are often used as a means to assess the patient's condition, the utility of on-site vital signs largely depends on the interpretation of experts. Artificial prediction of whether a patient needs intensive care mainly depends on vital signs, which may delay treatment.

Computer and machine learning can continuously, timely, and accurately process a large number of different data including vital signs and time trends. They were not only conducive to trauma diagnosis, but also accurately predicted the needs of patients for intensive care [17, 36]. It could even predict the needs of life-saving interventions, such as tracheal intubation, blood transfusion, cardiopulmonary resuscitation, etc. [18, 19, 20].

2.8 Predicting Outcomes

Patients are divided into different levels based on their vital signs, nature of injury, and Glasgow coma score, so that they are ranked before they arrive at the ED. The earlier the patient receives surgery or medication, the more likely the patient is to survive. If the mortality rate of patients can be accurately predicted, it will play a vital role in the classification of patients' priority treatment and the reasonable allocation of medical resources. Machine learning methods have always been the backbone of modeling in this field. A large number of diverse datasets are used to train and build accurate models, which will effectively predict patient risks and survival probabilities. Tsiklidis *et al.* [38] analyzed 32 key features of 799,233 complete patients' records, and finally incorporated eight features of systolic blood pressure, heart rate, respiratory rate, body temperature, oxygen saturation, gender, age and Glasgow coma score for model training. The accuracy rate of this model for predicting the outcome of trauma patients was 92.4%. Models established by other scholars also predicted and verified death, multiple organ failure, and blood transfusion, initially showed the value of machine learning and artificial intelligence in predicting prognosis [2, 3, 14, 18, 23, 24, 31, 34]. A recent meta-analysis evaluated the value of machine learning in predicting prognosis. A total of 64 studies, including 2,433,180 trauma patients, were included. The main predictive indicators included mortality, length of stay, etc., although the included studies proved that the models had different values and advantages in predicting prognosis, but they also had different algorithms and different evaluations of algorithm performance. The prediction sensitivity-specificity gap value varied greatly between 0.035 and 0.927. Earlier machine learning was widely used in practice, but it still needs further verification by prospective, randomized clinical trials [21].

3. Prospects

In the near future, Artificial Intelligence will undoubtedly become an indispensable part in the diagnosis and treatment of trauma, and Artificial Intelligence-based algorithms will help enhance our work processes. We can imagine unexpected traffic accident-caused multiple injuries. The Artificial Intelligence-based rescue system optimizes the deployment of rescue resources. Based on the continuously detected vital signs data and the injured part, the patient's injury severity is assessed and survival rate is predicted automatically pre-hospital. After the X-ray or CT examination, the diagnosis system based on machine learning will make an accurate diagnosis immediately. If liver rupture occurs, it will automatically determine the degree of liver rupture, calculate blood loss, judge the possibility of massive blood transfusion, predict whether it needs immediate embolization therapy, and provide the information of postoperative hospital stay and prognosis. In this hypothetical example, Artificial Intelligence involves many aspects of care and management of trauma patients, and will ultimately improve their diagnosis and treatment.

4. Conclusion

Machine learning and Artificial Intelligence have attracted considerable attention in the field of trauma. The initial research and application have also shown priority and advantages over traditional diagnostic approach. However, research on trauma is still relatively limited. Artificial Intelligence and machine learning applied to trauma systems are still in their infancy, and many unknowns still need to be resolved. More research is still expected on the accuracy and reliability of the diagnosis and prediction of trauma-related disorders.

References

1. Abujaber, A., Fadlalla, A., Gammoh, D., *et al.* (2020). Using trauma registry data to predict prolonged mechanical ventilation in patients with traumatic brain injury: Machine learning approach, *PLoS One,* 15: e0235231.

2. Abujaber, A., Fadlalla, A., Gammoh, D., *et al.* (2020). Prediction of in-hospital mortality in patients with post traumatic brain injury using National Trauma Registry and Machine Learning Approach, *Scand. J. Trauma Resusc. Emerg. Med.*, 28: 44.

3. Almaghrabi, F., Xu, D. and Yang, J. (2019). A new machine learning technique for predicting traumatic injuries outcomes based on the vital signs. *In:* Data Quality of a Wearable Vital Signs Monitor in the Pre-hospital and Emergency Departments for Enhancing Prediction of Needs for Life-saving Interventions in Trauma Patients. *2019 25th International Conference on Automation and Computing (ICAC)*, p. 1-5.

4. Bolourani, S., Thompson, D., Siskind, S., *et al.* (2021). Cleaning up the MESS: Can machine learning be used to predict lower extremity amputation after trauma-associated arterial injury? *J. Am. Coll. Surg.*, 232: 102-113 e4.

5. Cheng, C.T., Chen, C.C., Cheng, F.J., *et al.* (2020). A human-algorithm integration system for hip fracture detection on plain radiography: system development and validation study, *JMIR Med. Inform.*, 8: e19416.

6. Cheng, C.T., Wang, Y., Chen, H.W., *et al.* (2021). A scalable physician-level deep learning algorithm detects universal trauma on pelvic radiographs, *Nat. Commun.*, 12: 1066.

7. Chilamkurthy, S., Ghosh, R., Tanamala, S., *et al.* (2018). Deep learning algorithms for detection of critical findings in head CT scans: A retrospective study, *Lancet*, 392: 2388-2396.

8. Davis, M.A., Rao, B., Cedeno, P.A., *et al.* (2020). Machine learning and improved quality metrics in acute intracranial hemorrhage by noncontrast computed tomography, *Curr. Probl. Diagn. Radiol.* (In Press)

9. Dreizin, D., Chen, T., Liang, Y., *et al.* (2021). Added value of deep learning-based liver parenchymal CT volumetry for predicting major arterial injury after blunt hepatic trauma: A decision tree analysis, *Abdom. Radiol.* (NY), 46(6): 2556-2566.

10. Dreizin, D., Goldmann, F., LeBedis, C., *et al.* (2021). An Automated deep learning method for tile AO/OTA pelvic fracture severity grading from trauma whole-body CT, *J. Digit Imaging*, 34: 53-65.

11. Duron, L., Ducarouge, A., Gillibert, A., *et al.* (2021). Assessment of an AI aid in detection of adult appendicular skeletal fractures by emergency physicians and radiologists: A multicenter cross-sectional diagnostic study, *Radiology*, 203886.

12. England, J.R., Gross, J.S., White, E.A., *et al.* (2018). Detection of traumatic pediatric elbow joint effusion using a deep convolutional neural network, *AJR Am. J. Roentgenol.*, 211: 1361-1368.

13. Fleck, D.E., Ernest, N., Asch, R., *et al.* (2021). Predicting post-concussion symptom recovery in adolescents using a novel artificial intelligence, *J. Neurotrauma*, 38: 830-836.

14. Ginat, D.T. (2020). Analysis of head CT scans flagged by deep learning software for acute intracranial hemorrhage, *Neuroradiology*, 62: 335-340.

15. Gorczyca, ,M.T., Toscano, N.C. and Cheng, J.D. (2019). The trauma severity model: An ensemble machine learning approach to risk prediction, *Comput. Biol. Med.* , 108: 9-19.

16. Guy, S., Jacquet, C., Tsenkoff, D., *et al.* (2021). Deep learning for the radiographic diagnosis of proximal femur fractures: Limitations and programming issues, *Orthop. Traumatol. Surg. Res.*, 107: 102837.

17. Hale, A.T., Stonko, D.P., Lim, J., *et al.* (2018). Using an artificial neural network to predict traumatic brain injury, *J. Neurosurg. Pediatr.*, 23: 219-226.

18. He, L., Luo, L., Hou, X., *et al.* (2021). Predicting venous thromboembolism in hospitalized trauma patients: A combination of the Caprini score and data-driven machine learning model, *BMC Emerg. Med.,* 21: 60.

19. Kang, D.Y., Cho, K.J., Kwon, O., *et al.* (2020). Artificial Intelligence algorithm to predict the need for critical care in prehospital emergency medical services, *Scand. J. Trauma Resusc. Emerg. Med.*, 28: 17.

20. Kuo, P.J., Wu, S.C., Chien, P.C., *et al.* (2018). Derivation and validation of different machine-learning models in mortality prediction of trauma in motorcycle riders: A cross-sectional retrospective study in southern Taiwan, *BMJ Open*, 8: e018252.

21. Liu, N.T., Holcomb, J.B., Wade, C.E., *et al.* (2014). Utility of vital signs, heart rate variability and complexity, and machine learning for identifying the need for lifesaving interventions in trauma patients, *Shock*, 42: 108-114.

22. Liu, N.T., Holcomb, J.B., Wade, C.E., *et al.* (2015). Data quality of a wearable vital signs monitor in the pre-hospital and emergency departments for enhancing prediction of needs for life-saving interventions in trauma patients, *J. Med. Eng. Technol.*, 39: 316-321.

23. Liu, N.T., Holcomb, J.B., Wade, C.E., *et al.* (2014). Development and validation of a machine learning algorithm and hybrid system to predict the need for life-saving interventions in trauma patients, *Med. Biol. Eng. Comput.*, 52: 193-203.

24. Liu, N.T. and Salinas, J. (2017). Machine Learning for Predicting Outcomes in Trauma, *Shock*, 48: 504-510.

25. Maurer, L.R., Bertsimas, D., Bouardi, H.T., *et al.* (2021). Trauma Outcome Predictor (TOP): An Artificial-Intelligence (AI) Interactive Smartphone Tool to Predict Outcomes in Trauma Patients, *J. Trauma Acute Care Surg.*, 91(1): 93-99.

26. Lyu, W.H., Xia, F., Zhou, C.S., *et al.* (2021). Application of deep learning-based chest CT auxiliary diagnosis system in emergency trauma patients, *Zhonghua Yi Xue Za Zhi*, 101: 481-486.

27. Niggli, C., Pape, H.C., Niggli, P., *et al.* (2021). Validation of a visual-based analytics tool for outcome prediction in polytrauma patients (WATSON Trauma Pathway Explorer) and comparison with the predictive values of TRISS, *J. Clin. Med.*, 10.

28. Ojeda, P., Zawaideh, M., Mossa-Basha, M., *et al.* (2020). *The Utility of Deep Learning: Evaluation of a Convolutional Neural Network for Detection of Intracranial Bleeds on Non-Contrast Head-computed Tomography Studies*. 1 March, 2019, 109493J.

29. Olczak, J., Emilson, F., Razavian, A., *et al.* (2021). Ankle fracture classification using deep learning: Automating detailed AO Foundation/Orthopedic Trauma Association (AO/OTA) 2018 malleolar fracture identification reaches a high degree of correct classification, *Acta Orthop.* 92: 102-108.

30. Pennell, C., Polet, C., Arthur, L.G., *et al.* (2020). Risk assessment for intra-abdominal injury following blunt trauma in children: Derivation and validation of a machine learning model, *J. Trauma Acute Care Surg.*, 89: 153-159.

31. Perkins, Z.B., Yet, B., Sharrock, A., *et al.* (2020). Predicting the outcome of limb revascularization in patients with lower-extremity arterial trauma: Development and external validation of a supervised machine-learning algorithm to support surgical decisions, *Ann. Surg.*, 272: 564-572.

32. Ramos-Lima, L.F., Waikamp, V., Antonelli-Salgado, T., *et al.* (2020). The use of machine learning techniques in trauma-related disorders: A systematic review, *J. Psychiatr. Res.*, 121: 159-172.

33. Rau, C.S., Wu, S.C., Chuang, J.F., *et al.* (2019). Machine learning models of survival prediction in trauma patients, *J. Clin. Med.*, 8.

34. Rashidi, H.H., Sen, S., Palmieri, T.L., *et al.* (2020). Early recognition of burn- and trauma-related acute kidney injury: A pilot comparison of machine learning techniques, *Sci. Rep.*, 10: 205.

35. Rashidi, H.H., Makley, A., Palmieri, T.L., *et al.* (2021). Enhancing military burn- and trauma-related acute kidney injury prediction through an automated machine learning platform and point-of-care testing, *Arch. Pathol. Lab. Med.*, 145: 320-326.

36. Shafaf, N. and Malek, H. (2019). Applications of machine learning approaches in emergency medicine: A review article, *Arch. Acad. Emerg. Med.*, 7: 34.

37. Shahi, N., Shahi, A.K., Phillips, R., *et al.* (2021). Decision-making in pediatric blunt solid organ injury: A deep learning approach to predict massive transfusion, need for operative management, and mortality risk, *J. Pediatr. Surg.*, 56: 379-384.

38. Shahverdi Kondori, M. and Malek, H. (2021). Determining the need for computed tomography scan following blunt chest trauma through machine learning approaches, *Arch. Acad. Emerg. Med.*, 9: e15.

39. Sjogren, A.R., Leo, M.M., Feldman, J., *et al.* (2016). Image segmentation and machine learning for detection of abdominal free fluid in focused assessment with sonography for trauma examinations: A pilot study, *J. Ultrasound Med.*, 35: 2501-2509.

40. Staziaki, P.V., Wu, D., Rayan, J.C., *et al.* (2021). Machine learning combining CT findings and clinical parameters improves prediction of length of stay and ICU admission in torso trauma, *Eur. Radiol.*, 31(7): 5434-5441.

41. Stonko, D.P., Guillamondegui, O.D., Fischer, P.E., *et al.* (2021). Artificial Intelligence in trauma systems, *Surgery*, 169: 1295-1299.

42. Tsiklidis, E.J., Sims, C., Sinno, T., *et al.* (2020). Using the National Trauma Data Bank (NTDB) and machine learning to predict trauma patient mortality at admission, *PLoS One*, 15: e0242166.

43. Tsurumi, A., Flaherty, P.J., Que, Y.A., *et al.* (2020). Multi-Biomarker Prediction Models for Multiple Infection Episodes Following Blunt Trauma, *Science*, 23: 101659.

44. Weikert, T., Noordtzij, L.A., Bremerich, J., *et al.* (2020). Assessment of a deep learning algorithm for the detection of rib fractures on whole-body trauma computed tomography, *Korean J. Radiol.*, 21: 891-899.

45. Weisberg, E.M., Chu, L.C. and Fishman, E.K. (2020). The first use of Artificial Intelligence (AI) in the ER: Triage not diagnosis, *Emerg. Radiol.*, 27: 361-366.

46. Zeineddin, A., Hu, P., Yang, S., *et al.* (2020). Prehospital continuous vital signs predict need for reboa and resuscitative thoracotomy, *J. Trauma Acute Care Surg.*, 91(5): 798-802.

DCGAN-based Facial Expression Synthesis for Emotion Well-being Monitoring with Feature Extraction and Cluster Grouping

Eaby Kollonoor Babu[1], Kamlesh Mistry[1] and Li Zhang[2]

[1] Computer and Information Sciences, Northumbria University, NE1 8ST, UK
[2] Department of Computer Science, Royal Holloway, University of London, UK

1. Introduction

The generative adversarial network (GAN) is a deep learning framework that is capable of generating diverse images according to the set criteria. The GAN model is first proposed by Goodfellow *et al.* [5] in 2014. It is able to autonomously generate images for diverse application domains. In this research, we employ the GAN model in combination with predefined facial expression labels, for realistic synthetic facial expression image generation. Specifically, we use Deep Convolutional Generative Adversarial Network (DCGAN) to generate facial expression images. Two datasets, i.e. Bosphorus3D [17] and BU_3DFE [23] datasets are used in this research. These datasets contain facial images of individuals showing emotions, like anger, disgust, fear, joy, surprise, neutral, and sadness. The DCGAN model is trained using the above datasets to generate realistic facial expression images. The generated images are subsequently evaluated using several evaluation criteria to measure the image quality. The main contributions of our work are listed as follows.

- An embedding and reshape layer for clustering of images is combined with the conventional DCGAN model to improve the robustness and capability of the proposed model for realistic synthetic image generation.
- A novel cluster grouping technique is used in the image generation process. It extracts the features from the facial images that are prominent in showcasing the expression of a human subject using a logical axiom.
- The proposed facial expression synthetic mechanism is able to generate realistic images for any given application domain.

The proposed work could be used to assist facial expression classification, facial distortion/cramp detection, facial changes associated with pain and depression/stress related to heart disease and seizure conditions. Therefore our work not only shows great potential in human robot interaction for healthcare and patient monitoring, but also can be used to conduct disease diagnosis, such as Parkinson's, facial paralysis, stroke and heart conditions as well as depression and general emotional well-being, where the data distributions are sparse. We discuss related studies, the proposed image generation model, and evaluation details and result analysis in the subsequent sections.

2. Related Work

Generation of facial expression images has been widely adopted in the field. Such techniques have been used in a variety of application domains, such as security, healthcare, forensic science, and many more. Rossler *et al.* [16] showcased a study to address image forgery, which becomes an increasing security threat these days. They proposed a method that conducted facial manipulation detection and performed benchmark test on different facial images. Their work indicated the importance of such systems for preventing face representation attack. Thies *et al.* [19] proposed a well-known facial expression manipulation model, namely Face2Face. It transferred a facial expression of one person to another person in real-time in resources-constrained settings. In the last 20 years, a large amount of research has been carried out in this domain due to its growing interest, which has contributed to a rapid growth of related deep learning methods. Karras *et al.* [7] produced a hyper realistic face generator system. It was capable of generating facial images with significant enhancement of resolution and clarity. Their research showcased the achievement of advancement in facial image generation techniques. An overview of other facial image generation methodologies using deep neural networks is presented in Lu *et al.* [11]. Also, a fader network was proposed by Lample *et al.* [9] for use in the training method to disentangle attribute-related features from the latent space. In addition, a face de-morphing method was proposed by Ferrara *et al.* [2]. It restored the facial images from morphed images with the assistance of face morphing operation in the morphing tool. Otberdout *et al.* [13] conducted facial image generation by utilizing the facial geometry modelling. Their work modelled the changes of facial landmarks as curves encoded in a spherical shape. Isola *et al.* [6] proposed the pix2pix network for facial image generation. However, this algorithm was too computationally intensive. A deterministic one-shot learning mechanism encoding the target class was developed by Zhang *et al.* [24]. Such strategies enabled the generation of facial expression images that were conditioned on discrete emotional states. Ververas and Zafeiriou [20] proposed an image-to-image transformation algorithm, called SliderGAN. It was able to transform a facial image to another, using different settings of statistical blend shapes. Kollias *et al.* [8] fitted a 3D Morphable Model (3DMM) to a target neutral image. An effect was then used in conjunction with the 3DMM image for the target image generation with a new face. The generation of facial features was affected either by different types of expressions, or the valence (positive or negative of an emotion) and arousal (power of the emotion activation) intensities prescribed by the users.

3. The Proposed Facial Image Generation System

In this research, we propose a variant of the DCGAN model for facial expression image generation. It incorporates embedding and reshape layers on both Generator and Discriminator models in conjunction with a novel cluster grouping technique to improve image generation performance. We introduce each key system component below.

3.1 Pre-training Procedures

We first conduct feature extraction from the images. Each of the original images shown in Fig. 1(a) is clustered into three regions, i.e. the eye line cluster (*see* Fig. 1(b)), the nose and mouth cluster (*see* Fig. 1(c)) and the combined eye-line, nose and mouth (E-N-M) region (*see* Fig. 1(d)).

This is achieved by using an eye detection technique on the image. The eye detection is performed by using a gradient function which compares the nearby pixels with the set gradient value. Then all small dark spots are found. If the found dark spots are in the same row, then it has a higher probability of being the eye. Then the centroid is determined, using the formula below:

$$X' = \frac{\int_{X_{min}}^{X_{max}} dA * X}{A} \tag{1}$$

where A is the total area of the dark spot and dA is the spot at a given value of X. In addition, X_{min} and X_{max} are the left and right extreme boundary values of the dark spot.

After centroid coordinates are obtained, an upper and lower eye cluster bounder limit of ±35% pixels from the center of the eye-line with respect to x-axis is determined. This step helps in generating a rectangular image with only the eye area of the subject extracted as shown in Fig. 1(b). After the centroids from both the eyes are identified, a left and right boundary limit of ±65% pixels for the nose and mouth cluster with respect to y-axis from the lower eye cluster boundary is determined. This step produces another rectangular image which contains only the nose and mouth

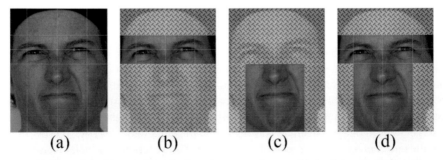

$$\text{(a)} \qquad\qquad \text{(b)} \qquad\qquad \text{(c)} \qquad\qquad \text{(d)}$$

Fig. 1: (a) The original image from the Bosphorus 3D dataset, (b) the detected eye line cluster, (c) the detected nose and mouth cluster, and (d) the detected combined E-N-M cluster

of the subject, as illustrated in Fig. 1(c). These two images are then combined to form the third cluster which is named as E-N-M cluster, as indicated in Fig. 1(d), which comprises the facial expression feature containing the eyes, nose, and mouth of the subject. Then these images are used in training the DCGAN model. Since the feature extracted images are used for the training, we are able to achieve a greater improvement in the training process with reduced costs. A training time comparison study was conducted to understand the processing time. The computational costs for the training stage of this study are shown in Table 1 below.

Table 1: Training computational cost comparison

Dataset/Epoch Setting	Training time	
	Using E-N-M Cluster Images (hh:mm:ss)	Using Direct Images from Dataset (hh:mm:ss)
BU_3DFE (100 epochs)	02:12:44	03:47:02
Bosphorus3D (2500 epochs)	66:01:50	72:19:22
BU_3DFE + Bosphorus3D (100 epochs)	02:17:02	02:54:34
BU_3DFE + Bosphorus3D (2500 epochs)	81:45:31	96:12:26

3.2 The Variant of the DCGAN Model

We use the concept of DCGAN to train our image generation system based on two facial image datasets with seven expressions, i.e. BU_3DFE and Bosphorus3Ds datasets. Specifically, we conduct four experiments as indicated below: Experiment 1 uses the BU_3DFE dataset with 100 epochs.

- Experiment 2 uses the Bosphorus3Ds dataset with 2500 epochs.
- Experiment 3 uses the combination of BU_3DFE and Bosphorus3Ds with 100 training epochs.
- Experiment 4 uses the combination of BU_3DFE and Bosphorus3Ds with 2500 epochs.

3.2.1 The Proposed Generator Model

The Generator model consists of a newly proposed embedding and reshape layer to accommodate the image labels in the image generation process. The main goal of using this layer is to compress the input feature space into a smaller one. Also, as we have different categories of expressions to be generated, this layer acts as a look-up table for the selected categories, thereby acting as a low dimensional vector representation of discrete values. A project and reshape layer takes 1-by-1-by-100 arrays of random noise inputs and converts it into 7-by-7-by-128 arrays. Then, the above two layers are concatenated and passed to the subsequent layers, i.e. transposed convolution, batch normalization, and ReLU layers, for upscaling the arrays to 64-by-64-by-3 ones. Next, single transposed convolution with a three

5-by-5 filters corresponding to the three RGB channels of the generated images is performed. Finally a tanh operation is performed to produce a symmetric value close to zero. These values are the output from the generator. The structure of the Generator is shown in Fig. 2 below.

The problem we encounter with the Generator and Discriminator in any DCGAN

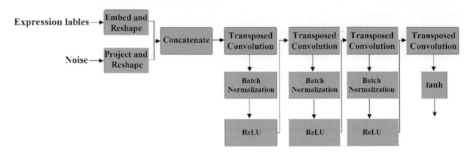

Fig. 2: The network structure of the proposed Generator model

model is to find the correct Nash equilibrium point. So, in our proposed algorithm, we use a technique, i.e. the bilateral tolerance search technique (BTS), to tackle the problem. In this method, we equip a search loop that constantly monitors the difference between previous and present output values of the Generator in correspondence with changes in present and previous output values of the Discriminator. A differential derivative function is used where the present and previous values for Discriminator and Generator are compared. When the change is less than 0.01%, the system is regarded as performing at its best potential. Equations (2)-(3) [18] below indicate the relation between the present and previous differential derivate functions of the Discriminator.

$$D_k(x, y) = \frac{1}{k} \sum\nolimits_{i \le k}^{n} [f(x_k, y_k)dx_k + f(x_k, y_k)dy_k] \qquad (2)$$

$$D_{(k-1)}(x, y) = \frac{1}{(k-1)} \sum\nolimits_{i \le (k-1)}^{n} [f(x_{(k-1)}, y_{(k-1)})dx_{(k-1)} + f(x_{(k-1)}, y_{(k-1)})dy_{(k-1)}] \qquad (3)$$

In a similar manner, the Generator is also developed using the proposed bilateral tolerance search technique (BTS). Equations (4)-(5) [18] denote the functional formulae used for the differential derivative functions of the Generator.

$$G_k(x, y) = \frac{1}{k} \sum\nolimits_{i \le k}^{n} [f(x_k, y_k)dx_k + f(x_k, y_k)dy_k] \qquad (4)$$

$$G_{(k-1)}(x, y) = \frac{1}{(k-1)} \sum\nolimits_{i \le (k-1)}^{n} [f(x_{(k-1)}, y_{(k-1)})dx_{(k-1)} + f(x_{(k-1)}, y_{(k-1)})dy_{(k-1)}] \qquad (5)$$

In Equations (2)-(5) [18], we have computed the present and previous values of both Generator and Discriminator in an iterative manner for comparison. Finally, a function that produces the Discriminator's validation accuracy is formulated in Equation 6 [18]. This function is designed to keep validating the score of each of the categories as the iteration progresses for the Discriminator.

$$\nabla D = \left[\begin{array}{cc} \nabla\left(\dfrac{D_k}{D_{(k-1)}}\right) \leq 0.01; & \text{Validate category score} \\ 0; & \text{Elsewhere} \end{array} \right] \tag{6}$$

The disadvantage of putting the threshold value to 0.01% is the high computational cost at the training stage but resulting in promising performances using limited resources such as a normal PC. This method can be further enhanced using Pareto optimal strategies or other similar optimization techniques.

3.2.2 The Proposed Discriminator Model

With respect to the Discriminator, images from the dataset are passed to a dropout function with a dropout probability of 0.5. The labels from the dataset are used as input to an embedding and reshape function which marks the images with their corresponding labels. By concatenating this layer, we get a discrete low dimensional vector look-up table for obtaining a comparison score to assess the feature map with the facial expression category set. This method drastically improves the performance of the Discriminator as the training progresses. Then the output from the above blocks is concatenated and passed on to a convolution layer, which features a 5-by-5 filter with an increasing number of filters for each subsequent layer. After batch normalization in each layer, a leaky ReLU with a scale factor of 0.2 is provided in the Discriminator network. Finally, a convolution layer with 4-by-4 filter is used at the output stage. To obtain the output probability ranging between 0 and 1, a sigmoid function is used. The structure of the Discriminator is shown in Fig. 3 below.

The Generator labels the image and stores it in its corresponding class only if the value of ΔG is below the set threshold. The detailed function is defined in Equation (7).

$$\nabla G = \left[\begin{array}{cc} \nabla\left(\dfrac{G_k}{G_{(k-1)}}\right) \leq 0.01; & \text{Select to category score} \\ 0; & \text{Elsewhere} \end{array} \right] \tag{7}$$

3.2.3 The Input and Output of the Proposed Model

Our model employs an input image with any of the seven facial expressions as inputs. It stores the provided expression label and then passes the image to the

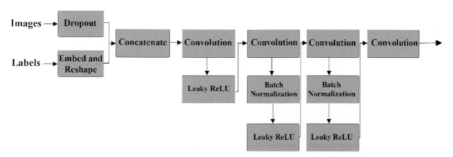

Fig. 3: The network structure of the proposed Discriminator model

trained DCGAN model comprising the previously introduced proposed Generator and Discriminator networks for generating new images with other six expressions.

4. Evaluation

4.1 Classification Performance

We perform a classification performance comparison study with varying batch sizes at the training stage. The results indicate that a batch size of 64 achieves a better performance, but with a significantly increased processing time as compared with those using a batch size of 32. As the system is able to produce better images with a batch size of 64, we use this batch size setting for generating the output images shown in the following subsections. The graph for the performance study is shown in Fig. 4.

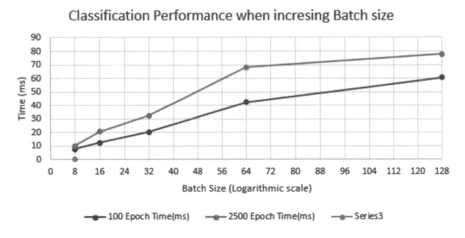

Fig. 4: Network classification performance comparison using different batch size settings.

4.2 Image Quality Measures Using SSIM, PSNR, and Sharpness Difference

To measure the quality of the generated images, we employ three measurement techniques. First, a PSNR metric is used. A better result is achieved when the model is trained using the Bosphorus3D dataset with 2500 epochs. When the SSIM measurement is used, the models trained using the combined dataset and the Bosphorus3D dataset both with 2500 epochs obtain similar performances. However, the output generated using the model trained on the Bosphorus3D dataset has a slightly higher edge over the other experimental setting. Finally, a sharpness measurement test is also conducted. The results indicate that the output images generated using the model trained on the Bosphorus3D dataset with 2500 epochs have better sharpness compared to those generated from other training processes. Table 2 shows the detailed results pertaining to image quality measurement.

Table 2: Results for image quality measurements using different metrics

	PSNR	SSIM	Sharpness Rate
Bosphorus3D (2500 epochs)	18.86	0.72	3.28
BU_3DFE + Bosphorus3D (2500 epochs)	16.32	0.68	3.06
BU_3DFE (100 epochs)	14.18	0.61	2.49
BU_3DFE + Bosphorus3D (100 epochs)	10.49	0.54	1.52

Figure 5 illustrates real facial expression training images with six expressions from the Bosphorus3D dataset, while Fig. 6 shows the training images extracted from the BU_3DFE dataset.

The proposed modified DCGAN model is able to generate realistic facial images with reasonable expressional accuracy. Our model was trained in using four different settings produces four sets of output images. These are produced by our proposed algorithm reported in Section III.

Figure 7 shows example of images from the model trained using the Bosphorus3D dataset with 2500 epochs. As indicated in Fig. 7, the images with anger, disgust, joy, and neutral emotions are comparatively more accurate; however, the output for the sad expressions is not up to the standard while the expression of fear is the least effective.

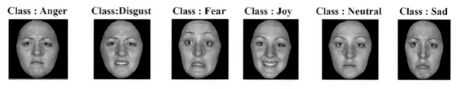

Fig. 5: Real facial expression images from the Bosphorus3Ds dataset

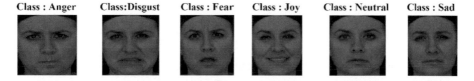

Fig. 6: Real facial expression images from the BU_3DFE dataset

Fig. 7: Output from the model trained using the Bosphorus3Ds dataset with 2500 epochs

We subsequently illustrate the output images generated by the model t rained using the combination of both datasets with 2500 epochs in Fig. 8. It shows that the generated images for expressions, such as joy, anger, and neutral show expected emotions, although the clarity of the generated images is not as good as those in the previous experiment.

Figure 9 contains the images generated by the models that were trained on BU_3DFE and the combined (Bosphorus3Ds + BU_3DFE) datasets, each with 100 epochs, respectively. As shown in Fig. 9, it is clear that the expressions of joy, neutral, and sadness are observable as expected, but the images generated for the other three expressions, i.e. anger, disgust and fear, do not relate well to the target expression labels, owing to the adoption of a comparatively smaller number of training epochs (i.e. 100).

As indicated in the existing studies, it was a significant challenge to run the DCGAN models using limited resources. The proposed system was able to perform on resources-constrained devices with reasonable performances. We will also further improve the proposed model by using other facial expression datasets and optimization strategies for model training in future directions.

Class : Anger   Class:Disgust   Class : Fear   Class : Joy   Class : Neutral   Class : Sad

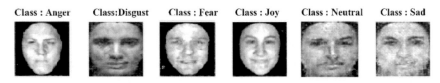

Fig. 8: Output from the model training using both Bosphorus3Ds and BU_3DFE with 2500 epochs

Class : Anger   Class:Disgust   Class : Fear   Class : Joy   Class : Neutral   Class : Sad

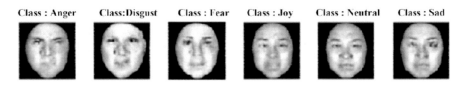

Fig. 9: Output from the model trained using the BU_3DFE dataset with 100 epochs

5. Conclusion

In this research, we have proposed an automatic synthetic facial expression generation system, which segments expression-dominant regions of facial images. It embeds a variant of the DCGAN model with additional layers, i.e. embedding and reshape layers, for the generation of diverse emotional synthesized facial images. The system is trained and tested, using two public datasets, i.e. Bosphorus3Ds and BU_3DFE. On evaluation by using several experimental settings, the overall system shows promising performances for facial expression image synthesis. For future work, we aim to incorporate diverse deep architectures for Generator and Discriminator models in conjunction with saliency detection mechanisms to further enhance the performance. We also aim to use the proposed model to generate facial or medical images [1, 3, 4, 10, 12, 14, 15, 21, 22, 25] to tackle the challenge of sparse data distributions to assist effective disease diagnosis, such as facial paralysis, stroke, and Parkinson's disease.

References

1. Bourouis, A., Feham, M., Hossain, M.A. and Zhang, L. (2014). An intelligent mobile based decision support system for retinal disease diagnosis, *Decision Support Systems*, Elsevier, 59: 341-350.
2. Ferrara, M., Franco, A. and Maltoni, D. (2018). Face demorphing, *IEEE Trans. Inf. Forensics Security*, 13(4): 1008-1017.
3. Fielding, B. and Zhang, L. (2020). Evolving deep DenseBlock architecture ensembles for image classification, *Electronics*, 9(11): 1880.
4. Frid-Adar, M., Diamant, I., Klang, E., Amitai, M., Goldberger, J. and Greenspan, H. (2018). GAN-based synthetic medical image augmentation for increased CNN performance in liver lesion classification, *Neurocomputing*, 321: 321-331.
5. Goodfellow, I., Pouget-Abadie, J., Mirza, M., Xu, B., Warde-Farley, D., Ozair, S., Courville, A. and Bengio, Y. (2014). Generative adversarial nets, *Advances in Neural Information Processing Systems*, 27.
6. Isola, P., Zhu, J.-Y., Zhou, T. and Efros, A.A. (2017). Image-to-image translation with conditional adversarial networks, *Proceedings of the IEEE Conference on Computer Vision and Pattern Recognition (CVPR)*, pp. 1125-1134.
7. Karras, T., Aila, T., Laine, S. and Lehtinen, J. 2017. Progressive growing of GANs for improved quality, stability, and variation. arXiv preprint arXiv: 1710.10196.
8. Kollias, D., Cheng, S., Ververas, E., Kotsia, I. and Zafeiriou, S. (2020). Deep neural network augmentation: Generating faces for affect analysis, *Int. Journal of Computer Vision*, 1-30.
9. Lample, G., Zeghidour, N., Usunier, N., Bordes, A., Denoyer, L. and Ranzato, M.A. (2017). Fader networks: Manipulating images by sliding attributes, *Advances in Neural Information Processing Systems*, 30.
10. Lawrence, T. and Zhang, L. (2019). IoTNet: An efficient and accurate convolutional neural network for IoT devices, *Sensors*, 19(24): 5541.
11. Lu, Z., Li, Z., Cao, J., He, R. and Sun, Z. (2017). Recent progress of face image synthesis. *In: IAPR Asian Conference on Pattern Recognition.*
12. Mistry, K., Zhang, L., Neoh, S.C., Lim, C.P. and Fielding, B. (2017). A micro-GA embedded PSO feature selection approach to intelligent facial emotion recognition, *IEEE Transactions on Cybernetics*, 47(6): 1496-1509.
13. Otberdout, N., Daoudi, M., Kacem, A., Ballihi, L. and Berretti, S. (2020). Dynamic Facial Expression Generation on Hilbert Hypersphere with Conditional Wasserstein Generative Adversarial Nets. *In: IEEE Transactions on Pattern Analysis and Machine Intelligence.*
14. Pandit, D., Zhang, L., Liu, C., Chattopadhyay, S., Aslam, N. and Lim, C.P. (2017). A lightweight QRS detector for single lead ECG signals using a max-min difference algorithm, *Computer Methods and Programs in Biomedicine*, 144: 61-75.
15. Raj, S. and Ray, K.C. (2018). Sparse representation of ECG signals for automated recognition of cardiac arrhythmias, *Expert Systems with Applications*, 105: 49-64.
16. Rossler, A., Cozzolino, D., Verdoliva, L., Riess, C., Thies, J. and Niessner, M. (2019).

FaceForensics++: Learning to detect manipulated facial images. *In: Proceedings of the IEEE/CVF International Conference on Computer Vision (ICCV)*, pp. 1-11.

17. Savran, A., Alyüz, N., Dibeklioğlu, H., Çeliktutan, O., Gökberk, B., Sankur, B. and Akarun, L. Bosphorus database for 3D face analysis. *In: European Workshop on Biometrics and Identity Management*, pp. 47-56, Springer, Berlin, Heidelberg.

18. Srivastava, A., Valkov, L., Russell, C., Gutmann, M.U. and Sutton, C. (2017). Veegan: Reducing mode collapse in GANs using implicit variational learning. *In: Proceedings of the 31st International Conference on Neural Information Processing Systems,* pp. 3310-3320.

19. Thies, J., Zollhofer, M., Stamminger, M., Theobalt, C. and Nießner, M. (2016). Face2Face: Real-time face capture and reenactment of RGB videos. *In: IEEE Conference on Computer Vision and Pattern Recognition (CVPR)*, pp. 2387-2395.

20. Ververas, E. and Zafeiriou, S. (2019), Slidergan: Synthesizing expressive face images by sliding 3d blendshape parameters, arXiv preprint arXiv: 1908.09638.

21. Wall, C., Young, F., Zhang, L., Phillips, E.J., Jiang, R. and Yu, Y. (2020). Deep learning based melanoma diagnosis using dermoscopic images. *In: Developments of Artificial Intelligence Technologies in Computation and Robotics: Proceedings of the 14th International FLINS Conference* (FLINS 2020), 907-914.

22. Wang, Q., Yu, Y., Gao, H., Zhang, L., Cao, Y., Mao, L., Dou, K. and Ni, W. (2019). Network representation learning enhanced recommendation algorithm, *IEEE Access*, 7: 61388-61399.

23. Yin, L., Wei, X., Sun, Y., Wang, J. and Rosato, M.J. (2006). A 3D facial expression database for facial behavior research. *In: 7th International Conference on Automatic Face and Gesture Recognition (FGR06)*, pp. 211-216, IEEE.

24. Zhang, F., Zhang, T., Mao, Q. and Xu, C. (2018). Joint pose and expression modeling for facial expression recognition. *In: IEEE Conf. on Computer Vision and Pattern Recognition (CVPR)*, 3359-3368.

25. Zhang, L., Mistry, K., Lim, C.P. and Neoh, S.C. (2018). Feature selection using firefly optimization for classification and regression models. *Decision Support Systems*, 106: 64-85.

A Hybrid-DE for Automatic Retinal Image-based Blood Vessel Segmentation

Colin Paul Joy[1], Kamlesh Mistry[2], Gobind Pillai[1] and Li Zhang[3]

[1] School of Computing, Teesside University, Middlesbrough TS1 3BX, UK
[2] Computer and Information Sciences, Northumbria University, NE1 8ST, UK
[3] Department of Computer Science, Royal Holloway, University of London, UK

1. Introduction

Retinal vein segmentation and portrayal of basic properties of veins, like length, width, convolution, and repercussion design are utilized for retinal determination, investigation, treatment, and evaluation of diverse visual ailments, like diabetes, hypertension, and so forth. The changes in retinal veins can aid the medical diagnosis of diseases, such as diabetic retinopathy, glaucoma, retinal artery occlusion, branch vein occlusion, central retinal vein occlusion, diabetes, etc. In current situation, fast development of diabetes is one of the major difficulties of well-being assurance. If it is not treated in time, then it may lead to visual deficiency in individuals before middle age that cause vision misfortune. In order to conduct an efficient and accurate diabetic retinopathy, a system is required that can segment retinal vein from retinal images and analyze it. Medical people acknowledge that manual vessel segmentation can provide accurate segmentation but at the same time, significant amount of time and domain knowledge is required. This urges a need for an automated retinal vein segmentation system. In biomedical, robotized retinal vein segmentation is becoming progressively popular for detecting retinal pathologies. Retinal veins can be seen in different thickness levels, but thick veins are easily identifiable but segmenting the thin veins is very tedious and challenging for both automated and manual systems. Over the time many researchers have worked on automated retinal vein segmentation methods, such as matched filter, multi-scale strategy, vessel tracking, and pattern classification-based methodology. In the present situation, matched filter is the most popular method for retinal vein segmentation due to its accuracy.

In order to address the above-mentioned challenges and constraints, this paper proposes a novel variant of DE to segment the retinal vein images. The main contributions of the proposed system are pointed below:

1. A novel DE variant is proposed for retinal vein segmentation, which employs three population approach to identify each category of vein, i.e. thick veins, thin veins, and non-vein areas.
2. Micro-FireFly Algorithm (mFA) is embedded with the multi-populational DE algorithm, in order to mitigate the premature convergence and local optimum problems of conventional DE.
3. The proposed DE variant also applies the diversity maintenance strategy of micro-Genetic Algorithm (mGA) to keep all the original populations in a non-replaceable memory, which remains intact during the lifetime of the algorithm, in order to reduce the probability of premature convergence.
4. Our proposed system is evaluated with DRIVE, STARE, and HRF datasets and is also compared with other existing methods reported in the literature.

This paper explores the discrimination capabilities in the texture of fundus to differentiate between healthy and un-healthy images, where the focus is to examine the performance of an extended variant of differential evolution (DE) to carry out automatic retinal blood vessel segmentation. An overall system architecture diagram is presented in Fig. 1. The rest of the paper is structured as follows: Section 2 focuses on related work, Section 3 on retinal vein segmentation system including the details on proposed DE variant, Section 4 on evaluation of the proposed system and finally, Section 5 concludes the work while addressing future directions.

2. Related Work

The authors [1] proposed an automated retinal blood vessel segmentation using artificial bee colony optimization and fuzzy c-means clustering. In their work artificial bee colony optimization is used as global search method to find cluster centers of the fuzzy c-means objective function. In order to localize small vessels with a different fitness function, a pattern search approach was used for optimization. Hassan *et al.* [2] used blood vessel segmentation approach to extract the vasculature on retinal fundus images. Their work involved particle swarm optimization (PSO) to determine the n-1 optimal n-level thresholds on retinal fundus images. Their work was tested on the DRIVE datasets and its efficiency compared with alternative methods. The approach proposed by Sreejini and Govindan [3] made use of improved noise suppression features of multi-scale Gaussian matched filter. The parameter values of the filter were obtained through particle swarm optimization and hence the accuracy of retina vessel segmentation was improved. Arnay *et al.* [4] worked on the optic cup segmentation in retinal fundus images using Ant Colony Optimization approach. In their approach, artificial agents produced solutions through a heuristic that used the intensity gradient of the optic disc area and the curvature of the vessels. The exploration capabilities of the agents were limited on their own, but by sharing the experience of the entire colony, they obtained accurate cup segmentations, even in images with a weak or non-obvious pallor.

In the paper by Wang *et al.* [5], a supervised method involving two superior classifiers, namely convolutional neural network (CNN) and random forest (RF) was used to deal with retinal blood vessel segmentation. The CNN was first used as a

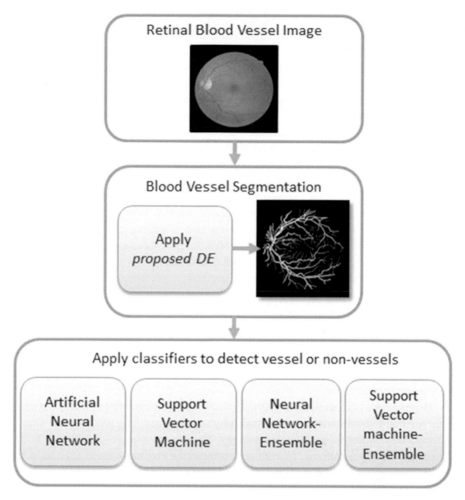

Fig. 1: System architecture of the proposed system

trainable hierarchical feature extractor and then ensemble RFs worked as a trainable classifier. Their approach combined the merits of feature learning and traditional classifier and was able to automatically learn features from the raw images and predict the patterns. Morales *et al*. [6] worked on differentiating between the texture of fundus images in case of pathological and healthy images. The authors used LBP as a texture descriptor tool for retinal images and compared their work with other descriptors, such as LBP filtering and local phase quantization. Hatami and Goldbaum [7] proposed a novel LBP method and showed that it was robust against low-contrast and low-quality fundus images, and helped in image classification by including additional AV texture and shape information. Fraz *et al*. [8] presented a survey of the blood vessel segmentation methods in two-dimensional, fundus camera acquired retinal images. Their paper has details about review, analysis, and classification of the retinal vessel extraction algorithms and methodologies along with highlights of the key points and the performance measures.

Staal *et al.* [9] proposed a ridge-based vessel segmentation method in colored images of the retina. The system extracted the image ridges based on the vessel centerlines. The ridges were then used to compose primitive line elements which helped an image to be partitioned into patches. Each image pixel was then assigned to the closest line element, which formed a local coordinate frame for its corresponding patch. The feature vector thereby computed for every pixel utilized the properties of the patches and the line elements. Sequential forward feature selection was used for feature selection and kappaNN-classifier to classify the feature vectors. Accuracy of 0.944 versus 0.947 for a second observer was reported in their work.

Marin *et al.* [10] used a new supervised method for detection of blood vessels in digital retinal images. They used a neural network (NN) for pixel classification and computed a 7-D vector composed of grey-level and moment invariants-based features for pixel representation. The authors showed that their algorithm was effective and robust for automated screening and detection in patients with early diabetic retinopathy.

In the work by You *et al.* [11], novel extracting of the retinal vessels based on radial projection and semi-supervised method is presented. Any segmentation method would need to consider two different processes to detect different types of vessels, namely, thin, and wide vessels. The radial projection method locates the vessel centerlines, and the steerable complex wavelet provides better capability of enhancing vessels under different scales. The feature vector then represents the vessel pixel line strength. The major structures of vessels are then identified using semi-supervised self-training and the union of the thin and wide vessels produces the final image segmentation.

Budai *et al.* [12] showed a method which reduced calculation time, obtained high accuracy, and increased the sensitivity as compared to the original Frangi method. With improvements in technology, the quality and resolution of fundus images are rapidly increasing, hence the segmentation methods would need to tackle the new challenges of high resolutions. The results in their work showed an average accuracy over 94% with the benefit of low computational burden. Hannink *et al.* [13] proposed a new method using scale–orientation scores that performed much better at enhancing vessels throughout crossings and bifurcations than the multi–scale Frangi filter which is an established tool in retinal vascular imaging. The authors presented results using both methods on a public dataset.

Chakraborti *et al.* [14] used self-adaptive matched filter method for blood vessel detection in the retinal fundus images. They used orientation histogram and presented a new combination of vesselness filter giving high sensitivity with the matched filter giving high specificity.

3. Proposed System

In this section, we introduce the proposed retinal image analysis system. The overall system consists of two key steps, i.e. an *hvn*LBP-based feature extraction and proposed DE-based segmentation. Each step is introduced in detail in the following sub-sections.

3.1   The Horizontal Vertical Neighborhood LBP

Ojala *et al.* [15] proposed the conventional LBP which thresholds each of the 3x3 neighboring pixels with a center pixel value. The conventional LBP was further extended to use various numbers of circular neighboring pixels. The LBP operator $LBP_{p,r}$ can produce 2p different binary patterns, where p denotes the number of neighborhood pixels and r denotes the radius of the circular pattern. The equation for calculating the $LBP_{p,r}$ operator can be given as follows:

$$\text{LBP} = \sum_{p=0}^{p-1} S(g_p - g_c)2^p, \ S(x) = \begin{cases} 1 \text{ if } x \geq 0 \\ 0 \end{cases} \tag{1}$$

where g_p denotes the neighborhood pixel at location p and gc is the center pixel. The important information, such as edges, corners, spot, and flat area can be detected by using the LBP. The conventional LBP is robust to illumination and scaling variations but fails to deal with rotation variations. The gradient images contain enhanced edge information and are more stable than raw pixel intensities and can benefit to deal with rotation and illumination variations.

In order to improve the feature extraction quality in terms of low contrast ration, Mistry *et al.* [16] proposed horizontal vertical neighborhood LBP (*hvn*LBP). The *hvn*LBP operator can be calculated by using the following equation:

$$\begin{aligned} hvnLBP_{p,r} = \{ & S(\max(p_0,p_1,p_2)), \ S(\max(p_7,p_3)), \\ & S(\max(p_6,p_5,p_4)), \ S(\max(p_0,p_7,p_6)), \\ & S(\max(p_1,p_5)), \ S(\max(p_2,p_3,p_4)) \} \end{aligned} \tag{2}$$

where p_i denotes the pixel intensity of neighborhood pixels at the ith location, r is the radius, and S denotes the comparison operation, as follows:

$$S(\max(p_j, p_k, p_m)) = \begin{cases} 1 & \text{if maximum} \\ 0 & \text{if not maximum} \end{cases} \tag{3}$$

where p_j, p_k, and p_m represent the neighborhood pixels in a row or column. Note that p_k is removed if it represents the center pixel. In comparison to conventional LBP, the proposed extended *hvn*LBP operator captures more discriminative contrast information and can achieve better retinal vein representation. The feature extracted by *hvn*LBP is further processed by proposed DE.

3.2   Conventional DE

Storn and Price [17] proposed DE to deal with global optimization in continuous spaces. DE algorithm employs the scaling factor between two individuals and is also called the mutation factor. DE algorithm starts its search with the random initialization of vectors and tries to improve them to further obtain optimal solution. In DE the population with Np vectors in g generations is denoted as $P = \{X_1, X_1, \ldots, X_{np}\}$ where, $X_i = (X_{i,a}, \ldots, X_{i,D})$. Conventional DE consists of three important steps:

3.2.1   Mutation

In this step, three vectors from the population are selected randomly and the following mutation equation is applied:

$$V_i = X_{i_1} + F\left(X_{i_2} - X_{i_3}\right) \tag{4}$$

where $F \in [0, 2]$ it further controls the augmentation of the differential vector of $(X_{i2} - X_{i3})$. In DE, the F value plays a very important role in controlling the exploration ability. Higher the F value the higher the exploration ability and vise-versa.

3.2.2 Crossover

This step is applied to improve the diversity of the population by crossing the mutant and parent vector as follows:

$$U_{i,d} = \begin{cases} V_{i,d}, \text{if } r \text{ and } d(0,1) \le \text{ or } d_{\text{rand}} = d \\ x_{i,d}, \text{otherwise} \end{cases} \tag{5}$$

where d is the dimension, C_r is the cross rate parameter, and the trial vector U_i can be generated as given below:

$$U_i = (U_{i,1}, \ldots, U_{i,d}) \tag{6}$$

3.2.3 Selection

This is the final step in each iteration, where U_i or X_i vector is selected based on their fitness value. The best fitness value vector is selected and sent for next iteration until the end condition is met.

3.3 Proposed DE

The conventional differential evolution only employs one set of population to explore the search space [17]. The single set of population restricts the search in only one direction or one set of features. The application of retinal blood vessel segmentation can lead to multiple clusters of features, which makes conventional DE less useful in this scenario.

Algorithm 1: The Pseudo Code of the Proposed DE Variant

Step-1: Initialize three sets of populations
Step-2: Evaluate the fitness value for each population using the Equation 4, with separate criteria for each population.
Step-3: Store the current version of population in non-replaceable memory.
Step-4: While (satisfying termination criteria), perform standard DE steps as follows:
  Mutation using Equation 4.
  Crossover using Equation 5 and 6.
  Perform selection.
Re-evaluate the fitness value for everyone in the current population.
 If (iterations $>=30$)
  Select best individual from each population and compare it with members from non-replaceable memory.

Select the three most co-relating members and add them to micro population along with the three global bests.

Run FA for 100 iterations using micro population

Evaluate fitness for everyone in micro population using fitness criteria of the thin veins.

Swap global bests if the newly generated members have higher fitness value.

End If

Re-evaluate the fitness value for everyone in all populations.

End While

Step-4

In this paper, we employ a multi-population DE to cluster the retinal veins in three categories. First initializing the number of populations with the same size of 30 members. Each population has its own fitness criterion, i.e. the populations will look for thick blood vessels pixels, thin blood vessels pixels, and non-blood vessel pixels. However, it simply means that three DE's will run parallel to generate three sets of features.

In order to further diversify the populations and reduce the risk of pre-mature convergence, we employ mGA's diversity maintenance strategy of non-replaceable memory [16]. During the first iteration we store all the three populations in one set of non-replaceable memory. These stored members are further used to aid the micro population generation. After the first iteration, we apply DE on all three populations and when the system reaches 30 iterations (selected, based on experiments), a set of combined micro population will be created. This micro population will consist of six members, including the global best member from each population and three most correlating members from the non-replaceable memory. A FA with three mutation strategies will be applied to the micro population and this will generate new offspring, based on thin vein fitness criteria over. If the newly generated members are better than the previous global bests, then they will be replaced. In this version of FA, we have employed Gaussian mutation [16], Cauchy mutation [36] and Levy's flight mutations [36]. This allows the algorithm to find the best candidates to search the thin veins, as it is one of the challenging tasks. At the same time, it will improve diversity and balance the global and local exploration. The overall structure of the proposed DE is illustrated in Algorithm 1. The fitness function used to evaluate each individual is given as follows:

$$F(x) = w_a * acc_x + w_f * (number - feature_x)^{-1} \qquad (7)$$

where w_a and w_f are two predefined constant weights for acc_x (classification accuracy) and $number - feature_x$ (the number of features), respectively. In this paper, $w_a = 0.8$ and $w_f = 0.2$, with $w_a > w_f$ to represent the fact that classification accuracy is more important than the number of selected features. Fig. 2 shows the output results generated by using the proposed system. The classification accuracy will be evaluated differently for each population.

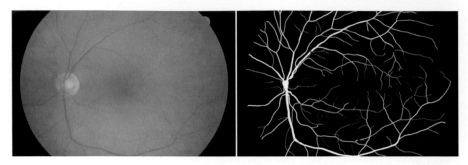

Fig. 2: Segmentation results generated by proposed system for image from HRF dataset

Diverse classifiers have been employed to further identify vein and non-vein categories. The selected classifiers are popular in image analysis and have proved to show an impressive performance, as in previous works [16] and [18]. NN, SVM, and the SVM-based and NN-based ensembles with SVM and NN as base classifiers respectively carry out the vessel and non-vessel classification. The input layer for NN is assigned to a number of features extracted from the proposed model. The NN classifier contains a secret layer along with an output layer with two nodes denoting vessel or non-vessel. Furthermore, the grid-search mechanism is availed to generate the most favorable parameter settings for the SVM classifier with the aim of achieving optimal performance. The most favorable settings generated for every single model NN and SVM previously described are also applied to the settings of each base classifier within each ensemble. Both ensembles utilize three base classifiers and a weighted majority voting collaboration method to generate the final categorization. Overall, the NN and SVM-based ensembles achieve the best accuracy when subjected to the images from the databases used as described in the next section.

4. Evaluation

We have proposed an automatic retinal blood vessel segmentation system, which segments blood vessels from fundus image and outperforms most of the existing methodologies. The proposed system is evaluated using three datasets i.e. DRIVE [9], STARE [19], and HRF [10]. The STARE (Structured Analysis of the Retina) project was conceived and initiated in 1975 by Michael Goldbaum, M.D., at the University of California, San Diego, USA. The STARE database has a set of around 400 raw retinal images. The DRIVE (Digital Retinal Images for Vessel Extraction) database was obtained from a diabetic retinopathy screening program in the Netherlands. The screening population consisted of 400 diabetic subjects, between 25-90 years of age. Forty photographs have been randomly selected, where 33 do not show any sign of diabetic retinopathy while seven show signs of mild early diabetic retinopathy. The set of 40 images was divided into a training and a test set, both containing 20 images. The HRF (High-Resolution Fundus) dataset has 15 retinal images of healthy patients, 15 images of patients with diabetic retinopathy, and 15 images of glaucomatous patients gathered through a collaborative research group in Germany.

The proposed system is implemented from scratch using C++ and OpenCV library under Ubuntu operating system. The learning algorithms, such as NN, SVM, NN- and SVM-based ensemble are imported from OpenCV and LibSVM library. The ground truth of the matching image is used to evaluate the performance of the proposed methodology on segmenting vessels from a fundus image. In order to measure the performance of the proposed system, we use accuracy, sensitivity, and specificity value. To calculate the accuracy, sensitivity, and specificity, we have to consider four measures, i.e. true positives, false positives, false negatives, and true negatives. The correctly categorized vessel pixels as vessels are denoted as true positive (TP) and correctly categorized non-vessel pixels as non-vessels are denoted as true negative (TN). Wrongly categorized non-vessels pixels as vessels are denoted as false positive (FP) and wrongly categorized vessels pixels as non-vessels are denoted as false negative (FN). The equations used to calculate accuracy, sensitivity, and specificity value are as follows:

$$\text{Accuracy} = (TP + TN)/(TP + TN + FP + FN) \tag{8}$$

$$\text{Sensitivity} = TP/(TP + FN) \tag{9}$$

$$\text{Specificity} = TN/(TN + FP) \tag{10}$$

In order to evaluate the performance of the proposed DE, it is compared with the existing state-of-the-art evolutionary algorithms. The selected evolutionary algorithms include, GA [20], DE [17], FA [11], PSO [21], Moth Flame Optimization (MFO) [22], and Whale Optimization algorithm (WOA) [23]. Moreover, this paper presents three sets of results using NN, SVM, ensemble NN and ensemble NN.

The first set of results presents the results using DRIVE dataset for training and testing. This set is illustrated in Tables 1 to 4 and each table shows the average accuracy, sensitivity, and specificity of all the images in DRIVE datasets over the 30 runs. The results obtained clearly show that the proposed DE outperforms the selected evolutionary algorithms while WOA is very close and second best. Overall Table 4 shows that SVM-based ensemble classifier can obtain best results for all the selected algorithms with proposed DE achieving the best results.

The second set of results use STARE dataset for training and testing. This set is illustrated in Tables 5 to 8 and each table shows the average accuracy, sensitivity, and specificity of all the images in STARE datasets over the 30 runs. The results obtained clearly show that the proposed DE outperforms the selected evolutionary algorithms. The performance results obtained by W OA are significantly lower than the proposed DE compared to results obtained in the first set. Overall Table 8 shows that SVM-based ensemble classifier can obtain best results for all the selected algorithms with proposed DE achieving 99.2% average accuracy.

Table 1: Performance of proposed system using NN classifier on DRIVE dataset

	Accuracy	Sensitivity	Specificity
GA	91.1	79.3	90.5
DE	92.5	81.7	91.3
FA	92.0	82.4	93.6
PSO	90.4	80.5	91.7
MFO	93.8	83.2	94.0
WOA	95.1	84.3	94.8
Proposed DE	**95.5**	**85.1**	**96.3**

Table 2: Performance of proposed system using SVM classifier on DRIVE dataset

	Accuracy	Sensitivity	Specificity
GA	91.5	80.0	89.7
DE	91.8	81.6	92.4
FA	93.2	83.0	92.9
PSO	92.6	82.8	93.0
MFO	94.4	85.5	95.1
WOA	96.0	86.6	95.7
Proposed DE	**96.0**	**87.0**	**96.2**

Table 3: Performance of proposed system using Ensemble NN classifier on DRIVE dataset

	Accuracy	Sensitivity	Specificity
GA	91.5	80.0	91.5
DE	92.2	81.9	94.3
FA	94.8	84.0	93.7
PSO	93.5	84.3	94.8
MFO	95.6	86.4	95.1
WOA	97.4	86.9	96.4
Proposed DE	**96.7**	**88.1**	**97.7**

Table 4: Performance of proposed system using Ensemble SVM classifier
on DRIVE dataset

	Accuracy	Sensitivity	Specificity
GA	93.2	80.0	91.0
DE	93.6	82.6	93.9
FA	94.4	83.8	93.6
PSO	93.0	84.6	94.2
MFO	96.4	85.8	96.1
WOA	96.4	86.6	95.9
Proposed DE	**97.9**	**87.8**	**96.2**

Table 5: Performance of proposed system using NN classifier on STARE dataset

	Accuracy	Sensitivity	Specificity
GA	91.4	78.9	87.0
DE	90.6	79.6	87.1
FA	91.8	80.4	88.2
PSO	91.4	80.8	87.1
MFO	91.9	82.4	90.0
WOA	92.9	83.3	91.8
Proposed DE	94.0	84.9	92.7

Table 6: Performance of proposed system using SVM classifier on STARE dataset

	Accuracy	Sensitivity	Specificity
GA	90.8	78.1	86.3
DE	90.2	79.8	86.9
FA	91.4	80.9	88.0
PSO	91.1	81.3	87.3
MFO	91.4	82.3	90.2
WOA	92.8	82.9	92.1
Proposed DE	93.1	85.2	93.0

Table 7: Performance of proposed system using Ensemble NN classifier on STARE dataset

	Accuracy	Sensitivity	Specificity
GA	92.5	79.0	86.6
DE	90.6	80.7	88.2
FA	93.6	82.7	90.1
PSO	92.0	82.1	89.1
MFO	92.5	82.9	91.8
WOA	95.6	85.5	92.6
Proposed DE	96.0	84.8	93.0

Table 8: Performance of proposed system using Ensemble SVM classifier on STARE dataset

	Accuracy	Sensitivity	Specificity
GA	90.6	78.7	86.3
DE	91.0	81.1	89.1
FA	92.7	81.9	89.4
PSO	90.9	82.3	88.3
MFO	93.1	83.5	90.6
WOA	97.5	85.3	93.8
Proposed DE	99.2	86.1	96.4

Table 9: Performance of proposed system using NN classifier on HRF dataset

	Accuracy	Sensitivity	Specificity
GA	86.6	75.9	81.9
DE	86.5	76.2	82.8
FA	88.4	77.5	83.6
PSO	88.6	76.3	83.8
MFO	89.2	80.4	85.4
WOA	90.3	80.0	86.5
Proposed DE	91.6	81.1	88.4

The third set of results use HRF dataset for training and testing. This set is illustrated in Tables 9 to 12 and each table shows the average accuracy, sensitivity, and specificity of all the images in HRF datasets over the 30 runs. The results obtained clearly show that the proposed DE outperforms the selected evolutionary algorithms while WOA is the second best. The performance results obtained by WOA are significantly lower than the proposed DE compared to results obtained in first set. Overall Table 4 shows that SVM-based ensemble classifier can obtain the best results for all the selected algorithms with proposed DE achieving 98.3% average accuracy.

Table 10: Performance of proposed system using SVM classifier on HRF dataset

	Accuracy	Sensitivity	Specificity
GA	86.9	75.2	81.5
DE	86.9	76.3	82.5
FA	89.0	78.0	83.4
PSO	88.6	76.3	83.5
MFO	89.7	80.3	85.8
WOA	90.9	80.6	86.3
Proposed DE	91.0	81.2	88.6

Table 11: Performance of proposed system using Ensemble NN classifier on HRF dataset

	Accuracy	**Sensitivity**	**Specificity**
GA	87.0	76.0	83.6
DE	87.4	76.3	84.6
FA	89.7	78.1	83.2
PSO	88.5	78.8	85.1
MFO	92.2	83.6	89.9
WOA	93.3	84.8	90.8
Proposed DE	97.2	86.1	93.8

Table 12: Performance of proposed system using Ensemble SVM classifier on HRF dataset

	Accuracy	**Sensitivity**	**Specificity**
GA	88.3	75.5	83.4
DE	88.5	76.5	84.0
FA	88.1	78.1	84.0
PSO	90.0	77.9	84.4
MFO	94.5	84.7	86.6
WOA	95.4	85.1	91.4
Proposed DE	98.3	87.0	94.9

A comparison of the proposed DE-based system using SVM-based ensemble classifier with other existing systems is illustrated in Table 5. The proposed system shows a better performance than other methodologies when evaluated with three publicly available datasets.

Table 13 shows that the proposed system outperforms specifically the works of Wang *et al.*, Moghimirad *et al.*, Geetha Ramani *et al.*, Imani *et al.*, Franklin *et al.*, Cheng *et al.*, and others when evaluated with DRIVE, STARE, and HRF datasets. For the DRIVE dataset, the new proposed algorithm achieved 97.9% accuracy, for STARE dataset, 99.2% accuracy and for HRF dataset, 98.3% accuracy, giving an average accuracy across three datasets to be 98.4%, along with 86.9% average sensitivity and 95.8% average specificity respectively. The above results show that the proposed system can accurately segment the blood vessels from the retinal fundus images. A further analysis of segmented vessels using the proposed system can lead to automatic disease diagnosis, like diabetic retinopathy, artery and vein occlusion, hypertensive retinopathy. Thus, the proposed system can benefit ophthalmologists in the screening of retinal diseases more efficiently.

5. Discussion and Conclusion

In this paper, we have proposed an automatic retinal blood vessel segmentation system, which segments blood vessels from fundus image while outperforming the

Table 13: Performance comparison with existing methodologies

Dataset	Methodology	Accuracy	Sensitivity	Specificity
DRIVE	Proposed work	97.9	87.8	96.2
	Wang *et al.* [24]	97.7	81.7	97.3
	Moghimirad *et al.* [25]	96.6	78.5	99.4
	Geetha Ramani *et al.* [26]	95.4	70.8	97.8
	Imani *et al.* [27]	95.2	75.2	97.5
	Franklin and Rajan [29]	95.0	68.7	98.2
	Roychowdhury *et al.* [29]	95.2	72.5	98.3
	Liu *et al.* [30]	94.7	73.5	97.7
STARE	Proposed work	99.2	86.1	96.4
	Wang *et al.* [24]	98.1	81.0	97.9
	Moghimirad *et al.* [25]	97.6	81.3	99.1
	Imani *et al.* [27]	95.9	75.0	97.5
	Annunziata *et al.* [31]	95.6	71.3	98.4
	Roychowdhury *et al.* [29]	95.2	77.2	97.3
	Liu *et al.* [30]	95.7	76.3	97.1
HRF	Proposed work	98.3	87.0	94.9
	Christodoulidis *et al.* [32]	94.8	85.1	95.8
	Cheng *et al.* [33]	96.1	70.4	98.6
	Annunziata *et al.* [31]	95.8	71.3	98.4
	Lázár *et al.* [34]	95.3	71.0	98.3
	Odstrcilik *et al.* [35]	94.9	77.4	96.7

other state-of-the-art systems. The overall system shows significant improvement in the segmentation and classification for efficient retinal image analysis. The proposed system uses an mFA embedded multi-population differential evolution (DE) algorithm in order to carry out automatic retinal blood vessel segmentation on retinal images. Multiple classifiers and other tools [36] are used to identify whether the retinal blood vessel images are healthy or unhealthy (i.e. with retinal disease). The proposed system is evaluated using three publicly available datasets, such as DRIVE, STARE, and HRF. Upon evaluation, the proposed system achieves a high average accuracy of 98.4% on three datasets and is on par with most of the systems reported in the literature. Due to its noticeable accuracy, this system could benefit ophthalmologists in retinal image analysis. It can further help in diagnosing the retinal diseases at an earlier stage, and thus lead to successful treatment of the eye patients. As far as we know, no one has attempted a differential evolution variant for retinal image analysis and diagnosis and hence, the work is novel.

References

1. Emary, E., Zawbaa, H.M., Hassanien, A.E., Schaefer, G. and Azar, A.T. (2014). Retinal blood vessel segmentation using bee colony optimisation and pattern search, *2014 International Joint Conference on Neural Networks (IJCNN)*, pp. 1001-1006. Beijing.
2. Hassan, G., Hassanien, A.E., El-bendary, N. and Fahmy, A. (2015). Blood vessel segmentation approach for extracting the vasculature on retinal fundus images using Particle Swarm Optimization. *11th International Computer Engineering Conference (ICENCO)*, Cairo, pp. 290-296.
3. Sreejini, K.S. and Govindan, V.K. (2015). Improved multiscale matched filter for retinal vessel segmentation using PSO algorithm, *Egyptian Informatics Journal*, 16(3): 253-260.
4. Arnay, R., Fumero, F. and Sigut, J. (2017). Ant colony optimization-based method for optic cup segmentation in retinal images, *Applied Soft Computing*, 52: 409-417.
5. Wang, S., Yin, Y., Cao, G., Wei, B., Zheng, Y. and Yang, G. (2015). Hierarchical retinal blood vessel segmentation based on feature and ensemble learning, *Neurocomputing*, 149(B): 708-717.
6. Morales, S., Engan, K., Naranjo, V. and Colomer, A. (2017). Retinal Disease Screening Through Local Binary Patterns', *IEEE Journal of Biomedical and Health Informatics*, 21(1): 184-192
7. Hatami, N. and Goldbaum, M. (2016). Automatic identification of retinal arteries and veins in fundus images using local binary patterns, *Investigative Ophthalmology and Visual Science*, 55(5): 232.
8. Fraz, M.M., Remagnino, P., Hoppe, A., Uyyanonvara, B., Rudnicka, A.R., Owen, C.G. and Barman, S.A. (2012). Blood vessel segmentation methodologies in retinal images – A survey, *Comput. Methods Programs Biomed.*, 108(1): 407-433.
9. Staal, J., Abramoff, M.D., Niemeijer, M., Viergever, M.A. and Ginneken, B.V. (2004). Ridge-based vessel segmentation in color images of the retina, *IEEE Trans. Med. Imaging*, 23(4): 501-509.
10. Marin, D., Aquino, A., Gegundez-Arias, M.E. and Bravo, J.M. (2011). A new supervised method for blood vessel segmentation in retinal images by using gray- level and moment invariants-based features, *IEEE Trans. Med. Imaging*, 30(1): 146-158.
11. You, X., Peng, Q., Yuan, Y., Cheung, Y-M. and Lei, J. (2011). Segmentation of retinal blood vessels using the radial projection and semi-supervised approach, *Pattern Recognition*, 44: 2314-2324
12. Budai, A., Bock, R., Maier, A., Hornegger, J. and Michelson, G. (2013). Robust vessel segmentation in fundus images, *International Journal of Biomedical Imaging*, 11.
13. Hannink, J., Duits, R. and Bekkers, E. (2014). Crossing-preserving multi-scale vesselness. *In: Medical Image Computing and Computer-Assisted Intervention*, International Conference on Medical Image Computing and Computer-assisted Intervention, Springer, 8674.
14. Chakraborti, T.K., Jha, D.K., Chowdhury, A.S. and Jiang, X. (2015). A self- adaptive matched filter for retinal blood vessel detection, *Machine Vision and Applications*, 26:55. Berlin Heidelberg, Springer Verlag,
15. Ojala, T., Pietikainen, M. and Harwood, D. (1996). A comparative study of texture measures with classification based on featured distribution, *Pattern Recognit.*, 29(1): 51-59.
16. Mistry, K., Zhang, L., Neoh, S.C., Lim, C.P. and Fielding, B. (2016). A micro-GA embedded PSO feature selection approach to intelligent facial emotion recognition', *IEEE Transactions on Cybernetics*, 47(6): 1496–1509.
17. Storn, R. and Price, K. (1997). Differential Evolution – A Simple and Efficient Heuristic for global Optimization over Continuous Spaces. *Journal of Global Optimization*, 11: 341.

18. Neoh, S.C., Zhang, L., Mistry, K., Hossain, M.A., Lim, C.P., N. Aslam and P. Kinghorn (2015). Intelligent facial emotion recognition using a layered encoding cascade optimization model, *Appl. Soft Comput.*, 34: 72-93.
19. Hoover, A.D., Kouznetsova, V. and Goldbaum, M. (2000). Locating blood vessels in retinal images by piece-wise threshold probing of a matched filter response, *IEEE Trans. Med. Imaging*, 19: 203-210.
20. Holland, J.H. (1992). Genetic algorithms. *Sci. Am.*, 267: 66-72.
21. Eberhart, R.C. and Kennedy, J. (1995). A new optimizer using particle swarm theory. *In: Proceedings of the Sixth International Symposium on Micro Machine and Human Science*, pp. 39-43.
22. Mirjalili, S. (2015). Moth-flame optimization algorithm: A novel nature- inspired heuristic paradigm, *Knowl.-based Syst.*, 89: 228-249.
23. Mirjalili, S. and Lewis, A. (2016). The whale optimization algorithm, *Adv. Eng.Softw.*,95: 51-67.
24. Wang, S., Yin, Y., Cao, G., Wei, B., Zheng, Y. and Yang, G. (2015). Hierarchical retinal blood vessel segmentation based on feature and ensemble learning, *Neurocomputing*, 149(B): 708-717.
25. Moghimirad, E., Rezatofighi, S.H. and Soltanian-Zadeh, H. (2012). Retinal vessel segmentation using a multi-scale medialness function, *Comput. Biol. Med.*, 42: 50-60.
26. Geetha Ramani, R. and Balasubramanian, L. (2016). Retinal blood vessel segmentation employing image processing and data mining techniques for computerized retinal image analysis, *Biocybernetics and Biomedical Engineering*, 36(1): 102-118.
27. Imani, E., Javidi, M. and Pourreza, H. (2015). Improvement of retinal blood vessel detection using morphological component analysis, *Comput. Methods Progr. Biomed.*, 118: 263-279.
28. Franklin, S.W. and Rajan, E. (2014). Computerized screening of diabetic retinopathy employing blood vessel segmentation in retinal images, *Biocybern Biomed Eng.*, 34: 117-124.
29. Roychowdhury, S., Koozekanani, D.D. and Parhi, K.K. (2015). Blood vessel segmentation of fundus images by major vessel extraction and subimage classification, *IEEE J. Biomed. Health Inf.*, 19: 1118-1128.
30. Liu, X., Zeng, Z. and Wang, X. (2014). Vessel segmentation in retinal images with a multiple kernel learning based method. *In: 2014 International Joint Conference on Neural Networks (IJCNN)*, pp. 507-511. IEEE.
31. Annunziata, R., Garzelli, A., Ballerini, L., Mecocci, A. and Trucco, E. (2017).Leveraging multiscale hessian-based enhancement with a novel exudate in painting technique for retinal vessel segmentation, *IEEE Journal Biomed Health Inform*, 20(4): 1129-1138.
32. Christodoulidis, A., Hurtut, T., Tahar, H.B. and Cheriet, F. (2016). A multi-scale tensor voting approach for small retinal vessel segmentation in high-resolution fundus images, *Comput. Med. Imaging Graph*, 52: 28-43
33. Cheng, E., Du, L., Wu, Y., Zhu, Y.J., Megalooikonomou, V. and Ling, H. (2014). Discriminative vessel segmentation in retinal images by fusing context- aware hybrid features, *Mach. Vis. Appl.*, 25: 1779-1792.
34. Lázár, I. and Hajdu, A. (2015). Segmentation of retinal vessels by means of directional response vector similarity and region growing, *Comput. Biol. Med.*, 66: 209-221.
35. Odstrcilik, J., Kolar, R., Budai, A., Hornegger, J., Jan, J., Gazarek, J., Kubena, T., Cernosek, P., Svoboda, O. and Angelopoulou, E. (2013). Retinal vessel segmentation by improved matched filtering: Evaluation on a new high- resolution fundus image database, *IET Image Process*, 7: 373-383.
36. Yang, X.S. (2010). Firefly algorithm, Levy's flight and global optimization, *Research and Development in Intelligent Systems*, 26: 209-218.

Artificial Intelligence for Accurate Detection and Analysis of Freezing of Gait in Parkinson's Disease

Debin Huang[1], Wenting Yang[1], Simeng Li[1], Hantao Li[1], Lipeng Wang[1],
Wei Zhang[2,3]*, Yuzhu Guo[1]*

[1] Department of Automation Science and Electrical Engineering, Beihang University,
    Beijing 100191, China
[2] Department of Neurology, Neurobiology and Geriatrics, Xuanwu Hospital of Capital
    Medical University, Beijing Institute of Geriatrics, Beijing 100053, China
[3] Department of Neurology, The Affiliated Hospital of Xuzhou Medical University,
    Xuzhou, Jiangsu 221006, China

1. Introduction

As the world's second prevalent neurodegenerative disease, Parkinson's disease
(PD) affects more than ten million people [60] worldwide and this number is
expected to double by 2050 [52], especially among the elderly. Because of the loss
of dopaminergic [10], PD symptoms are manifested as: slowness of motion, muscle
tremor and rigidity, freezing of gait (FoG) and other impaired motor functions [32,
33]. Results of a survey of 6,620 patients with PD showed that about half have the
experience of regular gait freeze [48]. FoG commonly occurs in gait initiation,
turning, passing through narrow space or approaching obstacles in the patient's daily
life, which significantly increases the risk of falling during walking. As the most
serious disability motor symptom of PD, FoG can often have a significant impact
on the quality of life of advanced PD patients [44, 83]. The latent pathology of
FoG is still unclear; however, there are some common characteristics in the gait,
such as sharp decrease in stride length, increase of cadence, and high-frequency leg
movements [62]. That FoG often happens suddenly, asymmetrically, and with a short
duration [64, 69, 83] makes the clinical detection, tracking, and evaluation of the
onset of FoG a challenging task. Freezing of gait in PD is common and debilitating,
thus increasing the demands on supportive caregivers' stress and non-motor illness
burden, such as anxiety and depression. Anxiety often occurs during 'off' periods; it
improves with better control of motor symptoms but can be a major source of distress
for patients even during the 'on' state. On-going assessment and punctual and proper

supportive care becomes increasingly important in advanced PD. Fall prevention is essential to avoid serious fracture or injuries. FoG is often associated with end-stage disease and is typically difficult to handle. With the reduction of the efficacy of medication, non-pharmacologic treatments, such as auditory cueing and visual cueing may eliminate or diminish the freezing episodes [3].

To clinically assess the FoG events, at present, designated self-report questionnaires from patients and manual video analysis [76] of in-lab activities by clinicians are utilized. The standard questionnaires include the Unified Parkinson's Disease Rating Scale (UPDRS) [18] and the new FoG-Questionnaire (NFoG-Q) [63]. As for manual video analysis method, time up and go task (TUGT) is a typical experimental pipeline to diagnose and assess PD and FoG [67]. These two schemes depend on test environment, experimental design, experience of clinicians, and the description ability of patients. Consequently, the results may be subjective and cannot provide reliable clinical guidance. Therefore, an accurate and automatic FoG detection approach is desired to objectively detect or even predict the onset of FoG in advance for neurologists to assess the disease status, progression, and make essential intervention to improve the patient's quality of life. Automatic detection of FoG in everyday life is crucially important for increasing patient's mobility and avoiding falls, which can improve the patient's life quality, increase self-confidence, and delay the progression of PD.

Despite their common occurrences in living conditions, the underlying mechanisms of FoG are still poorly understood [45] and accurately and timely detection of FoGs in a living condition is still a challenging task. Wearable inertial sensors are the commonest solution for detecting FoG in both laboratory and home environments [34, 49, 75] due to their cheap, lightweight, and unobtrusive features. Accelerations and gyroscopes from locomotion-related body locations, such as waist, chest, upper and lower limbs, have been utilized in the detection of FoG [58, 59]. Often used discriminative inertial features include Fourier spectrum [15, 58, 59], freezing of gait criterion [4], wavelet power [2, 17], k-index [36, 46] and freezing index (FI) [58], etc.

Moore *et al.* [58] proposed freeze index (FI) to characterize FoG using frequency domain information. The vertical accelerations at left ankle were recorded and transformed into frequency domain. The obtained spectra were divided into a normal locomotion band (0.5-3 Hz) and freeze band (3-8 Hz). FI was defined as the ratio of power spectral densities in freeze band and locomotion band. A simple threshold was determined to discriminate FoG from normal locomotion because more high frequency components were observed in FoG spectra. The FI is time-variant and the instant FI is calculated by using a short-time Fourier transform with a moving window of six seconds. Some amelioration and optimization to FI-based FoG automatic detection using wearable devices were investigated to improve the classification accuracy such as in [4, 34, 71, 80, 89]. However, in these studies, a FoG event is marked as correctly detected when the FoG duration is touched with the marked FoG duration. That is, the FoG episodes have not been accurately located. This limits the application of the FoG detection in the real time intervention. Moreover, prediction of the FoG before its occurrence is highly expected in fall prevention [61]. Therefore, an improvement of FI to improve the time-accuracy is important for the detection/ prediction of FoG. A new FI calculation method based on a time-frequency spectrum estimation is investigated to improve the accurate detection of the FoG.

Meanwhile, emerging findings about brain activity shed light on the pathophysiology of FoG. Recent studies have shown that EEG offers important information for understanding and forecasting FoG [27, 82]. An increase of power in θ and α sub-bands and a decrease of β and γ power have been observed in the dynamics of EEG of PD patients [15]. Event-related potentials (ERP), which remain latent in the EEG signals of PD patients, were found capable of reflecting cognitive impairments and a promising biomarker for motor disorders [45, 55, 78, 82].

Besides focusing on unimodal features, attention has also been drawn to multimodal feature fusion to improve FoG detection performance [86, 85, 28, 70]. The multimodal features include accelerations, gyroscopes, gait features (e.g. stride length and step velocity), brain and heart activity, eye movements [28, 70], EMG [86, 85], etc. Significant features were selected or extracted from the multimodal data with statistical methods [15, 70, 37] or machine learning-based methods [66]. These researches presented encouraging results of FoG detection but still have room for improvement. In a study [78], 90% sensitivity and 92% specificity of FoG detection were achieved with multimodal features. However, the generalization ability of the results may be limited due to the small number of patients. Using the R index which merged the gyroscope and EMG information, 95% sensitivity and 98% specificity were achieved in the study [55], but this method may have strong reliance on the self-designed R threshold due to the effects of interpersonal variability. Wang *et al.* utilized multimodal features from accelerations and EEG signals [85], which were the same as in our study, but obtained much lower performance on their data. In their recent study [86], using more multimodal features yielded 97% sensitivity and 73% precision in LOOCV. However, the acquisition of EEG and additional EOG signals can be problematic in real applications.

The advancement of wearable technologies makes it possible for researchers to accurately detect FoG in daily life, because light weight, unobtrusive and comfortable, wearable devices are suitable for long-term monitoring of PD in daily life conditions. However, when the applications of multimodality-based detection of FoG significantly increase the detection accuracy, the acquisition of multimodal signals can be problematic. Multimodal data are usually acquired with separate subsystems, which result in high cost, poor wearability, and inconvenience in practical use. For example, the collection of EEG signals needs a long-time preparation; the acquisition equipment is often expensive; and the EEG signal processing is a complicated and skilled job. This makes the continuous monitoring of brain activity infeasible in practical applications. Though single electrode EEG has been designed to improve the wearability [43], limited information restricts the applications of the method. On the flip side, most multimodality-based FoG studies are simple stacks of multimodal physiological features, neglecting dynamic interactions among different modes which highly relate to the occurrence of FoG [19, 24, 28]. For example, Günther *et al.* found that an increase in EEG-EMG coupling at the beginning of stop and FoG episodes provides a better understanding of the pathophysiology of FoG [19]. Therefore, how to enhance users' wearing experience while preserving the high performance of multimodal data remains a challenging task. In this study, we will use the idea of proxy measurement [24, 26] to overcome the shortcomings of EEG acquisition, and proxy represents the expensive, time-consuming, and less-wearable EEG features from cheap, easy-obtained inertial signals. The new proxy features which integrate both EEG and gait modal information will be proved surprisingly powerful in the detection of FoG.

Diverse but complementary methodologies are required to uncover the complex determinants and pathophysiology of freezing of gait. To develop future therapeutic avenues, we need a deeper understanding of the disseminated functional-anatomic network and its associated dynamic processes. Most methods characterize the brain activities from frequency, space or time domain. A new representation which gives a complete description of the brain dynamics is essential to reveal the underlying pathophysiology of FoG. The final task of this study is to use the dynamic mode decomposition (DMD) to extract the spatial-temporal-frequency coupled dynamic modes to provide a possible path for the prediction of FoG.

2. Experimental Design and Data Preparation

In order to comprehensively study the mechanisms underlying FoG and develop a highly accurate detection method, an experiment is designed and multimodal data were acquired.

2.1 Participants

In order to conduct the experiments safely and obtain valid data which include sufficient FoG episodes, the participants were selected, based on the following inclusion criteria:

(1)   FoG occurs during the off time.
(2)   Being able to walk independently during the off time.
(3)   No severe vision or hearing loss, dementia, or other neurological/orthopedic diseases.

2.2 Data Collection

Some patients only wear two inertial sensors (mounted on left tibia and left wrist, respectively). TP9, TP10 (signal of the mastoid process of the temporary bone) were used as reference in data preprocessing. IO (electrooculogram) was given in the dataset without preprocessing.

The multimodal sensoring platform acquires EEG, EMG, ACC, and SC. The locations of each sensor is given in Table 1. Among them, the 25 channel EEG and three-channel EMG signals were acquired through a 32-channel wireless MOVE system (BRAIN PRODUCTS, Germany). ACC and SC were acquired using self-designed hardware subsystems based on TDK MPU6050 6-DoF accelerometer and gyro, and STMicroelectronics STM32 processor, see Table 1 and Fig. 1.

Studies have shown that FoG is related to the brain activities in the frontal, parietal, and occipital lobes [1, 69, 84]. Accordingly, EEG signals were acquired at

Table 1: Hardware configuration and location of the sensoring system

Sensing Type	System	Sensor Quantity	Sensor Location
28D-EEG	The wireless MOVE	28	FP1, FP2, F3, F4, C3, C4, P3, P4, O1, O2, F7, F8, P7, P8, Fz, Cz, Pz, FC1, FC2, CP1, CP2, FC5, FC6, CP5, CP6, TP9, TP10, IO
3D-EMG		3	Gastrocnemius muscle of right leg; tibialis anterior muscle of left and right legs
3D-accelerometer	MPU6050	4	Lateral tibia of left and right legs
3D-Gyro		4	Fifth lumbar spine; wrist
1D-SC	LM324	2	The second belly of the index finger and middle finger of the left hand

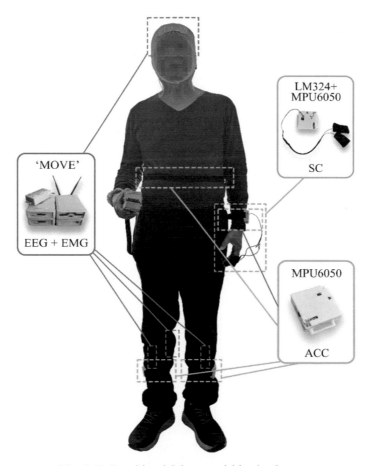

Fig. 1: FoG multimodal data acquisition hardware system

28 channels of the international 10-20 EEG electrode systems shown in Table 1 and Fig. 2 and artifacts were removed using EEGLAB [14].

Three channels of the 'MOVE' system were used to collect EMG signals. According to Alice Nieuwboer *et al.*'s results [83], the EMG signals were collected at the gastrocnemius muscle (GS) of the right leg and tibias anterior muscle (TA) of the left and right legs, respectively, which are shown in Fig. 3.

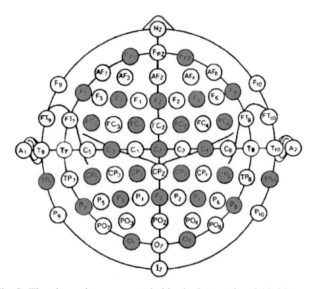

Fig. 2: The channels were recorded in the international 10-20 system

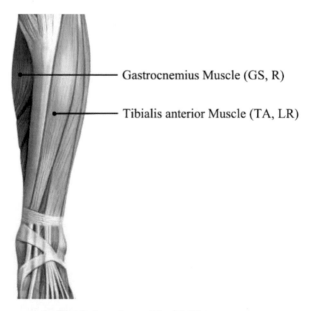

Gastrocnemius Muscle (GS, R)

Tibialis anterior Muscle (TA, LR)

Fig. 3: Locations of the EMG sensor

The SC acquisition was integrated in one of the ACC acquisition subsystems mounted on the left arm. The other three accelerators were mounted at the lateral tibia of the left and right legs, and fifth lumbar spine of the waist (L5), respectively. Both SC and ACCs were sampled at 100Hz and stored on a TF memory card. SC was recorded at the second belly of the left index finger and middle finger.

2.3 Protocol

The occurrence of FoG can be affected by the environment and the patient's emotional state and chiefly materializes in living circumstances in daily life. The tasks which may trigger FoG have been well reported in the literature, such as walking through narrow spaces, approaching obstacles, and so on [76]. Based on this knowledge, a procedure including two types of task were carefully designed to trigger FoG episodes.

The data acquisition were done under the supervision of the Ethics Committee of Beijing Xuanwu Hospital. Data were collected in the off-medication state of patients. Four specialists taking different roles in the process, included an accompanying doctor, a system operator, a video recorder and a stopwatch time controller.

Experiments were conducted as per the following procedures:

(1) All participants read and signed the informed consent.
(2) The participants were asked to take a physical examination and fill in medical history and Unified PD Rating Scale (UPDRS) questionnaires to confirm the participants meet the inclusion criteria. The participants did not take medicine within two hours before collecting data to ensure that they were in off time.
(3) Participants wore the multimodal sensing equipment under the help of professional technicians, including EEG caps, EMG electrodes, ACC, and SC sensors mounted at the specified locations as discussed before.
(4) Completed the tasks according to the experimental paradigm as discussed below. The video was recorded during the whole experiment.
(5) Checked the saved data to avoid any faults in the experimental process; the data that did not meet the requirements were to be discarded, and the task was redone after a two-minute rest.

Four tasks were conducted by each participant, which were defined as:

(1) *Task 1*: This walking task was conducted in a setting as shown in Fig. 4. Participants started from a sitting state. When a participant is ready, he rises from a chair at point *A* and marches to the junction *B* between the room and a narrow corridor. Turn right and walk into the corridor. Bypass the obstacle 1 (can be a chair or a square region on the floor) by turning their bodies. Continue going straight along the narrow corridor until point *C*. Make a U-turn at the end of the corridor, and go along an opposite direction. Bypass the three obstacles 1, 2 and 3 by turning their body. When they reach the left end of the corridor, point *D*, make another U-turn, bypassing obstacles 3 and 2, and reach the door of the room. Enter the room, and walk back to the chair, and sit down.
(2) *Task 2*: Repeat the above task one more time.
(3) *Task 3*: This task was conducted in a setting as shown in Fig. 5, where a square

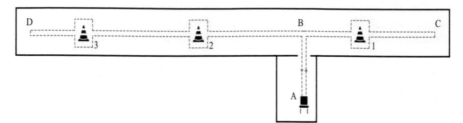

Fig. 4: Experimental settings of task 1-2

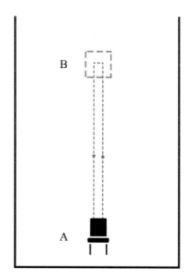

Fig. 5: Experimental settings of task 3-4

was drawn on the ground for patients to make a turn in a limited space. When the patient is ready, stand up from the chair at the end *A* in the room and march to the pre-pasted square mark at the end *B* in the room. The participant makes a U-turn in the narrow square region, and then walks straight back to the chair and sits down.

(4) *Task 4*: Repeat Task 3. Then end the experiment.

3. Detection of FoG Based on Freezing Index

Due to their cheap, lightweight and unobtrusive features, wearable inertial sensors are the commonest solution for detecting FoG in both laboratory and home environments. Based on pure acceleration information, freezing index has been proposed as an easy to use indicator for the freezing of gait. However, the short-time-Fourier-transform based calculation of FI cannot well balance the temporal and frequency resolution [35] and leads to poor accuracy in the detection of FoG. A new in indirect time-frequency spectral estimation (TFFS) method has been proposed to improve the performance of the FI.

3.1 Asymmetric Basis Function TV-ARMA Model and Identification

In this study a new indirect non-parametric time-frequency spectral estimation method is proposed [25]. The new method essentially includes two steps: the system identification of TV-ARMA model from non-stationary time series and the estimation of time-frequency spectrum from the obtained time-domain model. The first step is crucially important because the TV-ARMA is needed to be capable of not only fitting the measurements but also accurately characterizing the time-varying system dynamics so that the associated frequency domain representation, namely, the frequency response function, will agree with the time-frequency spectrum of the non-stationary process.

Consider the general TV-ARMA model [22, 23, 57]:

$$y(k) = \sum_{i=1}^{p} a_i(k) y(k-i) + \sum_{j=1}^{q} b_j(k) e(k-j) + e(k) \tag{1}$$

where $k = 1, 2, \ldots N$ (N is the length of samples) is the time index of observed data; $y(k)$ and $e(k)$ denote observed system output (time series) and noise, respectively, with maximum delays p and q respectively, $e(k)$ is the model error that is an i.i.d. white noise sequence with zero mean and variance δ_e^2. In the model, $a_i(k)$, and $b_j(k)$ stand for the corresponding time-varying coefficients which may change smoothly or abruptly in the process.

In the multi-basis function time-varying dynamic system identification algorithms, the time-varying coefficients $a_i(k)$ and $b_j(k)$ are approximated as superposition of a set of over completed basis functions. Our resent results [39] showed that a sparser representation can be obtained when postsynaptic currents, like asymmetric basis functions were used.

$$\begin{cases} a_i(k) = \sum_{m=1}^{M} \alpha_{i,m} \varphi_m(k) \\ b_j(k) = \sum_{l=1}^{L} \beta_{j,l} \varphi_l(k) \end{cases} \tag{2}$$

where M and L are the size of the two sets. $\alpha_{i,m}, \beta_{j,l}$ denotes the final time-independent parameters to be estimated after decomposition using asymmetric alpha basis function. $\Phi_{M(L)} = \left\{ \varphi_{m(l)} \left(\dfrac{k-\mu}{\sigma} \right) \right\}_{m(l)=1}^{M(L)}$ is a set of alpha basis function shaped by:

$$\varphi(k \mid a,b) = \frac{\Gamma(a+b)}{\Gamma(a)\Gamma(b)} (k-0)^{a-1} (1-k)^{b-1} \tag{3}$$

where $1 \leq a \leq b$, $a, b \in \mathbb{N}$ are the parameters control the shape of the wavelet function. $\Gamma(\cdot)$ denotes the generalized factorial function of Euler. μ and σ are the translation and scale parameters of alpha basis function.

Substituting (2) into (1), the TV-ARMA model can thus be rewritten as:

$$y(k) = \sum_{i=1}^{p}\sum_{m=1}^{M}\alpha_{i,m}\varphi_m(k)y(k-i) + \sum_{j=1}^{q}\sum_{l=1}^{L}\beta_{j,l}\varphi_l(k)e(k-j) + e(k) \qquad (4)$$

Model (4) can be rearranged as a simple form:

$$y(k) = \left[\sum_{i=1}^{p}\sum_{m=1}^{M}\alpha_{i,m}\right]\left[\sum_{m}^{M}\varphi_m(k)\sum_{i}^{p}y(k-i)\right]$$

$$+ \left[\sum_{j=1}^{q}\sum_{l=1}^{L}\beta_{j,l}\right]\left[\sum_{l}^{L}\varphi_m(k)\sum_{j}^{q}e(k-j)\right] + e(k) \qquad (5)$$

A matrix form for the formula (5) is given as:

$$y = \Psi\theta^T + e \qquad (6)$$

where the new variables $\Psi = \left\{[y(k-1),\cdots,y(k-p)]_{k=1}^{N}\otimes\Phi_M, [e(k-1),\cdots,e(k\text{-}q)]_{k=1}^{N}\otimes\Phi_L\right\}$ denotes the candidate dictionary, in which $\otimes$ is the Kronecker product, and θ^T indicates the transpose of time-invariant parameter matrix $\theta = [\alpha_{1,1},\alpha_{1,2}\cdots,\alpha_{1,M},\cdots\alpha_{p,M},\beta_{1,1},\beta_{1,2}\cdots,\beta_{1,M},\cdots\beta_{q,M}]$, y and e are the model output data and noise error respectively.

As discussed above, the initial TV-ARMA model (1) can be treated as the system identification of time-invariant linear-in-the-parameter model (6) with model terms Ψ and the associated constant coefficients θ. However, the over-complete basis functions may lead to numerically ill-posed problems. A local regularization-assisted orthogonal forward regression (LROFR) algorithm is used to select the most significant asymmetric basis function and obtain a sparse representation of the non-stationary process. For the details about LROFR, refer to paper [39].

In order to identify the moving average noise model, an iterative process is employed because the noise terms $e(k-1)$, ..., $e(k-q)$ are not known. In the process, a TV-AR model is firstly identified using LROFR algorithm and the residual series $\xi^{(1)}(k-1)$, ..., $\xi^{(1)}(k-q)$ are used to replace the noise terms $e(k-1)$, ..., $e(k-q)$, respectively. A TV-ARMA model can then be identified and a new residual series $\xi^{(2)}(k-1)$, ..., $\xi^{(2)}(k-q)$ is then obtained. Repeat the process until the model is converged. A similar identification techniques has been used in the identification of time invariant NARMAX model [8].

3.2 Time-frequency Spectral Estimation

Once the TV-ARMA structure and the associated time-varying coefficients are determined using the LROFR algorithm, the time frequency spectrum can then be estimated using the following rational spectral estimation formula [38, 39]:

$$P_{TDS}(t,f) = \delta_e^2 \frac{\left|1 + \sum_{j=1}^{q}\hat{b}_j(t)e^{-k2\pi jf/f_s}\right|^2}{\left|1 - \sum_{i=1}^{p}\hat{a}_i(t)e^{-k2\pi if/f_s}\right|} \qquad (7)$$

where $P_{TDS}(t, f)$ is the TFSE value at time t and frequency f, $\hat{a}_i(t)$ and $\hat{b}_j(t)$ is the estimation of TV-ARMA model parameters $a_i(k)$, $b_j(k)$, δ_e^2 is the variance of model residual, f_s is the sampling frequency and $j = \sqrt{-1}$ denotes the imaginary of a complex number.

Formula (7) represents the frequency response function (FRF) of a time-varying ARMA process, which is a frequency domain description of the original time series. This description is sufficiently accurate when the time-varying coefficients change at a relatively slow rate [65]. It can be observed that poles and zeros of the FRF (7) correspond to peaks and valleys in the TFS respectively. The frequency spectrum, therefore, changes with the changes of the poles and zeros, which are determined by the time-varying coefficients. The non-stationary dynamic process is characterized by the time-varying ARMA model and then by the time-varying FRF. There are several significant advantages in the indirect parametric TFSE method than the direct transforming method, such as, short-time-Fourier transforms and wavelet transforms. Firstly, the obtained TFS is robust to the effects of noise because the noise pollution has automatically been eliminated in the modeling process and only the most important system modals (frequency components) remain in the simple model structure. Secondly, the obtained TFS is less widespread and the most important frequency components play a prominent role. Finally, smooth coefficients, which can change abruptly but still continue over time, can be obtained when proper regularization techniques are used in the modeling process. A smoothly changing spectrum can then be obtained. These features can be observed from the results in Section III.

3.3 Calculation of FI

The freeze index (FI) at time t was defined as the ratio of the area under power spectra in the 'freeze' band (3-8 Hz) and the area under the spectra in the 'locomotion' band (0.5-3 Hz) [57]. The instant power spectra is traditionally calculated using short time Fourier transform with a fixed window (centered at time t) [57, 5, 30, 90]. The calculated FI may be sensitive to the window size. For example, the power spectra cannot be accurately estimated based on a limited number of data in a narrow window whilst the power spectra can be blurred when a long window is used. To reduce the effect of the window size, the instant FI is calculated employing the $P_{TDS}(t, f)$ as

$$FI(t) = \frac{\int_3^8 P_{TDS}(t, f)df}{\int_{0.5}^3 P_{TDS}(t, f)df} \tag{8}$$

where the time as well as frequency are continuous independent variables.

The whole proposed processes for the calculation of FI and the automatical detection of FoG can be summarized as follows:

Step 1: Data preprocess: Preprocess the raw accelerometer data using low pass and zero lag filter with a 16 Hz cutoff frequency, then discard the data that was not conformable with protocol.

Step 2: *Model construction and asymmetric basis function expansion*: Construct the dictionary of TV-ARMA (p, q) model terms $[y(k-1), \cdots, y(k-p), e(k-1), \cdots, e(k-q)]_{k=1}^{N}$ and the over-complete alpha basis function basis $\Phi_{M(L)} = \left\{ \varphi_{m(l)} \right\}_{m(l)=1}^{M(L)}$; then combine the two dictionaries by $\otimes$ to generate the candidate term dictionary Ψ, and the order of model p and q is determined by APRESS criterion.

Step 3: *Model terms selection and parameter estimation*: Use the LROFR algorithm to select the model terms and approximate the corresponding time-varying parameter to obtain the best fitness of data.

Step 4: *System identification of the noise model*: Firstly, apply LROFR algorithm to identify a TV-AR model, then replace the noise terms in TV-ARMA model with the residual series obtained in TV-AR model. Repeat the process until the model is converged.

Step 5: *Time frequency spectral estimation*: Estimate the high resolution time frequency spectral using (7).

Step 6: *Calculation of FI*: Calculate the high resolution TFSE and FI defined in (8).

Step 7: *Automatic detection of FoG*: Select an optimal threshold to classify the FoG and free-FoG automatically.

3.4 Results

The performance of three different TFSEs and FIs in terms of accuracy and specificity under the effects of inter and intra-subject variability, an example of 20s data fragment are illustrated in Fig. 6.

Based on the advanced system identification and time-frequency spectral estimation techniques, the FI is significantly improved in the robustness and temporal accuracy. Results have shown that the new FI can be used for detecting different FoG patterns and a preselected threshold is applicable to other patients. Combined with a cheap wearable sensor, FoG can easily be detected and even predicted, using just vertical accelerations measured at shank level. The new FI offers a promising application of wearable sensors in continuous FoG monitoring and automatic management in home conditions to improve the patient;s living quality and reduce the increasing caregiver stress.

4. Detection of FoG Based on Multimodal Data

To our knowledge, only a few studies have employed multimodal features to detect FoG [86, 85, 28, 70]. Among these studies, a wide range of multimodal signals, including accelerations, gyroscopes, EMG, EEG, EOG, and ECG signals, were analyzed. In a study [78], 90% sensitivity and 92% specificity of FoG detection were achieved with multimodal features. Using the R index which merged the gyroscope and EMG information, 95% sensitivity and 98% specificity were achieved in the study [55]. Wang *et al.* utilized multimodal features from accelerations and EEG

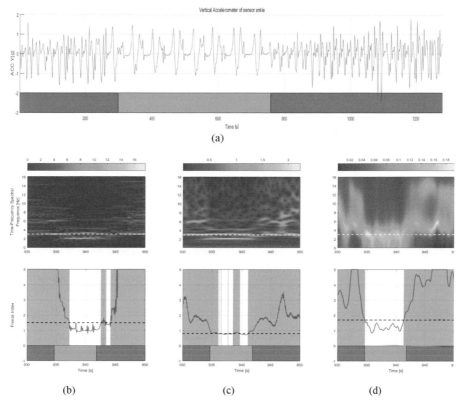

Fig. 6: Another sequence with a 20s window extracted from subject 02 (a), and TFSE and FI-based STFT (b), CWT (c), asymmetric basis function TV-ARMA (d) [25]

signals [85]. In their recent study [86], use of more multimodal features yielded 97% sensitivity and 73% precision in LOOCV. However, the acquisition of EEG and additional EOG signals can be problematic in real applications.

In order to further verify the usefulness of multimodal data, the FoG detection was studied based on our own multimodal data.

4.1 Feature Extraction

According to the references [27, 28, 52, 79] as many as 16 statistical features were utilized in this study for FoG classification, to verify the data quality. The features, along with the associated brief description, are listed in Table 2. Each modal data was verified individually, and the multimodal data were verified by the combined features. With these features, the multimodal data were verified, based on intra-subject validation error and inter-subject validation errors.

For the EEG data, using the method proposed in Handojoseno *et al.* in 2012 [27, 28, 29, 52], 5-scale discrete wavelet transforms (DWT) for original EEG signal of each 25 rest channels to obtain five rhythms, that is, the δ wave, in 0-3.9 Hz; θ wave, in 3.9-7.8 Hz; α wave, in 7.8-15.6 Hz; β wave, in 15.6-31.3 Hz; and γ wave, in 31.3-62.5 Hz. The single-channel of EEG extracted as its wavelet energy (WE) in δ, θ, α-bands and its total wavelet entropy (TWE), denoted as WE_δ, WE_θ, WE_α, and TWE, respectively, as features of EEG.

Table 2: Features and brief description

Data	Channel Quantity	Feature	Description
EEG	25	WE_δ	Represents changes in energy of FoG and
		WE_θ	locomotion period of PD patients' EEG signal
		WE_α	
		TWE	Represents changes in energy complexity
EMG	3	MAV	Estimation of the STD of EMG signals
		ZC	Related to the frequency of EMG signals
		SSC	
		WL	Directly related to the EMG signals
ACC	3	SE	Evaluate the repeatability of the waveform
		STD	Standard Deviation
		TP	Detection algorithm of FoG proposed by
		FI	Moore *et al.* [60]

According to the Parseval Theorem and Shannon Information Entropy Theory, the features WE and TWE are calculated as follows (1) and (2). Among them, WE_j is the WE of the jth component in the single-channel EEG after the 5-scale DWT; y_j is the corresponding jth EEG signal; N is the length of the data.

$$WE_j = \sum_{k=1}^{N} |y_j|^2 \tag{9}$$

$$TWE = -\sum_j \frac{WE_j}{\sum_j WE_j} \log \frac{WE_j}{\sum_j WE_j} \tag{10}$$

For the EMG data, four features of single-channel of EMG were extracted as its Mean Absolute Value (MAV), Zeros Crossing (ZC), Slope Sign Change (SSC), and Wave Length (WL), respectively.

For the ACC data, the accelerations in three directions at lateral tibias of the left or right leg were extracted as the associated sample entropy (SE), standard deviation (STD), sum (Total Power, TP) and ratio (Freezing Index [58], FI) of the frequency components of freezing band (3-8 Hz) and locomotion band (0.5-3 Hz).

For the label, FoG data were segmented using a sliding window method with a window length to three samples and a sliding step size of 0.3 sample intervals (500 Hz). Each segment was assigned a common label based on the proportion of FoG sample, that is, the percentage of FoG (PFG) defined as (11).

$$PFG = \frac{N_{FoG}}{N_{FoG} + N_N} \times 100\% \tag{11}$$

where N_{FoG} is the number of FoG samples in the segment, while *NN* is the number of FoG-free samples. The label of the segment is determined by (12), where *T* is the appropriate threshold selected by the researcher, which is usually around 0.75-0.85.

$$NewLabel = \begin{cases} 1, & if \ PFG \geq T \\ -1, & if \ PFG < T \end{cases} \qquad (12)$$

The labeling threshold was set to 80% in the verification, which would indicate FoG's appearance in the data segment if the percentage of the old positive label in the segment is over 80%.

4.2 FoG Detection and Evaluation

Based on the extracted features, a SVM model with radial basis function kernel is used to classify FoG from normal locomotion [47, 78]. Four types of training are based on different feature combinations, including individual EEG features, individual EMG features, individual ACC features, and all features together. Simultaneously, divide the dataset into the training set and test set with a quantity ratio of 0.25 randomly, and the method of cross-validation and grid search is used to determine the hyper-parameters of SVM automatically. Replicate each experiment 20 times, and the average value of evaluation indicators, including accuracy, sensitivity, specificity, precision, F1 value, area under curve (AUC) are used to evaluate the classification performance.

The data were evaluated by its discrimination ability of FoG from normal locomotion. The performance of the different mode data was compared under two

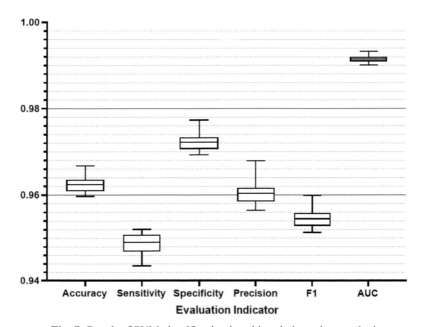

Fig. 7: Result of SVM classification in subject-independent method

different settings, namely, subject-dependent and subject-independent verifications. In the subject-dependent case, the data from each subject were divided into training and test sets. The performance of the multimodal data was evaluated individually for each subject. On the contrary, the subject-independent verification mixed data from all subjects and evaluated the performance of the dataset considering the inter-subject variability.

4.2.1   Subject-dependent Analysis

For the subject-dependent verification of individual patient data quality, the SVM classifier is used to train the data. Results were shown in Table 2.

In addition, the duration of FoG events in the data has a significant impact on the classification results. Either the patient with many FoG (ID:03) or the patient (ID:09) with few FoG events produce a relatively poor performance, which may be caused by the unbalanced data for the training of the classifier.

4.2.2   Subject-independent Analysis

Figure 9 is a box chart of the value of evaluation indicators, while the upper and lower lines of it are the maximum and minimum values, and the mid-line show the average value.

5.   High-wearable Detection of FoG

Freezing of gait is the most serious symptom of Parkinson's disease and prompt detection of freezing of gait (FoG) is crucial to fall prevention and effective intervention. Although multimodal data combining electroencephalography (EEG) benefits accurate FoG detection, the preparation, acquisition, and analysis of EEG signals are delicate, time-consuming and costly, which impedes the application of multimodal information in FoG detection. By proxy measuring [24, 26] the EEG information from a highly-wearable inertial sensor, a new accurate FoG detection method which possess both the advantages of rich multimodal information and of high wearability, were proposed. EEG features were reconstructed from inertial accelerations (ACC) using a long short-term memory (LSTM) network and the proxy EEG features, instead of the real EEG features, were used to accurately detect FoG, avoiding complicated EEG acquisition. Combined with traditional ACC features, the proxy EEG features performed surprisingly well and even better than the original multimodal features did. Creatively using the proxy modal data as sources of information and the target modal data as the supervision of the cross-sensory feature transfer, the new method is capable of extracting effective features from simple sensory system and enables an accurate detection of FoG with cheap, wearable sensors. We anticipate that this new method will provide a promising path for the long-term monitoring of FoG and greatly improve the healthcare management of Parkinson's disease in living environments.

Due to the complex physiological processes of human locomotion, which involve intrinsic nonlinear, neural dynamics in gait initiation, automatic rhythmicity, posture preparation and execution of gait [51, 31, 81], etc., a static linear model with a short memory is insufficient to characterize the cross-sensory information interactions [21]. In this study, a LSTM network [88] was utilized to build the nonlinear proxy measurement model (named LSTM-PM model) from ACC to EEG sub-band components. The structure of the LSTM-PM model is shown in Fig. 10. Tri-axis ACC signals from left leg and their associated time delays were used as the input of the LSTM-PM model and the δ, θ, α and β sub-band waves of Fz, Cz and O1 channels were taken as the target variables to be proxy measured. There are two primary considerations for reconstructing the sub-band components rather than original EEG signals. On one hand, due to the very rich information in the EEG signals, direct reconstruction of the EEG signals can be very challenging. Additionally, the high frequency components of EEG made little contribution to FoG detection and prediction [37], and, therefore, it is unnecessary for the PM model to represent all the details in EEG data. On the other hand, extraction of the abstract WE features, based on an inaccuracy reconstruction of the EEG signals, may introduce further error and corrupt the subsequent FoG detection.

As a result, the LSTM-PM model fused the EEG features into the ACC data and realized the cross-sensory feature transfer. Based on the cross-sensory reconstructed pmEEG sub-band waves, proxy brain electrophysiological features, WEs, RWEs, and TWEs can be calculated as defined in Subsection 2.4. The proxy EEG features were then used for the subsequent FoG/non-FoG classification.

5.1 FoG Detection

The whole scheme of the proposed method is presented as a two-stage feature extraction and FoG classification. In the first stage, the most efficient multimodal features of each mode, namely, the EEG and ACC features, were extracted, respectively, to identify FoG events. The second stage trained the LSTM-PM networks and implemented the cross-sensory feature transfer from EEG to abstract ACC transforms, so that the EEG sub-waves (e.g. δ, θ, α, β waves) were reconstructed as the proxy EEG signals directly from ACC through the well-trained LSTM-PM regressor. Now, the feature extraction of multimodal data becomes feature extraction of pure ACC data, including the traditional ACC features and the pmEEG features.

After multimodal feature extraction, support vector machine (SVM) with RBF kernel was employed to discriminate FoG from non-FoG events. The optimal hyperparameters, penalty parameter c and Gaussian kernel width g, of SVM were determined by a grid-search strategy based on the performance of a three-fold cross-validation. The grid-search were conducted in a logarithmic space ranging from 2^{-8} to 2^8 with an interval of 0.25.

5.2 Subject-dependent Validation

5.2.1 Intra-session FoG Detection

In the subject-dependent experiments, both training and test sets are taken from the same participant. In the intra-session detection, the average metrics of five-fold-cross-validation are reported in Table 3. It can be observed that models based on multimodal features surpassed unimodal feature models at almost all five metrics. However, the pseudo-unimodal model based on the reconstructed pmEEG features produced even better results in many metrics than the traditional multimodal data did. In multimodal feature models, pmEEG-ACC combination produced better performance than original multimodal feature combination EEG-ACC models did with increments of 3.7%, 2.2% and 5.4% in (subject-averaged) accuracy, specificity and F1-score, respectively. Particularly, the FoG detection of subject S6 based on EEG modal, ACC modal and EEG-ACC multimodal features without using the proxy information fusion had poor sensitivity with scores of 0.579, 0.272 and 0.696, respectively, which implied a big portion of FoG cases were misclassified as normal gait by these models. The large false negative error indicates that the models fail to detect FoG and increase the fall risks of patients in practices. In contrast, the performance has been significantly improved in our pmEEG-ACC feature combination and even in the pseudo-unimodal pmEEG models, where the sensitivity increased to an acceptable level of 0.844 and 0.842, respectively. These inspiring results can be explained as the pmEEG-ACC features merged information from both accelerations and EEG sub-band components and promoted the deep fusion of multimodal information. Compared with models based on pure EEG or ACC features, the model based on pseudo-unimodal pmEEG features also achieved encouraging performance on average metrics. This significant improvement in unimodal settings further demonstrated the effectiveness of the novel cross-sensory feature transfer method.

It is worthy to emphasize that both pmEEG and pmEEG-ACC-based models used only the inertial accelerations once the LSTM-PM model had been established for each participant. Their subject-averaged metrics improved by 10% to 20% as compared to the pure ACC-based models and some metrics increased over 30%, for example, the precision and sensitivity of S6. This means, using the simple inertial data collection settings, the new pmEEG-ACC cross-sensory features can significantly improve the performance of FoG detection and reach an even better performance than the true multimodal data where costly, time-consuming, and complicated data collection are essential. More importantly, the new method enables the accurate detection of FoG under a high-wearable setting using only inertial sensors. The cost of the data acquisition equipment can be reduced from tens of thousands of dollars to a few dozen dollars using our self-designed inertial acquisition system. This makes the high accurate detection of FoG possible in the daily-life environment and greatly promotes the practical applications.

Note that S7 and S8 achieved appreciable performance at some metrics on single EEG or ACC features, which is owing to their extremely unbalanced datasets. Careful checking of the data showed that the whole experiment of S7 was almost manifested as FoG while S8 had nearly no FoG during tasks. This explained the very low specificity of S7 and low sensitivity of S8 in all five settings. Hence, the metrics of S7 and S8 were excluded when computing the subject-averaged metrics in Table 4.

Table 3: Results of four types of SVM classification in subject-dependent analysis

Patient ID	Features	Accuracy	Sensitivity	Specificity	Precision	F1 Value	AUC
01	EEG	0.9709	0.9420	0.9833	0.9602	0.9510	0.9953
	EMG	0.8866	0.8028	0.9224	0.8162	0.8092	0.9335
	ACC	0.8983	0.8617	0.9136	0.8094	0.8345	0.9613
	ALL	0.9701	0.9380	0.9838	0.9608	0.9492	0.9939
02[N]	/	/	/	/	/	/	/
03	EEG	0.9651	0.9855	0.8825	0.9714	0.9784	0.9923
	EMG	0.8303	0.9589	0.3093	0.8492	0.9007	0.7828
	ACC	0.9374	0.1154	0.9953	0.7438	0.1743	0.9007
	ALL	0.9781	0.7570	0.9938	0.8989	0.8204	0.9817
04	EEG	0.9784	0.9462	0.9850	0.9292	0.9375	0.9941
	EMG	0.9448	0.8057	0.9729	0.8570	0.8301	0.9616
	ACC	0.9400	0.7500	0.9799	0.8867	0.8117	0.9622
	ALL	0.9770	0.9304	0.9864	0.9326	0.9314	0.9958
05[N]	/	/	/	/	/	/	/
06	EEG	0.9564	0.9136	0.9736	0.9331	0.9232	0.9847
	EMG	0.8816	0.7853	0.9196	0.7938	0.7894	0.9269
	ACC	0.8575	0.7846	0.8863	0.7313	0.7566	0.9164
	ALL	0.9653	0.9497	0.9714	0.9293	0.9393	0.9907
07	EEG	0.8789	0.8646	0.8895	0.8525	0.8585	0.9423
	EMG	0.8555	0.7897	0.9042	0.8591	0.8229	0.9220
	ACC	0.8709	0.8283	0.9024	0.8624	0.8450	0.9349
	ALL	0.9252	0.9484	0.9080	0.8845	0.9151	0.9693
08(1)	EEG	0.9600	0.9439	0.9719	0.9613	0.9525	0.9878
	EMG	0.8872	0.8694	0.9005	0.8661	0.8677	0.9501
	ACC	0.9353	0.9289	0.9400	0.9198	0.9242	0.9809
	ALL	0.9715	0.9608	0.9794	0.9718	0.9663	0.9895
08(2)	EEG	0.9459	0.9487	0.9432	0.9404	0.9445	0.9766
	EMG	0.8874	0.8743	0.8999	0.8932	0.8836	0.9368
	ACC	0.9318	0.9533	0.9111	0.9114	0.9318	0.9799
	ALL	0.9560	0.9587	0.9534	0.9516	0.9551	0.9857
09	EEG	0.9762	0.7429	0.9928	0.8814	0.8047	0.9801
	EMG	0.9357	0.0545	0.9980	0.7123	0.1013	0.7856
	ACC	0.9238	0.9733	0.7198	0.9348	0.9536	0.9519
	ALL	0.9718	0.9863	0.9130	0.9787	0.9825	0.9945
10	EEG	0.9422	0.9042	0.9720	0.9620	0.9322	0.9827
	EMG	0.8772	0.8414	0.9054	0.8744	0.8575	0.9372
	ACC	0.8327	0.8507	0.8186	0.7860	0.8170	0.9033
	ALL	0.9585	0.9301	0.9806	0.9741	0.9516	0.9889

Table 3: *(Contd.)*

Patient ID	Features	Accuracy	Sensitivity	Specificity	Precision	F1 Value	AUC
11	EEG	0.9639	0.9284	0.9826	0.9657	0.9466	0.9940
	EMG	0.8600	0.7292	0.9288	0.8444	0.7821	0.9146
	ACC	0.8490	0.7699	0.8907	0.7879	0.7787	0.9173
	ALL	0.9666	0.9337	0.9839	0.9684	0.9507	0.9944
12	EEG	0.9852	0.9566	0.9927	0.9718	0.9640	0.9947
	EMG	0.8888	0.5809	0.9693	0.8333	0.6836	0.8905
	ACC	0.9125	0.7349	0.9589	0.8240	0.7767	0.9427
	ALL	0.9857	0.9637	0.9915	0.9674	0.9655	0.9982
Average	EEG	0.95664	0.91606	0.96081	0.93900	0.92666	0.98404
	EMG	0.88501	0.73564	0.87547	0.83626	0.75709	0.90378
	ACC	0.89901	0.77738	0.90150	0.83613	0.78219	0.94106
	ALL	0.96597	0.93244	0.96774	0.94709	0.93882	0.98932

An N next to the patient ID indicates that almost no FoG appeared during the experiment. ALL = All features together.

5.2.2 *Inter-session FoG Detection*

In inter-session FoG detection setting, the training and testing data are taken from the same subject but at different sessions. That is, the well-trained LSTM-PM models based on one session data were employed to proxy reconstruct EEG sub-band components of other sessions which were collected from the same subject. In the inter-session FoG detection, the subject-averaged metrics were calculated with seven subjects because session-2 of S7 was better balanced with 1749 FoG and 689

Table 4: Model performance in single FoG detection

Metrics	#No	Single Modal Features			Multimodal Features	
		EEG-modality	ACC-modality	pmEEG modality	EEG-ACC multimodality	pmEEG-ACC multimodality
Accuracy	S1	0.837	0.864	0.914	0.936	**0.950**
	S2	0.857	0.858	0.920	0.935	**0.951**
	S3	0.749	0.768	0.930	0.856	**0.937**
	S4	0.879	0.816	0.920	0.916	**0.928**
	S5	0.817	0.785	0.908	0.904	**0.944**
	S6	0.828	0.736	0.924	0.871	**0.931**
	S7[un]	0.931	0.916	**0.957**	0.946	0.955
	S8[un]	0.970	0.963	0.974	0.974	**0.979**
	Ave	0.828	0.805	0.919	0.903	**0.940**
Precision	S1	0.826	0.773	0.902	0.904	**0.927**
	S2	0.859	0.831	0.917	0.933	**0.934**
	S3	0.709	0.696	0.919	0.830	**0.925**

Metric	Subject					
	S4	0.875	0.790	0.925	0.920	**0.929**
	S5	0.791	0.790	0.886	0.884	**0.926**
	S6	0.755	0.578	0.885	0.819	**0.901**
	S7^{un}	0.937	0.924	0.964	0.962	**0.967**
	S8^{un}	*Nan*	*Nan*	*Nan*	**0.793**	0.767
	Ave	0.803	0.743	0.906	0.882	**0.924**
Specificity	S1	0.925	0.866	0.948	0.949	**0.960**
	S2	0.880	0.843	0.928	**0.941**	0.940
	S3	0.829	0.784	0.947	0.890	**0.951**
	S4	0.887	0.802	0.934	0.929	**0.937**
	S5	0.841	0.856	0.909	0.910	**0.942**
	S6	0.927	0.919	0.956	0.940	**0.964**
	S7^{un}	0.399	0.254	0.660	0.647	**0.695**
	S8^{un}	**0.999**	0.996	0.997	0.994	0.996
	Ave	0.882	0.845	0.937	0.927	**0.949**
Sensitivity	S1	0.673	0.864	0.849	0.909	**0.930**
	S2	0.830	0.877	0.911	0.928	**0.964**
	S3	0.629	0.745	0.903	0.805	**0.915**
	S4	0.870	0.832	0.904	0.900	**0.919**
	S5	0.786	0.694	0.908	0.895	**0.944**
	S6	0.579	0.272	0.842	0.696	**0.844**
	S7^{un}	0.989	0.988	**0.989**	0.979	0.983
	S8^{un}	0.267	0.167	0.367	0.500	**0.627**
	Ave	0.728	0.714	0.886	0.856	**0.919**
F1-score	S1	0.741	0.815	0.873	0.907	**0.928**
	S2	0.843	0.853	0.914	0.930	**0.948**
	S3	0.665	0.719	0.911	0.817	**0.920**
	S4	0.872	0.810	0.914	0.910	**0.924**
	S5	0.788	0.736	0.896	0.890	**0.935**
	S6	0.653	0.365	0.863	0.750	**0.871**
	S7^{un}	0.963	0.955	0.976	0.970	**0.975**
	S8^{un}	*Nan*	*Nan*	*Nan*	0.601	**0.622**
	Ave	0.760	0.716	0.895	0.867	**0.921**

^{un} Means an unbalanced data for the subject. *Nan* means not a number because TP and FP in the confusion matrix are zero.

non-FoG events and the result was credible. Comprehensive comparisons of the performance of the five different feature combinations are presented in Table 5.

In inter-session evaluation, each feature extraction method worked well and each feature combination produced similar performances as they did in the intra-session evaluation. More importantly, the performances of the pmEEG and pmEEG-ACC combinations did not get worse in inter-session settings. This illustrated that the PM cross-sensory method works well when the LSTM-PM model was trained in a data session but applied to another data session.

Based on the results in Table 5, the same conclusion can be drawn that the proxy pseudo-multimodal features showed better performances than the original multimodal features did. The five subject-averaged metrics were promoted by 3.6%, 3.4%, 1.6%, 7.4%, and 5.7% in pmEEG-ACC models, compared with their original EEG-ACC counterparties. The proxy pseudo-unimodal pmEEG performed even better than the original multimodal combination EEG-ACC in many metrics. Using only the inertial information, the proxy features can significantly improve the detection of the FoG from all the criteria, which is in line with intra-session evaluation. This means that we can use the same LSTM-PM models for different FoG detection sessions.

5.3 Subject-independent Validation

It has been shown that the proxy cross-sensory feature extraction method can significantly promote the accurate detection of FoG with a much simpler data acquisition. In this section, we will show that the proxy cross-sensory feature extraction method works well in a subject-independent setting. For traditional feature extractions in the ACC, EEG, and ACC-EEG combination, there were no differences

Table 5: Model performance in inter-session FoG detection

Metrics	#No	Single Modal Features			Multimodal Features	
		EEG-modality	ACC-modality	pmEEG modality	EEG-ACC multimodality	pmEEG-ACC multimodality
Accuracy	S1	0.811	0.888	0.931	0.923	**0.952**
	S2	0.876	0.881	0.917	0.935	**0.962**
	S3	0.807	0.834	0.914	0.885	**0.941**
	S4	0.828	0.748	0.883	0.863	**0.897**
	S5	0.847	0.875	0.922	0.916	**0.951**
	S6	0.800	0.754	0.888	0.863	**0.926**
	S7	0.861	0.853	0.892	0.915	**0.922**
	S8[un]	0.950	0.935	0.948	0.950	**0.960**
	Ave	0.833	0.833	0.907	0.900	**0.936**
Precision	S1	0.767	0.821	0.906	0.887	**0.918**
	S2	0.882	0.888	0.924	0.943	**0.971**
	S3	0.735	0.697	0.855	0.826	**0.887**

	S4	0.841	0.719	0.873	0.863	**0.891**
	S5	0.814	0.853	0.901	0.912	**0.936**
	S6	0.712	0.631	0.894	0.834	**0.898**
	S7	0.877	0.866	0.907	**0.932**	0.929
	S8^{un}	0.817	*Nan*	0.570	**0.833**	0.824
	Ave	0.804	0.782	0.894	0.885	**0.919**
Specificity	S1	0.876	0.887	0.945	0.934	**0.948**
	S2	0.887	0.891	0.926	0.945	**0.971**
	S3	0.928	0.874	0.946	0.939	**0.955**
	S4	0.852	0.693	0.875	0.868	**0.892**
	S5	0.877	0.904	0.934	0.943	**0.958**
	S6	0.885	0.857	**0.959**	0.935	0.955
	S7	0.667	0.630	0.754	**0.824**	0.814
	S8^{un}	**0.996**	0.993	0.986	**0.996**	0.993
	Ave	0.853	0.819	0.905	0.912	**0.928**
Sensitivity	S1	0.703	0.888	0.907	0.902	**0.955**
	S2	0.867	0.873	0.909	0.927	**0.953**
	S3	0.502	0.736	0.836	0.751	**0.909**
	S4	0.805	0.806	0.892	0.858	**0.902**
	S5	0.805	0.831	0.903	0.876	**0.941**
	S6	0.614	0.533	0.734	0.705	**0.863**
	S7	0.938	0.940	0.946	0.951	**0.964**
	S8^{un}	0.228	0.047	0.326	0.250	**0.453**
	Ave	0.748	0.801	0.875	0.853	**0.927**
F1-score	S1	0.731	0.853	0.905	0.894	**0.936**
	S2	0.874	0.880	0.916	0.934	**0.962**
	S3	0.593	0.712	0.843	0.785	**0.897**
	S4	0.822	0.759	0.882	0.860	**0.896**
	S5	0.809	0.841	0.902	0.893	**0.938**
	S6	0.658	0.576	0.805	0.763	**0.879**
	S7	0.907	0.901	0.926	0.941	**0.946**
	S8^{un}	0.350	*Nan*	0.411	0.355	**0.542**
	Ave	0.771	0.789	0.883	0.867	**0.922**

^{un} Means an unbalanced data for the subject and *Nan* means not a number.

between the subject- dependent and independent cases and they produced exactly same results as those in the intra-subject cases which were used as the baselines for the following comparisons.

In subject-independent experiment setting, all eight subjects were used to evaluate the performance of the proposed proxy cross-sensory feature extraction method using a leave one out cross-validation (LOOCV) method, that is, build the LSTM-PM models with the data of seven subjects and evaluate the model performance on the eighth subject which did not involve the model training. Considering the unbalanced data in session-1 of S7 and S8, the average metrics of LOOCV were calculated only based on the results of six subjects, except S7 and S8 although we still reported the inter-subject classification metrics on the two unbalanced datasets. The complete classification metrics of each subject and the average metrics of all subjects are presented in Table 6 where only the results of pmEEG-ACC features were reported.

Table 6: Model performance based on pmEEG-ACC features in the subject-independent setting

Metrics	Subjects Index								
	S1	S2	S3	S4	S5	S6	S7^{un}	S8^{un}	Average
Accuracy	0.946	0.946	0.849	0.916	0.922	0.927	0.971	0.975	0.918
Precision	0.910	0.939	0.784	0.895	0.904	0.894	0.978	0.720	0.888
Specificity	0.955	0.930	0.841	0.920	0.941	0.965	0.658	0.993	0.925
Sensitivity	0.930	0.960	0.863	0.912	0.896	0.821	0.992	0.523	0.897
F1-score	0.920	0.950	0.820	0.903	0.899	0.854	0.985	0.594	0.891

Compared with the results in Subsection 5.2, the performance of the pmEEG-ACC features in the inter-subject setting was slightly worse than that in the intra-subject setting. However, the performance of the pmEEG-ACC in the inter-subject setting is still better than that of the EEG-ACC multimodal features in the intra-subject setting without cross-sensory feature extraction. The subject-averaged accuracy, specificity and F1-score of pmEEG-ACC features in inter-subject setting are 9.0%, 4.3% and 13.1% higher than those of pure EEG features, and 11.3%, 8.0% and 17.5% higher than those of pure ACC features in intra-subject setting. In order to further illustrate the effectiveness of the new proposed method in different settings, Fig. 8 shows the performance comparison of subject S1 among original EEG-ACC, intra-subject pmEEG-ACC, and inter-subject pmEEG-ACC models. It can be observed that all five metrics of inter-subject cross-sensory features on S1 are slightly lower than those of intra-subject pmEEG-ACC features, but higher than those of original EEG-ACC features. Similar results were observed on the other participants.

Despite the studies on multimodal FoG detection, no study have explored the cross-modal interactions and exploit the inter-modal information complementary and redundancy in the detection of FoG so far. To the best of our knowledge, this is the first study to adopt proxy measurement method for cross-sensory feature transfer and accurate FoG detection. The new method achieved a surprisingly good performance

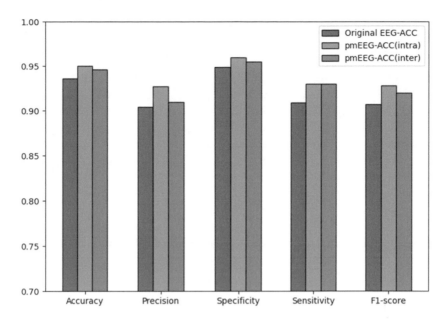

Fig. 8: The performance comparison of cross-sensory features in intra- and inter-subject settings

which was even better than the original multimodal features did. There are two potential reasons to explain the encouraging results. Firstly, the pmEEG features, which were derived from accelerations but integrated information from both gait and brain activity, realize deep fusion of gait and EEG information. This can be demonstrated by significant improvements (over 10% incensement) in the accuracy of FoG detection based on pmEEG features compared to pure ACC features in subject-dependent case. In other comparisons, the excellent results of pmEEG features in the inter-session and inter-subject FoG detection further showed the effectiveness and robustness of the new proposed method. The second reason lies in the observation that, in the pseudo-multimodal pmEEG-ACC combination, traditional gait features provided complementary information and enhanced the cross-modal interactions. The reduction in representative power of pmEEG features caused by information loss of the LSTM-PM process has been compensated by the traditional gait features in pmEEG-ACC combination. The PM process is essentially an enhanced feature extraction from pure ACC data. What makes the feature extraction different is that the supervision information was not from the classification labels but the features from another sensor, namely, the EEG sub-band components in this study. Therefore, the PM process extracted features which reflect the brain activities during the occurrence of FoG from ACC data in actual. Besides, the PM process is actually a new fine designed multimodal information fusion method. Different from traditional information fusion methods which treat multimodal information equally, the two sensory information plays different roles in the PM process. The proxy variable, ACC, is used as the sources for feature extraction while the target variable, EEG sub-bands, are used as the supervision information for abstract feature extraction.

6. Transient Brain Activities during the Occurrence of FoG

In the study of FoG, we are not only concerned about the accurate detection of the FoG but the prediction of FoG so that prompt intervention can be conducted to prevent the fall risk. Plenty of studies have showed that brain activities shown by the EEG can benefit the analysis and prediction of FoG. However, most of the current methods were based on either the spatial features, such as the common spatial patterns, frequency domain features, such as EEG rhythms, temporal features, such as the event-related potential information, or their combinations, such as the time-frequency spectra. Few methods characterize a complete picture of the brain activities, including all spatial, temporal, and frequency information in one representation. A recently developed dynamic mode decomposition method which was originally used in the study of fluid mechanism is used to characterize the dynamic activities of brain during the occurrence of FoG.

6.1 DMD Methodology

The DMD method provides a spatiotemporal decomposition of data into a set of dynamic modes that are derived from snapshots or measurements of a given system in time [81, 87]. The DMD method can be viewed as computing the eigenvalues and eigenvectors (low-dimensional modes) of a finite-dimensional linear model that approximates the infinite-dimensional Koopman operator. Since the operator is linear, the decomposition gives the growth rates and frequencies associated with each mode. If the underlying dynamics are linear, then the DMD method recovers the leading eigenvalues and eigenvectors normally computed, using standard solution methods for linear differential equations.

Let $x_i \in R^{25}$ be a column vector containing measurements from 25 channels at the time instant i in EEG signals and compose a data matrix:

$$X = [x_1, x_2, \cdots, x_m] \in R^{25 \times 125} \tag{13}$$

which represents a segmented EEG signal. In order to minimize the approximation error, we define two data matrices as follows:

$$X_1 = [x_1, x_2, \cdots, x_{m-1}] \in R^{25 \times 124}$$
$$X_2 = [x_2, x_3, \cdots, x_m] \in R^{25 \times 124} \tag{14}$$

where X_1 is the data matrix constructed by the measurement vectors for 124 consecutive time instances from the beginning, and X_2 is the data matrix temporally shifted by the time unit 1 from X_1. Then, the locally linear approximation can be written by:

$$X_2 = A X_1$$

and the best-fit A matrix is given by:

$$A = X_2 X_1^\dagger$$

where $X_1^\dagger$ denotes the Moore-Penrose pseudoinverse of X_1.

The DMD modes, also called dynamic modes, are the eigenvectors of A, and each DMD mode corresponds to a particular eigenvalue of A.

In practice, when the state dimension is large, the matrix A may be intractable to analyze directly. Instead, using the following DMD algorithm, a low-rank approximation $\widetilde{A}$ can be obtained.

Take the singular value decomposition (SVD) of X_1.

$$X_1 = U\Sigma V^*$$

where * denotes the conjugate transpose, U is 25×r, Σ is diagonal and r × r, V is 124 × r, Here r is the rank of the reduced SVD approximation to X_1.

Define the matrix

$$\widetilde{A} = U^* A U = U^* X_2 V\Sigma^{-1}$$

and compute the eigende composition of $\widetilde{A}$

$$\widetilde{A}W = W\Lambda$$

where columns of W are eigenvectors and Λ is a diagonal matrix containing the corresponding eigenvalues λ_k.

Reconstruct the eigende composition of A from W and Λ. In particular, the eigenvalues of A are given by Λ and the eigenvectors of A (DMD modes) are given by columns of Φ:

$$\Phi = UW$$

With the low-rank approximations of both the eigenvalues and the eigenvectors in hand, the projected future solution can be constructed for all time in the future.

$$\omega_k = \log(\lambda_k)/\Delta t$$

$$x(t) = \sum_{k=1}^{r} \phi_k \exp(\omega_k t)b_k = \Phi \exp(\Omega t)b \tag{16}$$

$$x(t) = \sum_{k=1}^{r} \phi_r \exp(\omega_k t)b_k \tag{17}$$

where b_k is the initial amplitude of each mode, Φ is the matrix whose columns are the DMD eigenvectors Φ_k, and $\Omega = \text{diag}(\omega)$ is a diagonal matrix whose entries are the eigenvalues ω_k. The eigenvectors ϕ_k are the same size as the state x, and b is a vector composed by all the coefficients b_k ($k = 1,2, ..., r$).

The phase of eigenvalues can be converted into frequency (Hz) by:

$$f = \frac{abs(imag(\omega))}{2\pi} \tag{18}$$

6.2 Dynamic Modes of FoG

In this subsection, an improved DMD method [77] was used to extract the dynamic modes of EEG data to give a complete description of the brain activities before the occurrence, pre-frizzing and during the FoG. Preliminary results are reported to show the transient brain activities and the promising potentials in the prediction of FoG.

Two typical class of FoGs were considered, including the gait initiation failure and failure to continue. For each case, data segments of 125 data points in sequential non-FoG, pre-freezing and FoG periods were used to extract the dynamic modes. A pair of complex dynamic mode includes the oscillations $e^{(\sigma \pm j\omega)t}$ and the corresponding a pair of conjugated eigenvectors ϕ and ϕ^*. The oscillations represent the rhythm of the brain activity and the amplitudes and phases of the eigenvectors represent the spatial distribution of rhythm power and the traveling (distribution) of the sub-band waves. A greater phase means the rhythm occurrence is earlier than the region with smaller phase values. The dynamic modes of the two FoGs are shown in Figs. 9 and 10, respectively.

Case I: Failure to Initiate

OMD methods find the first 10 main modes in the δ, θ, α, β and γ sub-bands, respectively. It can be found that OMD method finds very similar rhythms in non-FoG, pre-freezing and FoG periods. Namely, brain activities share the same rhythms. However, the spatial patterns, including amplitude and phase, are different, which reflects the changes of the brain activity during the occurrence of FoG.

Generally speaking, in the relatively lower frequency bands δ, θ, α, the brain oscillates with relatively standing patterns which have relatively small phase differences (in-phase) or fixed 180° degree differences (out-phase) among different brain areas. However, these brain activity patterns start to blur during pre-freezing which is manifested as the reduction in the amplitude and the emergence of more complex phase patterns. This may indicate that the clear pattern energies decline through a spatial diffusion where more travelling waves were observed. On the contrary, the brain activities in the higher frequency bands β, and γ were enhanced and new, clearer standing waves emerged.

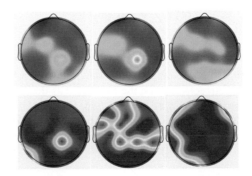

$(0.92 \pm 3.88*2\pi j)(-3.75 \pm 2.77*2\pi j)(-13.11 \pm 1.42*2\pi j)$
(a) The first mode in δ rhythm

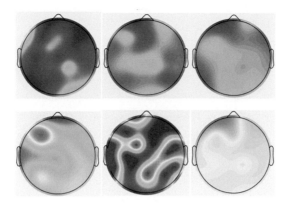

(10.80±6.67*2πj) (-4.60±7.56*2πj)(-6.19±7.41*2πj)
(b) The second mode in θ rhythm

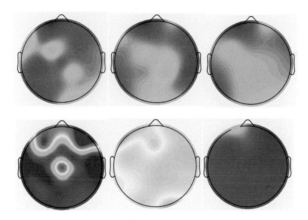

(-2.19±10.13*2πj) (-1.95±10.25*2πj) (-6.12±8.17*2πj)
(c) The third mode in α rhythm

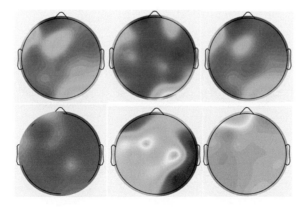

(-14.65±14.65*2πj) (-0.04±15.81*2πj) (-2.62±15.23*2πj)
(d) The fourth mode in α rhythm

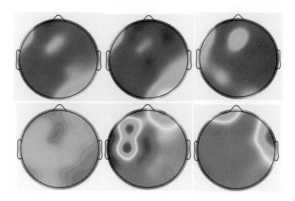

(5.76±17.49*2πj) (4.33±18.59*2πj) (11.90±19.78*2πj)
(e) The fifth mode in β rhythm

(6.46±22.69*2πj) (-5.90±22.46*2πj) (-5.25±19.96*2πj)
(f) The sixth mode in β rhythm

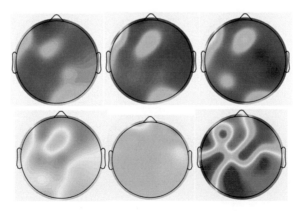

(-1.74±27.28*2πj)(-3.95±27.10*2πj)(-2.55±25.73*2πj)
(g) The seventh mode in β rhythm

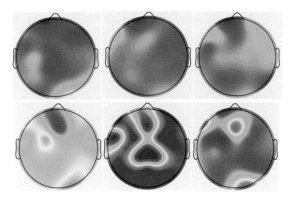

(2.89±30.35*2πj) (6.91±29.48*2πj) (-12.84±29.00*2πj)
(h) The eighth mode in β rhythm

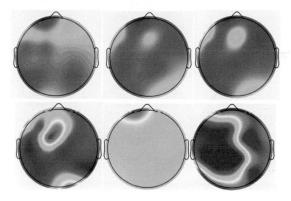

(-0.24±36.74*2πj)(7.59±35.89*2πj)(6.17±33.88*2πj)
(i) The ninth mode in γ rhythm

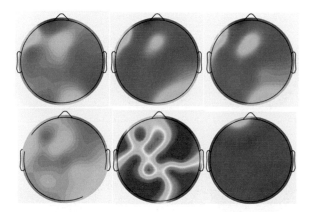

(2.26±42.16*2πj)(9.45±37.34*2πj)(0.63±39.25*2πj)
(j) The tenth mode in γ rhythm

Fig. 9: Dynamic modes of an inability to initiate FoG EEG

The upper row of each subplot shows the spatial distribution of the amplitudes and the lower row shows the distribution of the phases; three columns from left to right are before-freezing, pre-freezing and during freezing of gait. The complex numbers in the brackets indicate the corresponding eigenvalues of DMD modes.

Case II: Failure to Continue

For the inability to continue FoG, a clearer pattern was generated in δ, and γ rhythms on prefrontal lobe, indicating that more attention was paid to walking. Clear patterns in θ and β rhythms were weakened which may indicate the weakened locomotion function.

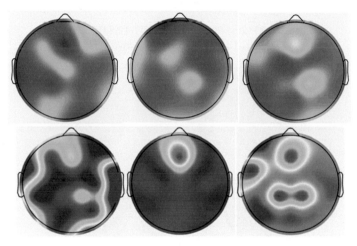

$(5.27\pm3.00*2\pi j)(9.38\pm3.71*2\pi j)(-3.62\pm3.17*2\pi j)$
(a) The first mode in δ rhythm

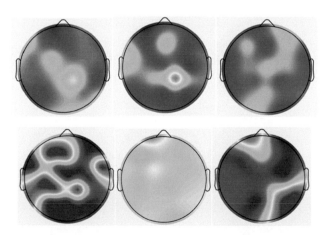

$(-6.68\pm7.23*2\pi j)(-2.39\pm6.38*2\pi j)(-6.06\pm6.43*2\pi j)$
(b) The second mode in θ rhythm

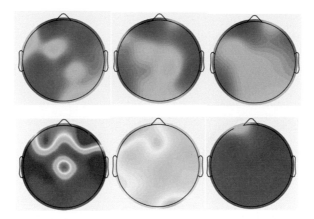

(-14.20±9.87*2πj)(3.47±11.52*2πj)(-1.04±9.96*2πj)
(c) The third mode in α rhythm

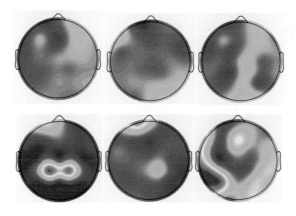

(4.18±12.10*2πj)(-2.12±15.32*2πj)(-1.30±14.64*2πj)
(d) The fourth mode in α rhythm

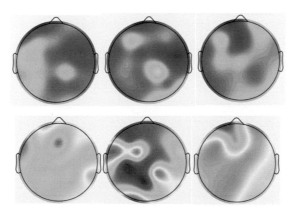

(2.48±19.77*2πj)(-0.88±18.71*2πj)(-6.03±20.01*2πj)
(e) The fifth mode in β rhythm

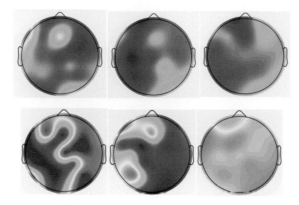

$(-20.78\pm20.15*2\pi j)(1.13\pm22.35*2\pi j)(-2.38\pm22.30*2\pi j)$
(f) The sixth mode in β rhythm

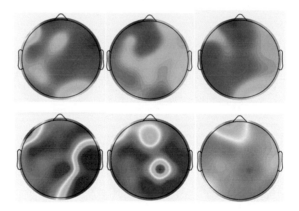

$(-3.82\pm23.93*2\pi j)(-4.79\pm27.27*2\pi j)(2.20\pm26.36*2\pi j)$
(g) The seventh mode in β rhythm

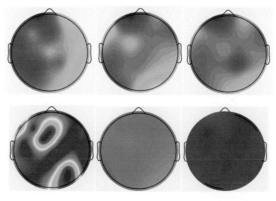

$(-1.29\pm27.61*2\pi j)(-8.57\pm31.74*2\pi j)(5.08\pm31.26*2\pi j)$
(h) The eighth mode in β rhythm

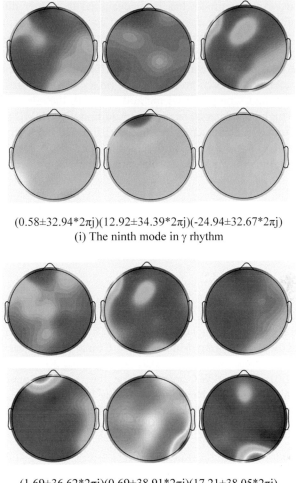

$(0.58\pm32.94*2\pi j)(12.92\pm34.39*2\pi j)(-24.94\pm32.67*2\pi j)$
(i) The ninth mode in γ rhythm

$(1.69\pm36.62*2\pi j)(0.69\pm38.91*2\pi j)(17.21\pm38.05*2\pi j)$
(j) The tenth mode in γ rhythm

Fig. 10: Dynamic modes of an inability to continue FoG EEG

The upper row of each subplot shows spatial distribution of the amplitudes and the lower row shows the distribution of phases; three columns from left to right are before freezing, pre-freezing, and during freezing of gait. The complex numbers in the brackets indicate the corresponding eigenvalues of DMD modes.

It is worth emphasizing that the dynamics modes provide not only the energy changes in brain lobes but also the specific functional regions involved in the occurrence of FoG, which may be used for the prediction of FoG.

7. Conclusion

This study provides a complete solution for accurate detection and prediction of FoG from the point of view of improving prediction accuracy, improving wearability and promptitude. An experiment has been designed to trigger different types of FoGs and acquire multimodal data. Using pure acceleration data, a new accurate calculation method for the FI index has been proposed to improve the estimation of FoG. In order to further improve the accuracy of the FoG detection but not to deteriorate the excellent wearability of the inertial sensor, a pseudo-multimodal FoG detection method has been proposed. The proxy EEG features were extracted from acceleration data. The proposed method has both the advantage of excellent wearability and high FoG detection accuracy. Additionally, a DMD method has been proposed for complete characterization of the EEG activity before, pre-freezing and during freezing of gait from spatial, temporal, and frequency domain. The full picture of EEG activity may provide a promising path to the prediction of FoG before its occurrence.

Data Availability

The datasets that support the findings of this study are publicly available in Mendeley Data [42] and research based on the dataset is allowed.

Acknowledgements

Authors WY, SL, YG, DH, and LW, gratefully acknowledge the support from the National Natural Science Foundation of China (Grant No.61876015), the Natural Science Foundation of Beijing Municipality, China (Grant No. 4202040), and Science and Technology Innovation 2030 Major Program of China (Grant No. 2018AAA001400). Authors WZ acknowledge the partial support from Beijing Municipal Administration of Hospitals' Mission Plan (No. SML20150803), the National Key R&D Program of China (No. 2018YFC1312001, 2017YFC0840105), and Beijing Municipal Science & Technology Commission (No. Z171100000117013).

References

1. Allen, D.P. and MacKinnon, C.D. (2010). Time-frequency analysis of movement-related spectral power in EEG during repetitive movements: A comparison of methods, *Journal of Neuroscience Methods*, 186(1): 107-115.
2. Assam, R. and Seidl, T. (2014). Prediction of freezing of gait from Parkinson's disease movement time series using conditional random fields. *In: Proceedings of the Third ACM SIGSPATIAL International Workshop on the Use of GIS in Public Health*, pp. 11-20.
3. Sweeney, D., Quinlan, L.R., Browne, P., Richardson, M., Meskell, P. and OLaighin, G. (2019). A technological review of wearable cueing devices addressing freezing of gait in Parkinson's disease. *Sensors*, 19(6): 1277.
4. Azevedo Coste, C., Sijobert, B., Pissard-Gibollet, R., Pasquier, M., Espiau, B. and Geny, C. (2014). Detection of freezing of gait in Parkinson's disease: Preliminary results, *Sensors*, 14: 6819-6827.
5. Bächlin, M., Hausdorff, J., Roggen, D., Giladi, N., Plotnik, M. and Tröster, G. (2009). Online detection of freezing of gait in Parkinson's disease patients: A performance

characterization. *In: Proceedings of the Fourth International Conference on Body Area Networks*, pp. 1-8.

6. Bachlin, M., Plotnik, M., Roggen, D., Maidan, I., Hausdorff, J.M., Giladi, N. and Troster, G. (2010). Wearable assistant for Parkinson's disease patients with the freezing of gait symptom, *IEEE Transactions on Information Technology in Biomedicine*, 14(2): 436-446.

7. Billings, S.A. and Wei, H.L. (2008). An adaptive orthogonal search algorithm for model subset selection and non-linear system identification, *International Journal of Control*, 81(5): 714-724.

8. Billings, S.A., Chen, S.A. and Korenberg, M. (1989). Identification of MIMO non-linear systems using a forward-regression orthogonal estimator, *International Journal of Control*, 49(6): 2157-2189.

9. Borzì, L. *et al.* (2021). Prediction of freezing of gait in Parkinson's disease using wearables and machine learning, *Sensors* (Basel), 21, doi: 10.3390/s21020614

10. Braak, H., Ghebremedhin, E., Rüb, U., Bratzke, H. and Del Tredici, K. (2004). Stages in the development of Parkinson's disease-related pathology, *Cell and Tissue Research*, 318(1): 121-134.

11. Butler, J.S. *et al.* (2017). Motor preparation rather than decision-making differentiates Parkinson's disease patients with and without freezing of gait, *Clinical Neurophysiology*, 128: 463-471.

12. Camps, J., Samà, A., Martín, M., Rodríguez-Martín, D., Pérez-López, C., Moreno Arostegui, J.M., Cabestany, J., Català, A., Alcaine, S., Mestre, B., Prats, A., Crespo-Maraver, M.C., Counihan, T.J., Browne, P., Quinlan, L.R., Laighin, G.Ó., Sweeney, D., Lewy, H., Vainstein, G., Costa, A., Annicchiarico, R., Bayés, À. and Rodríguez-Molinero, A. (2018). Deep learning for freezing of gait detection in Parkinson's disease patients in their homes using a waist-worn inertial measurement unit, *Knowledge-based Systems*, 139: 119-131.

13. Daubechies, J. Lu and Wu, H.-T. (2011). Synchrosqueezed wavelet transforms: An empirical mode decomposition-like tool, *Applied and Computational Harmonic Analysis*, 30(2): 2043-2061.

14. Delorme, A. and Makeig, S. (2004). EEGLAB: An open source toolbox for analysis of single-trial EEG dynamics including independent component analysis, *Journal of Neuroscience Methods*, 134: 9-21.

15. Delval, A. *et al.* (2010). Objective detection of subtle freezing of gait episodes in Parkinson's disease, *Movement Disorders*, 25: 1684-1693.

16. Delval, A. *et al.* (2018). Motor preparation of step initiation: Error-related cortical oscillations, *Neuroscience*, 393: 12-23.

17. El-Attar, A. *et al.* (2019). Hybrid DWT-FFT features for detecting freezing of gait in Parkinson's disease. *In: Information Technology and Intelligent Transportation Systems*, pp. 117-126. IOS Press.

18. Fahn, S. and Elton, R. (1987). Unified Parkinson's disease rating scale. *In:* Fahn, S., Marsden, C., Calne, D., Goldstein, M. (Eds.). *Recent Developments in Parkinson's Disease*, vol. 2.

19. Folstein, M.F., Folstein, S.E. and McHugh, P.R. (1975). Mini-mental state: A practical method for grading the cognitive state of patients for the clinician, *Journal of Psychiatric Research*, 12: 189-198; doi:10.1016/0022-3956(75)90026-6

20. Gálvez, G., Recuero, M., Canuet, L. and Del-Pozo, F. (2018). Short-term effects of binaural beats on EEG power, functional connectivity, cognition, gait and anxiety in Parkinson's disease, *International Journal of Neural Systems*, 28: 1750055.

21. Gratwicke, J., Jahanshahi, M. and Foltynie, T. (2015). Parkinson's disease dementia: A neural networks perspective, *Brain*, 138: 1454-1476. .

22. Guo, Y., Wang, L., Li, Y., Luo, J., Wang, K., Billings, S.A. and Guo, L. (2019). Neural activity inspired asymmetric basis function TV-NARX model for the identification of time-varying dynamic systems, *Neurocomputing*, 357: 188-202.

23. Guo, Y., Guo, L.Z., Billings, S.A., Coca, D. and Lang, Z. (2013). A parametric frequency response method for non-linear time-varying systems. *International Journal of Systems Science*, 45(10): 2133-2144.

24. Guo, Y. *et al.* (2017). A new proxy measurement algorithm with application to the estimation of vertical ground reaction forces using wearable sensors, *Sensors,* 17: 2181.

25. Guo, Y., Wang, L., Li, Y., Guo, L. and Meng, F. (2019). The detection of freezing of gait in Parkinson's disease using asymmetric basis function TV-ARMA time-frequency spectral estimation method, *IEEE Trans. Neural Syst. Rehabil. Eng.*, 27: 2077-2086; doi: 10.1109/TNSRE.2019.2938301

26. Günther, M. *et al.* (2019). Coupling between leg muscle activation and EEG during normal walking, intentional stops, and freezing of gait in Parkinson's disease, *Frontiers in Physiology*, 10; doi: 10.3389/fphys.2019.00870

27. Handojoseno, A.A. *et al.* (2013). Using EEG spatial correlation, cross frequency energy, and wavelet coefficients for the prediction of freezing of gait in Parkinson's disease patients, *35th Annual International Conference of the IEEE Engineering in Medicine and Biology Society (EMBC)*, pp. 4263-4266.

28. Handojoseno, A.A. *et al.* (2012). The detection of freezing of gait in Parkinson's disease patients using EEG signals based on wavelet decomposition, *Conference Proceedings: Annual International Conference of the IEEE Engineering in Medicine and Biology Society*, 69-72; doi: 10.1109/EMBC.2012.6345873

29. Handojoseno, A. *et al.* (2018). Prediction of Freezing of Gait in Patients with Parkinson's Disease Using EEG Signals, *Studies in Health Technology and Informatics*, 246: 124-131.

30. Herrmann, C.S., Grigutsch, M. and Busch, N.A. (2005). 11 EEG oscillations and wavelet analysis, *Event-related Potentials: A Methods Handbook*, 229.

31. Hochreiter, S. and Schmidhuber, J. (1997). Long short-term memory, *Neural Computation*, 9: 1735-1780.

32. Hoehn, M.M. and Yahr, M.D. (1967). Parkinsonism: Onset, progression, and mortality, *Neurology*, 17: 427-442.

33. Jankovic, J. (2008). Parkinson's disease: Clinical features and diagnosis, *Journal of Neurology, Neurosurgery & Amp; Psychiatry* 79: 368-376; doi:10.1136/jnnp.2007.131045

34. Jovanov, E., Wang, E., Verhagen, L., Fredrickson, M. and Fratangelo, R. (2009), *DeFOG — A Real-time System for Detection and Unfreezing of Gait of Parkinson's Patients*, pp. 5151-5154.

35. Kadado, T., Maulik, D. and Chakrabarti, S. (1994). Comparison of parametric and nonparametric spectral estimation of continuous Doppler ultrasound shift waveforms, *In: Proceedings of IEEE 6th Digital Signal Processing Workshop*, pp. 145-148.

36. Kita, A., Lorenzi, P., Rao, R. and Irrera, F. (2017). Reliable and robust detection of freezing of gait episodes with wearable electronic devices, *E Sensors Journal*, 17: 1899-1908.

37. Laport, F., Iglesia, D., Dapena, A., Castro, P.M. and Vazquez-Araujo, F.J. (2021). Proposals and comparisons from one-sensor EEG and EOG human-machine interfaces, *Sensors*, 21: 2220.

38. Li, Y., Wei, H. and Billings, S.A. (2011). Identification of time-varying systems using multi-wavelet basis functions, *IEEE Transactions on Control Systems Technology*, 19(3): 656-663.

39. Li, Y., Lei, M., Guo, Y., Hu, Z. and Wei, H. (2018). Time-varying nonlinear causality detection using regularized orthogonal least squares and multi-wavelets with applications to EEG, *IEEE Access*, 6: 17826-17840.

40. Li, Y., Cui, W., Guo, Y., Huang, T., Yang, X. and Wei, H. (2018). Time-varying system identification using an ultra-orthogonal forward regression and multiwavelet basis functions with applications to EEG, *IEEE Transactions on Neural Networks and Learning Systems*, 29(7): 2960-2972.

41. Li, Y., Luo, M.-L. and Li, K. (2016). A multiwavelet-based time-varying model identification approach for time-frequency analysis of EEG signals, *Neurocomputing*, 193: 106-114.

42. Li, H. (2021). Multimodal dataset of freezing of gait in Parkinson's disease, *Mendeley Data*, V3; doi: http://dx.doi.org/10.17632/r8gmbtv7w2.2

43. Lin, A., Liu, K.K., Bartsch, R.P. and Ivanov, P. (2016). Delay-correlation landscape reveals characteristic time delays of brain rhythms and heart interactions, *Philosophical Transactions. Series A, Mathematical, Physical, and Engineering Sciences*, 374; doi:10.1098/rsta.2015.0182

44. Lokk and Delbari, A. (2012). Clinical aspects of palliative care in advanced Parkinson's disease, *BMC Palliative Care*, 11: 20-20.

45. Lonini, L. *et al.* (2018). Wearable sensors for Parkinson's disease: Which data are worth collecting for training symptom detection models, *NPJ Digital Medicine*, 1: 64; doi:10.1038/s41746-018-0071-z

46. Lorenzi, P. *et al.* (2015). Smart sensing systems for the detection of human motion disorders, *Procedia Engineering*, 120: 324-327.

47. Ly, Q.T. *et al.* (2017). Detection of gait initiation failure in Parkinson's disease based on wavelet transform and support vector machine, *Annual International Conference of the IEEE Engineering in Medicine and Biology Society*, Annual International Conference 2017, 3048-3051; doi:10.1109/embc.2017.8037500.

48. Macht, M., Kaussner, Y., Möller, J.C., Stiasny-Kolster, K., Eggert, K.M., Krüger, H.P. and Ellgring, H. (2007). Predictors of freezing in Parkinson's disease: A survey of 6,620 patients, *Movement Disorders*, 22(7): 953-956.

49. Maetzler, W., Domingos, J., Srulijes, K., Ferreira, J.J. and Bloem, B.R. (2013). Quantitative wearable sensors for objective assessment of Parkinson's disease, *Movement Disorders*, 28: 1628-1637.

50. Marder, E. and Bucher, D. (2001). Central pattern generators and the control of rhythmic movements, *Current Biology*, 11: R986-R996.

51. Marquez, J.S. *et al.* (2020). Neural correlates of freezing of gait in Parkinson's disease: An electrophysiology mini-review, *Front Neurol.*, 11: 571086; doi:10.3389/fneur.2020.571086

52. Mazilu, S., Calatroni, A., Gazit, E., Mirelman, A., Hausdorff, J.M. and Tröster, G. (2015). Prediction of freezing of gait in Parkinson's from physiological wearables: An exploratory study, *IEEE Journal of Biomedical and Health Informatics*, 19(6): 1843-1854.

53. Mazilu, S., Hardegger, M., Zhu, Z., Roggen, D., Tröster, G., Plotnik, M. and Hausdorff, J.M. (2012), *Online Detection of Freezing of Gait with Smartphones and Machine Learning Techniques*, pp. 123-130.

54. Mazilu, S., Blanke, U., Hardegger, M., Tröster, G., Gazit, E., Dorfman, M. and Hausdorff, J.M. (2014). *GaitAssist: A Wearable Assistant for Gait Training and Rehabilitation in Parkinson's Disease*, pp. 135-137.

55. Mazzetta, I. *et al.* (2019). Wearable sensors system for an improved analysis of freezing of gait in Parkinson's disease using electromyography and inertial signals, *Sensors*, 19: 948.

56. Mikos, V., Heng, C.-H., Tay, A., N. Shuang Yu Chia, K. Mui Ling Koh, D. May Leng Tan, and W. Lok Au (2017). Real-time patient adaptivity for freezing of gait classification through semi-supervised neural networks, *In: 16th IEEE International Conference on Machine Learning and Applications (ICMLA)*, pp. 871-876.

57. Grenier, Y. (1983). Time-dependent ARMA modeling of nonstationary signals, *IEEE Transactions on Acoustics, Speech, and Signal Processing*, 31(4): 899-911.

58. Moore, S.T., MacDougall, H.G. and Ondo, W.G. (2008). Ambulatory monitoring of freezing of gait in Parkinson's disease, *J Neurosci. Methods*, 167: 340-348; doi:10.1016/j.jneumeth.2007.08.023.

59. Moore, S.T. *et al.* (2013). Autonomous identification of freezing of gait in Parkinson's disease from lower-body segmental accelerometry, *Journal of Neuroengineering and Rehabilitation*, 10.

60. Muangpaisan, W., Mathew, A., Hori, H. and Seidel, D. (2011). A systematic review of the worldwide prevalence and incidence of Parkinson's disease, *Journal of the Medical Association of Thailand*, 94(6): 749.

61. Naghavi, N., Miller, A. and Wade, E. (2019). Towards real-time prediction of freezing of gait in patients with Parkinson's disease: Addressing the class imbalance problem, *Sensors* (Basel), 19; doi: 10.3390/s19183898

62. Nieuwboer, and Giladi, N. (2013). Characterizing freezing of gait in Parkinson's disease: Models of an episodic phenomenon, *Movement Disorders*, 28(11): 1509-1519.

63. Nieuwboer, L. Rochester, Herman, T., Vandenberghe, W., Ehab Emil, G., Thomaes, T. and Giladi, N. (2009). Reliability of the new freezing of gait questionnaire: Agreement between patients with Parkinson's disease and their carers. *Gait and Posture*, 459-463.

64. Nutt, J.G., Bloem, B.R., Giladi, N., Hallett, M., Horak, F.B. and Nieuwboer, A. (2011). Freezing of gait: Moving forward on a mysterious clinical phenomenon, *The Lancet Neurology*, 10(8): 734-744.

65. Pham, T.T., Moore, S.T., Lewis, S.J.G., Nguyen, D.N., Dutkiewicz, E., Fuglevand, A.J., McEwan, A.L. and Leong, P.H.W. (2017). Freezing of gait detection in Parkinson's disease: A subject-independent detector using anomaly scores, *IEEE Transactions on Biomedical Engineering*, 64(11): 2719-2728.

66. Penzel, T. *et al.* (2016). Modulations of heart rate, ECG, and cardio-respiratory coupling observed in polysomnography, *Front. Physiol.*, 7: 460; doi:10.3389/fphys.2016.00460

67. Podsiadlo, D. and Richardson, S. (1991). The timed up & go: A test of basic functional mobility for frail elderly persons. *Journal of the American Geriatrics Society*, 39(2): 142-148.

68. Pozzi, N.G. *et al.* (2019). Freezing of gait in Parkinson's disease reflects a sudden derangement of locomotor network dynamics, *Brain*, 142: 2037-2050; doi:10.1093/brain/awz141

69. Rahman, S., Griffin, H.J., Quinn, N.P. and Jahanshahi, M. (2008). The factors that induce or overcome freezing of gait in Parkinson's disease, *Behavioral Neurology*, 19(3): 127-136.

70. Rehman, R.Z.U. *et al.* (2019). Selecting clinically relevant gait characteristics for classification of early Parkinson's disease: A comprehensive machine learning approach, *Scientific Reports*, 9: 17269; doi:10.1038/s41598-019-53656-7

71. Rezvanian, S. and Lockhart, E.T. (2016). Towards real-time detection of freezing of gait using wavelet transform on wireless accelerometer data, *Sensors*, 16(4).

72. Rodríguez-Martín, D., Samà, A., Pérez-López, C., Català, A., Moreno, J.M., Arostegui, Cabestany, J., Bayés, À., Alcaine, S., Mestre, B., Prats, A., Crespo, M.C., Counihan, T.J., Browne, P., Quinlan, L.R., ÓLaighin, G., Sweeney, D., Lewy, H., Azuri, J., Vainstein, G., Annicchiarico, R., Costa, A. and Rodríguez-Molinero, A. (2017). Home detection

of freezing of gait using support vector machines through a single waist-worn triaxial accelerometer, *PLoS One*, 12(2): e0171764.

73. Samà, D. Rodríguez-Martín, Pérez-López, C., Català, A., Alcaine, S., Mestre, B., Prats, A., Crespo, M.C. and Bayés, À. (2018). Determining the optimal features in freezing of gait detection through a single waist accelerometer in home environments, *Pattern Recognition Letters*, 105: 135-143.

74. Shine, J.M. *et al.* (2014). Abnormal patterns of theta frequency oscillations during the temporal evolution of freezing of gait in Parkinson's disease, *Clinical Neurophysiology*, 125: 569-576; doi: https://doi.org/10.1016/j.clinph.2013.09.006

75. Silva de Lima, A.L. *et al.* (2017). Freezing of gait and fall detection in Parkinson's disease using wearable sensors: A systematic review, *J. Neurol.*, 264: 1642-1654; doi:10.1007/s00415-017-8424-0

76. Snijders, H., Weerdesteyn, V., Hagen, Y.J., Duysens, J., Giladi, N. and Bloem, B.R. (2016). Obstacle avoidance to elicit freezing of gait during treadmill walking, *Movement Disorders*, 25(1): 57-63.

77. Wynn, A., Pearson, D.S., Ganapathisubramani, B. and Goulart, P.J. (2013). Optimal mode decomposition for unsteady flows, *Journal of Fluid Mechanics*, 733: 473-503. doi:10.1017/jfm.2013.426.

78. Tahafchi, P. *et al.* (2017). Freezing-of-Gait detection using temporal, spatial, and physiological features with a support-vector-machine classifier, *In: 39th Annual International Conference of the IEEE Engineering in Medicine and Biology Society (EMBC)*, pp. 2867-2870.

79. Tochigi, Y., Segal, N.A., Vaseenon, T. and Brown, T.D. (2012). Entropy analysis of tri-axial leg acceleration signal waveforms for measurement of decrease of physiological variability in human gait, *J. Orthop. Res.*, 30: 897-904; doi:10.1002/jor.22022

80. Tripoliti, E.E., Tzallas, A.T., Tsipouras, M.G., Rigas, G., Bougia, P., Leontiou, M., Konitsiotis, S., Chondrogiorgi, M., Tsouli, S. and Fotiadis, D.I. (2013). Automatic detection of freezing of gait events in patients with Parkinson's disease, *Computer Methods and Programs in Biomedicine*, 110(1): 12-26,

81. Tu, J.H., Rowley, C.W., Luchtenburg, D.M., Brunton, S.L. and Kutz, J.N. (2014). On dynamic mode decomposition: Theory and applications, *Journal of Computational Dynamics*, 1(2): 391-421; https://doi.org/10.3934/jcd.2014.1.391

82. Wagner, J., Martínez-Cancino, R. and Makeig, S. (2019). Trial-by-trial source-resolved EEG responses to gait task challenges predict subsequent step adaptation, *NeuroImage*, 199: 691-703; doi: https://doi.org/10.1016/j.neuroimage

83. Walton, C.C., Shine, J.M., Hall, J.M., O'Callaghan, C., Mowszowski, L., Gilat, M., Szeto, J.Y.Y., Naismith, S.L. and Lewis, S.J.G. (2015). The major impact of freezing of gait on quality of life in Parkinson's disease, *Journal of Neurology*, 262(1): 108-115.

84. Wang, Q., Meng, L., Pang, J., Zhu, X. and Ming, D. (2020) Characterization of EEG data revealing relationships with cognitive and motor symptoms in Parkinson's disease: A systematic review, *Frontiers in Aging Neuroscience*, 12: 373.

85. Wang, Y. *et al.* (2020). Freezing of gait detection in Parkinson's disease via multimodal analysis of EEG and accelerometer signals, *42nd Annual International Conference of the IEEE Engineering in Medicine & Biology Society (EMBC)*, 847-850.

86. Wang, Y. *et al.* (2020). Characterizing and Detecting Freezing of Gait using Multi-modal Physiological Signals; arXiv preprint arXiv: 2009: 12660.

87. Wei, H.L. and Billings, S.A. (2002). Identification of time-varying systems using multiresolution wavelet models, *International Journal of Systems Science*, 33(15): 1217-1228, [p. 1223, Equation (41)].

88. Wynn, A., Pearson, D.S., Ganapathisubramani, B. and Goulart, P.J. (2013). Optimal mode decomposition for unsteady flows, *Journal of Fluid Mechanics*, 733: 473-503; doi:10.1017/jfm.2013.426

89. Zach, H., Janssen, A.M., Snijders, A.H., Delval, A., Ferraye, M.U., Auff, E., Weerdesteyn, V., Bloem, B.R. and Nonnekes, J. (2015). Identifying freezing of gait in Parkinson's disease during freezing provoking tasks using waist-mounted accelerometry, *Parkinsonism & Related Disorders*, 21(11): 1362-1366.

90. Zhang, Z.G., Hung, Y.S. and Chan, S.C. (2011). Local polynomial modeling of time-varying autoregressive models with application to time-frequency analysis of event-related EEG, *IEEE Transactions on Biomedical Engineering*, 58(3): 557-566.

Sparse Model Identification for Nonstationary and Nonlinear Neural Dynamics Based on Multiwavelet Basis Expansion

Song Xu[1],*, Lina Wang[2],* and Jingjing Liu[3]

[1] National Key Laboratory of Science and Technology on Aerospace Intelligence Control, Beijing Aerospace Automatic Control Institute, Beijing, China
[2] National Key Laboratory of Science and Technology on Aerospace Intelligence Control, Beijing Aerospace Automatic Control Institute, Beijing, China
[3] National Key Laboratory of Science and Technology on Aerospace Intelligence Control, Beijing Aerospace Automatic Control Institute, Beijing, China
E-mail: xusong618@163.com, violina@126.com

1. Introduction

Neurobiological processes, including synaptic transmission, generally involve large-scale complex nonstationary, and non-linear nervous systems. Therefore it is a crucial challenge for modeling the information transmission between the brain regions. For instance, studies have shown that synaptic plasticity with exact forms of long-term depression (LTD) and long-term potentiation (LTP) occurs in response to specific input patterns, which are generally expressed as system instability. In order to better understand the mechanism of these nonlinear and non-stationary processes, three main types of time-varying nonlinear system identification methods, based on spiking activities, have been proposed, and can be briefly introduced as follows.

The first kind of time-varying nonlinear system identification method is based on artificial neural network (ANN), which is widely applied in various fields, including speech recognition [20], prediction [19], etc. However, the hidden nodes and layers existing in the neural network structure generally have poor interpretation. The second kind of parametric estimation method for time-varying nonlinear systems are adaptive filtering strategies, such as Kalman filter [2, 3], steepest descent algorithm [4, 30], and stochastic gradient algorithm [11, 8]. By using an adaptive filter algorithm, the model parameters are iteratively updated, based on the difference between the estimated spike firing rate and the real spike occurrence [6], which has advantages in online estimation for the best fitting model of an unknown neural system. Nevertheless, if the model parameters change suddenly, the adaptive filter may not be able to capture the transient characteristics of the time-varying nonlinear system effectively in real time, on account of the unavoidable convergence time required by the adaptive filter strategy [17]. The third kind is based on the basis function extension theory, where the unknown time-varying linear or non-linear coefficients can be estimated by projecting themselves onto a series of specific basis functions. Therefore, the identification problem for time-varying coefficients is converted into the estimation of time-invariant parameters based on a predefined basis function. The basic expansion scheme can greatly reduce the number of unknown coefficients. The basis functions which are generally used to implement this strategy include Walsh basis [40], Chebyshev function [14, 13], Laguerre series [31], Legendre polynomial [15], and wavelet basis [39, 38]. Considering the limited number of basic functions commonly utilized, it is necessary to consider each type of basic function family according to its characteristic fit to the kernel shape. For example, the Legendre polynomials and the Chebyshev basis functions perform well in approximating smooth or slow parameter changes [20], while Haar and Walsh basis functions are more suitable in tracking sharp or abrupt changing parameters [33]. For the power system with spiking input and output data, Chebyshev polynomial expansion technology has been applied to research and test [1]. However, the complex nervous system often involves time-varying dynamic processes, and there may be smooth changes and sharp changes at the same time. Chebyshev polynomial expansion is not enough to capture the different kinds of potential nonstationary characteristics effectively. Besides, wavelet basis functions can be used to effectively characterize the rapid or slow changes that disappear outside the short time interval [23, 12]. Researchers propose to combine the least mean square algorithm and the cardinal B-spline basis function, which can effectively track the smooth and abrupt changes of coefficients in linear time-varying system [37]. However, considering that biological systems usually involve large-scale nonlinear and non-stationary systems, how to model and identify the time-varying nonlinear systems based on spike train data using basis function expansion approach still needs further exploration.

In this paper, a sparse modeling framework using multiwavelet basis expansion theory is proposed to identify potential nonstationary and nonlinear sparse neuron connectivity only with the recorded input and output spike train data. To be specific, firstly, a complete time-varying nonlinear generalized Laguerre-Volterra (TVNGLV) model framework based on all the recorded input and output spiking signals is constructed to effectively characterize the nonlinear and nonstationary neuro-dynamical system; secondly, significant input spiking signals are selected from the large-scale inputs using a regularization method, i.e. group-LASSO, and global sparsity can thus be obtained; thirdly, the nonstationary parameters in the sparse TVNGLV model will be projected into the multiple wavelet basis functions and different kinds of changes for the parameters to be estimated can thus be effectively tracked at the same time; fourthly, the effective orthogonal matching pursuit (OMP) algorithm is utilized to determine the model structure, so as to filter out the false regression terms in the redundant complete regression model effectively; finally, a maximum likelihood algorithm is used to estimate the transformed generalized time-invariant linear parameters and reconstruct the original time-varying nonlinear kernel function. The obvious advantage of the proposed framework is that it can effectively identify the time-varying nonlinear dynamic correlations among large-scale multivariable complex systems, and accurately track the fast and slow time-varying process based on spike train data. Simulation results show that compared with the traditional adaptive filter method and a single set of basis expansion methods, the proposed sparse method based on multi-wavelet basis functions expansion can obtain much more accurate results in tracking transient changes of artificial time-varying nonlinear spike train data.

The remaining part of this paper is organized as follows. Section 2 describes the methodology utilized for this study. Section 2.1 presents the dynamical multiple-input single-output model based on the generalized time-varying nonlinear Volterra series. The Laguerre expansion of Volterra kernel is illustrated in Section 2.2, where the Volterra series are projected on to orthonormal Laguerre basis functions, and therefore the number of the unknown parameter is significantly reduced. In Section 2.3, the group-LASSO regularization method is proposed in order to achieve group sparsity among the large-scale spiking inputs. Section 2.4 describes the multi-wavelet-based TVNGLV framework. In Sections 2.5, we detail the method of model structure selection, using the OMP algorithm. The parameter estimation and kernel reconstruction are separately given in Sections 2.6 and 2.7. The simulation results are shown in Section 3. In conclusion, Section 4 summarizes our main work.

2. Methodology

2.1 Multi-input Single-output Dynamical Time-varying Nonlinear Generalized Volterra Model

The multi-input and multi-output (MIMO) time-varying dynamical model for describing the neural activities has been formulated in the previous work [34, 35, 36], which can be further decomposed into a set of multi-input and single-output (MISO) spiking neural models of plausible structure as (Fig. 1).

$$w = u(K, x) + a(H, y) + \varepsilon(\sigma^2) \tag{1}$$

$$y = \begin{cases} 0, & \text{when } w < \theta \\ 1, & \text{when } w \geq \theta \end{cases} \tag{2}$$

where x indicates the input spike train and y indicates the output spike train here. The synaptic potential triggered by input spike trains can be given as u, and the after-potential triggered by the output spike is characterized by the function a, which will be introduced later in this section. The Gaussian white noise with variance σ can be represented as ε, which is generally produced by the intrinsic noise from the output and the unmeasured input neurons. The hidden variable w can therefore be obtained by means of adding the synaptic potential u, the after-potential a, and the noise term ε. When w exceeds the threshold value θ, the output speak will be generated and the feedback after-potential a will be triggered, and then it will be added to w. Specifically, the feedforward kernel K and the feedback kernel H describe the conversion from x to u and the conversion from y to a, respectively.

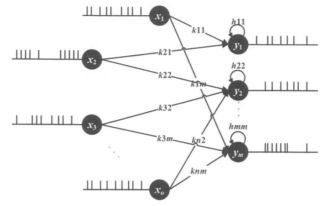

(a) Sparse connected MIMO neural network spike system

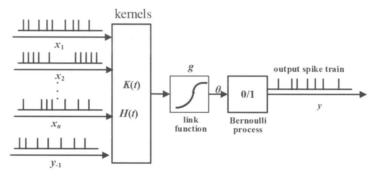

(b) Time-varying nonlinear MISO model from spike train data with similar structure.

Fig. 1: Time-varying multi-input and multi-output (MIMO) and multi-input and single-output (MISO) neural spiking structure. (a) Sparse connected MIMO neural network spike system. The solid lines between neural population indicate the existence of some causal connections. (b) Time-varying nonlinear MISO model from spike train data with similar structure.

Considering that the complex biological processes, such as the spiking activities of neural ensembles in the brain are inherently nonlinear; therefore, in this paper, the nonlinear Volterra series expansion of analytic functions are adopted for accounting for the point-process connection nature. Specifically, the synaptic potential u is defined with the nonlinear Volterra series of x as follows:

$$
\begin{aligned}
u(t) = k_0(t) &+ \sum_{n=1}^{N} \sum_{\tau=0}^{M_k} k_1^{(n)}(t,\tau) x_n(t-\tau) \\
&+ \sum_{n=1}^{N} \sum_{\tau_1=0}^{M_k} \sum_{\tau_2=0}^{M_k} k_{2s}^{(n)}(t,\tau_1,\tau_2) x_n(t-\tau_1) x_n(t-\tau_1) \\
&+ \sum_{n_1=1}^{N} \sum_{n_2=1}^{n_1-1} \sum_{\tau_1=0}^{M_k} \sum_{\tau_2=0}^{M_k} k_{2x}^{(n_1,n_2)}(t,\tau_1,\tau_2) x_{n_1}(t-\tau_1) x_{n_2}(t-\tau_2) \\
&+ \sum_{n=1}^{N} \sum_{\tau_1=1}^{M_k} \sum_{\tau_2=0}^{M_k} \sum_{\tau_3=0}^{M_k} k_{3s}^{(n)}(t,\tau_1,\tau_2,\tau_3) x_n(t-\tau_1) x_n(t-\tau_2) x_n(t-\tau_3) \\
&+ \dots
\end{aligned}
\tag{3}
$$

where $k_0(t)$ is the zeroth-order kernel representing the baseline value of u when the input equals zero; $k_1^{(n)}$ is the first-order kernels denoting the linear relation between the synaptic potential u and the n-th input spike train x_n, which are related with the time t and the time intervals τ between the past and the present time. The second-order and third-order nonlinear self-kernels can be given as $k_{2s}^{(n)}$ and $k_{3s}^{(n)}$ and involving the functions of the n-th input x_n and u respectively. $k_{2x}^{(n_1, n_2)}$ represent the second-order cross-kernel nonlinear interactions between each specific pair of inputs x_{n_1} and x_{n_2} as they affect u; the total number of inputs is N, and the memory length of the feedforward process can be denoted as M_k.

Similarly, feedback variable a is defined with the linear Volterra series as follows:

$$a(t) = \sum_{\tau=1}^{M_h} h(t, \tau) y\,(t - \tau) \tag{4}$$

where h represents the time-varying linear feedback kernel, and its memory length is denoted by M_h. Specifically, the time-varying nonlinear Volterra kernels in Equation (3) and (4) can be therefore given as $\{k_0, k_1^{(1)}, k_{2s}^{(n)}, k_{2x}^{(n_1, n_2)}, k_{3s}^{(n)}, ..., h\}$.

2.2 Laguerre Expansion of Volterra Kernels

In order to decrease the computation complexity of the time-varying nonlinear dynamical generalized Volterra model established in Equations (3) and (4), we utilized a Laguerre expansion technology of Volterra kernel (LEV) where the Volterra series is approximated by being projected on to a series of orthonormal Laguerre basis functions b:

$$
\begin{aligned}
u(t) = c_0(t) &+ \sum_{n=1}^{N} \sum_{j=1}^{L} c_1^{(n)}(t, j) v_j^{(n)}(t) \\
&+ \sum_{n=1}^{N} \sum_{j_1=1}^{L} \sum_{j_2=1}^{j_1} c_{2s}^{(n)}(t, j_1, j_2) v_{j_1}^{(n)}(t) v_{j_2}^{(n)}(t) \\
&+ \sum_{n_1=1}^{N} \sum_{n_2=1}^{n_1-1} \sum_{j_1=1}^{L} \sum_{j_2=1}^{L} c_{2x}^{(n_1, n_2)}(t, j_1, j_2) v_{j_1}^{(n_1)}(t) v_{j_2}^{(n_2)}(t) \\
&+ \sum_{n=1}^{N} \sum_{j_1=1}^{L} \sum_{j_2=1}^{j_1} \sum_{j_3=1}^{j_2} c_{3s}^{(n)}(t, j_1, j_2, j_3) v_{j_1}^{(n)}(t) v_{j_2}^{(n)}(t) v_{j_3}^{(n)}(t) \\
&+ ...
\end{aligned}
\tag{5}
$$

$$a(t) = \sum_{j=1}^{L} c_h(t, j) v_j^{(h)}(t) \tag{6}$$

where $v_j^{(n)}(t) = \sum_{\tau=0}^{M_k} b_j(\tau) x_n(t - \tau)$ and $v_j^{(h)}(t) = \sum_{\tau=1}^{M_h} b_j(\tau) y(t - \tau)$, $\{c_0, c_1^{(n)}, c_{2s}^{(n)}, c_{2x}^{(n_1, n_2)}, c_{3s}^{(n)}, ..., c_h\}$ represent the time-varying expansion parameters based on the LEV for different Volterra kernels, which are the functions of time. The total number of the Laguerre basis function is denoted as L. Previous publication described the formula of acquiring discrete Laguerre basis functions [25]. Considering that the Laguerre basis comprises the orthogonal function of exponential decay type (Fig. 2), it can fit various temporal processes effectively with a small number of basic functions. Specifically, Laguerre functions with order 3 is used in this study. The time-varying nonlinear generalized Laguerre-Volterra (TVNGLV) model is formed in equations (5) and (6). Considering that kernels are generally smooth, far fewer Laguerre basis functions than the memory length of M_k and M_h will be used, the LEV technology can largely reduce the number of unknown time-varying parameters.

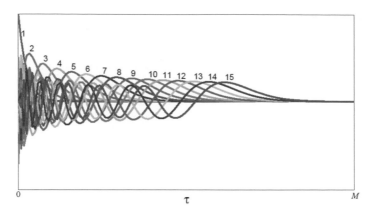

Fig. 2: Laguerre basis functions with orders from first to 15th

2.3 Group Sparsity Detection with Regularization Method

Studies have shown that the nodes in the biological neural network are not completely connected [24]. This is just a fraction of given input nodes can be connected to the output neuron node. However, in equations (5) and (6), the complete model based on Volterra kernels composed of all observed signals cannot effectively capture the sparsity of biological systems. Therefore, in this paper, we introduce the functional group sparsity for the sparse time-varying nonlinear generalized Laguerre Volterra model. Functional group sparsity selects valid inputs by forcing the kernel function of redundant input to zero by using the regularization method.

Regularization methods have been generally used in parametric estimation and variable selection, especially when the training data is sufficient, such as LASSO [27], SCAD [7], group-LASSO [21], and group-bridge [10]. LASSO and SCAD are usually used to select individual coefficients, while group-LASSO and group-bridge can achieve sparse coefficients at the group level. In the TVNGLV model constructed above, it can be grouped according to the Laguerre approximation coefficients of the corresponding input signals. For example, since $c_1^{(n)}(t, 1), c_1^{(n)}(t, 2),..., c_1^{(n)}(t, L)$ belong to the n-th input, they can be assigned to the same group. Given this group of settings, the sparsity of the functional group refers to the situation where the estimates of some groups of Laguerre expansion parameters are kept at zero. Thus, the group-LASSO regularization algorithm by means of minimizing the sum of square errors can be used for the sparse TVNGLV model. The penalty item in the group-LASSO approach can be composed of the mixture of L_2 and the L_1 penalty: within the group, it regularizes the shrinkage coefficient through the L_2 norm; across groups and elects the groups by the L_1 norm regularization method. In particular, the expression of the group-LASSO penalty used in this paper is expressed as follows:

$$p_2^1 = \sum_{n=1}^{N_0} \| \overline{c}_1^{(n)} \|_2^1 = \sum_{n=1}^{N_0} \left(\sum_{j=1}^{L} \left(\overline{c}_1^{(n)}(j) \right)^2 \right)^{\frac{1}{2}} \qquad (7)$$

where $\overline{c}_1^{(n)}(j)$ indicates the average of time-varying Laguerre expansion parameters over the entire data length. The regularization criterion for group-LASSO can therefore be given as:

$$C = \arg\min_{\lambda}\left\{\sum_{t=1}^{T}[u(t) - c_0 - \sum_{n=1}^{N_0}\sum_{j=1}^{L}\overline{c}_1^{(n)}(j)v_j^n(t)]^2 + \lambda P_2^1\right\} \tag{8}$$

where the regularization parameter $\lambda > 0$ determines the sparsity of the model, *i.e.*, the larger the value of λ, the more sparse the result of coefficients. The regularization parameter λ is explored in the range of 10^{-4} to 10^{-1}, and the deviance for each λ is achieved using the cross-validation theory. The spurious inputs can be found by using the minimum deviation λ and the corresponding C set. This is, if the Laguerre expansion coefficients of the group are all zero, the redundant input will therefore be identified.

Assuming that the selected Laguerre expansion coefficients can be given as $\left\{c_0(t), c_1^{(r_1)}(t, 1), ..., c_1^{(r_1)}(t, L),..., c_1^{(r_{N_0})}(t, 1), c_1^{(r_{N_0})}(t, L)\right\}$, the related inputs can therefore be represent as $x_{r_1}, x_{r_2}, ..., x_{r_{N_0}}$, where N_0 indicates the number of spurious inputs. According to Eq. (8), the synaptic potential in the sparse TVNGLV can be represented as follows:

$$
\begin{aligned}
u(t) = {} & c_0(t) + \sum_{n=1}^{N_0}\sum_{j=1}^{L} c_1^{(r_n)}(t, j)v_j^{(r_n)}(t) \\
& + \sum_{n=1}^{N_0}\sum_{j_1=1}^{L}\sum_{j_2=1}^{j_1} c_{2s}^{(r_n)}(t, j_1, j_2)v_{j_1}^{(r_n)}(t)v_{j_2}^{(r_n)}(t) \\
& + \sum_{n_1=1}^{N_0}\sum_{n_2=1}^{n_1-1}\sum_{j_1=1}^{L}\sum_{j_2=1}^{L} c_{2x}^{(r_{m}, r_{m})}(t, j_1, j_2)v_{j_1}^{(r_{m})}(t)v_{j_2}^{(r_{m})}(t) \\
& + \sum_{n=1}^{N_0}\sum_{j_1=1}^{L}\sum_{j_2=1}^{j_1}\sum_{j_3=1}^{j_2} c_{3s}^{(r_n)}(t, j_1, j_2, j_3)v_{j_1}^{(r_n)}(t)v_{j_2}^{(r_n)}(t)v_{j_3}^{(r_n)}(t) \\
& + ...
\end{aligned}
\tag{9}
$$

The sparsity of the function group can be achieved by using the group-LASSO regularization method, thereby accurately representing the connectivity of the neural network. In addition, time-varying nonlinear parameters can be effectively estimated, based on the following multiwavelet basis expansion scheme.

2.4 Multiwavelet-based Time-varying Nonlinear GLV Model

Previous work shows that the complexity of the model can be greatly reduced by extending the time-varying nonlinear Laguerre expansion parameters to a set of basic functions, since that the above TVNGLV model can be converted into an extended time-invariant regression model. Therefore, the identification of TVNGLV is simplified as deterministic model term selection and coefficients estimation, and the number of coefficients to be estimated is greatly reduced by involving unknown model coefficients into the basis expansion. Electrophysiological signals including spiking data generally contain both slow and fast changes. For reducing the number of parameters in the complex time-varying nonlinear spiking system, the multiwavelet basis function expansion method is proposed to approximate the time-varying parameter estimation process, and without considering the a priori knowledge of the nonstationary procedure, the trend of fast and slow changes can both be tracked. Compared with the conventional adaptive filtering method, like the SSPPF and basis expansion method using only one particular type of basis function, the novel TVNGLV framework based on multi-wavelet basis function expansion is more effective in terms of non-stationary spike train signal processing.

Time-varying   parameters $\left\{ c_0, c_1^{(r_n)}, c_{2s}^{(r_n)}, c_{2x}^{(r_{n_1}, r_{n_2})}, c_{3s}^{(r_n)}, ..., c_h \right\}$ in   Equations (9) and (6) can be expended with multi-wavelet basis functions, and therefore the following formula can be obtained:

$$c_0(t) = \sum_m \sum_{k \in \Gamma m} \alpha_{p,k}^{0,m} \psi_{p,k}^m \left(\frac{t}{T} \right)$$

$$c_1^{(r_n)}(t, j) = \sum_m \sum_{k \in \Gamma m} \alpha_{j,p,k}^{r_n,m} \psi_{p,k}^m \left(\frac{t}{T} \right)$$

$$c_{2s}^{(r_n)}(t, j_1, j_2) = \sum_m \sum_{k \in \Gamma m} \alpha_{j_1, j_2, p, k}^{r_n,m} \psi_{p,k}^m \left(\frac{t}{T} \right)$$

$$c_{2x}^{(r_{n_1}, r_{n_2})}(t, j_1, j_2) = \sum_m \sum_{k \in \Gamma m} \alpha_{j_1, j_2, p, k}^{r_{n_1}, r_{n_2} m} \psi_{p,k}^m \left(\frac{t}{T} \right) \qquad (10)$$

$$c_{3s}^{(r_n)}(t, j_1, j_2, j_3) = \sum_m \sum_{k \in \Gamma m} \alpha_{j_1, j_2, j_3, p, k}^{r_n,m} \psi_{p,k}^m \left(\frac{t}{T} \right)$$

$$\vdots$$

$$c_h(t, j) = \sum_m \sum_{k \in \Gamma m} \alpha_{j,p,k}^{h,m} \psi_{p,k}^m \left(\frac{t}{T} \right)$$

where the wavelet basis function with wavelet scale p and shift indices $k \in \Gamma_m$, $\Gamma_m = \{ k: \quad m \le k \le 2^p - 1 \}$ can be given as $\psi_{p,k}^m \left(\frac{t}{T} \right)$. The order of the wavelet basis functions is represented as the superscript m, and the length of observational samples is T. $\alpha_{p,k}^{0,m}$, $\alpha_{j,p,k}^{r_n,m}$, $\alpha_{j_1,j_2,p,k}^{r_n,m}$ and $\alpha_{j,p,k}^{h,m}$ denote the time-invariant coefficients to be estimated in the multiwavelet's expansion model. Multiple-wavelet basis functions of different orders can be utilized to approximate the time-varying parameters with both slow and fast changes. For instance, cardinal B-spline basis functions possess excellent approximation properties since they are completely supported. Based on this motivation, the mth order B-spline basis function (B_m) with the scale j and shift indices k can be used as $\alpha_{p,k}^m (\bullet)$ in equation (10) as follows:

$$\psi_{p,k}^m(x) = 2^{p/2} B_m (2^p x - k) \qquad (11)$$

where the variable $x = \frac{t}{T}$ is in the range of [0, 1] and k satisfies $- m \le k \le 2^p -$ 1. The higher the scale p, the more basis functions are utilized, and therefore the approximation resolution is improved; unfortunately, this will introduce more unknown parameters and further add computation complication. Generally, according to the criteria given in [9], the scale p is chosen to be 3 or greater. The first order cardinal B-spline function is a famous Haar function and can be represented as

$$B_1 = \chi_{[0,1]}(x) = \begin{cases} 1, \text{ if } x \in 0, 1 \\ 0, \text{ otherwise} \end{cases} \qquad (12)$$

The B-splines in common use with different orders are normally selected from $\{B_m(x): m = 2, 3, 4, 5\}$. Obviously, low order B-spline functions, such as first- and second-order, are piecewise non-smooth functions that can be applied to approximate signals with sharp or abrupt changes; while high order B-spline functions, *i.e.* 3, 4 and 5, often perform excellently on tracking steadily changing signals.

Therefore, multiwavelet basis functions of different orders can be employed to approximate the time-varying signals with both steady and sharp changes. A detailed discussion of the appropriate scale p and the selection of the order of the related wavelet can be seen in [18].

By substituting Equation (10) into Equations (6) and (9), the TVNGLV model can be obtained as follows:

$$u(t) = \sum_m \sum_{k \in \Gamma_m} \alpha_{p,k}^{0,m} \psi_{p,k}^m \left(\frac{t}{T}\right)$$

$$+ \sum_{n=1}^{N_0} \sum_{j=1}^{L} \sum_m \sum_{k \in \Gamma_m} \alpha_{j,p,k}^{r_n,m} \psi_{p,k}^m \left(\frac{t}{T}\right) v_j^{(r_n)}(t)$$

$$+ \sum_{n=1}^{N_0} \sum_{j_1=1}^{L} \sum_{j_2=1}^{j_1} \sum_m \sum_{k \in \Gamma_m} \alpha_{j_1,j_2,p,k}^{r_n,m} \psi_{p,k}^m \left(\frac{t}{T}\right) v_{j_1}^{(r_n)}(t) v_{j_2}^{(r_n)}(t) \qquad (13)$$

$$+ \sum_{n_1=1}^{N_0} \sum_{n_2=1}^{n_1-1} \sum_{j_1=1}^{L} \sum_{j_2=1}^{L} \sum_m \sum_{k \in \Gamma_m} \alpha_{j_1,j_2,p,k}^{r_{n_1},r_{n_2},m} \psi_{p,k}^m \left(\frac{t}{T}\right) v_{j_1}^{(r_{n_1})}(t) v_{j_2}^{(r_{n_2})}(t)$$

$$+ \sum_{n=1}^{N_0} \sum_{j_1=1}^{L} \sum_{j_2=1}^{j_1} \sum_{j_3=1}^{j_2} \sum_m \sum_{k \in \Gamma_m} \alpha_{j_1,j_2,j_3,p,k}^{r_n,m} \psi_{p,k}^m \left(\frac{t}{T}\right) v_{j_1}^{(r_n)}(t) v_{j_2}^{(r_n)}(t) v_{j_3}^{(r_n)}(t)$$

$$+ \ldots$$

$$a(t) = \sum_{l=1}^{L} \sum_m \sum_{k \in \Gamma_m} \alpha_{l,j,k}^{h,m} \psi_{j,k}^m \left(\frac{t}{T}\right) v_l^{(h)}(t) \qquad (14)$$

where $\left\{\alpha_{p,k}^{0,m}, \alpha_{j,p,k}^{r_n,m}, \alpha_{j_1,j_2,p,k}^{r_n,m}, \alpha_{j_1,j_2,p,k}^{r_{n_1},r_{n_2},m}, \alpha_{j_1,j_2,j_3,p,k}^{r_n,m}, \alpha_{j,p,k}^{h,m}\right\}$, $n = 1, 2, \ldots, N_0, j = 1, 2, \ldots$, L are the time invariant parameters. For simplicity, we can express the new variable as:

$$V_{p,k}^{0,m}(t) = \psi_{p,k}^m \left(\frac{t}{T}\right)$$

$$V_{j,p,k}^{r_n,m}(t) = \psi_{p,k}^m \left(\frac{t}{T}\right) v_j^{(r_n)}(t)$$

$$V_{j_1,j_2,p,k}^{r_n,m}(t) = \psi_{p,k}^m \left(\frac{t}{T}\right) v_j^{(r_n)}(t)$$

$$V_{j_1,j_2,p,k}^{r_{n_1},r_{n_2}m}(t) = \psi_{p,k}^m \left(\frac{t}{T}\right) v_{j_1}^{(r_n)}(t) v_{j_2}^{(r_n)}(t) \qquad (15)$$

$$V_{j_1,j_2,j_3,p,k}^{r_n,m}(t) = \psi_{p,k}^m \left(\frac{t}{T}\right) v_{j_1}^{(r_n)}(t) v_{j_2}^{(r_n)}(t) v_{j_3}^{(r_n)}(t)$$

$$V_{j,p,k}^{h,m}(t) = \psi_{p,k}^m \left(\frac{t}{T}\right) v_j^{(h)}(t)$$

Substituting Equation (15) into Equations (13) and (14) obtains

$$u(t) = \sum_m \sum_{k \in \Gamma_m} \alpha_{p,k}^{0,m} V_{p,k}^{0,m}(t)$$

$$+ \sum_{n=1}^{N_0} \sum_{j=1}^{L} \sum_m \sum_{k \in \Gamma_m} \alpha_{j,p,k}^{r_n,m} V_{j,p,k}^{r_n,m}(t)$$

$$+ \sum_{n=1}^{N_0} \sum_{j_1=1}^{L} \sum_{j_2=1}^{j_1} \sum_m \sum_{k \in \Gamma_m} \alpha_{j_1,j_2,p,k}^{r_n,m} V_{j_1,j_2,p,k}^{r_n,m}(t)$$

$$+ \sum_{n_1=1}^{N_0} \sum_{n_2=1}^{n_1-1} \sum_{j_1=1}^{L} \sum_{j_2=1}^{L} \sum_m \sum_{k \in \Gamma_m} \alpha_{j_1,j_2,p,k}^{r_{n_1},r_{n_2},m} V_{j_1,j_2,p,k}^{r_{n_1},r_{n_2},m}(t) \qquad (16)$$

$$+ \sum_{n=1}^{N_0} \sum_{j_1=1}^{L} \sum_{j_2=1}^{j_1} \sum_{j_3=1}^{j_2} \sum_m \sum_{k \in \Gamma_m} \alpha_{j_1,j_2,j_3,p,k}^{r_n,m} V_{j_1,j_2,j_3,p,k}^{r_n,m}(t)$$

$$+ \dots$$

$$a(t) = \sum_{j=1}^{L} \sum_m \sum_{k \in \Gamma_m} \alpha_{j,p,k}^{h,m} V_{j,p,k}^{h,m}(t) \qquad (17)$$

where $V_{p,k}^{0,m}(t), V_{j,p,k}^{r_n,m}(t), V_{j_1,j_2,p,k}^{r_n,m}(t), V_{j_1,j_2,p,k}^{r_{m_1},r_{m_2},m}(t), V_{j_1,j_2,j_3,p,k}^{r_n,m}(t)$ and $V_{j,p,k}^{h,m}(t)$ represent the model terms in the multiwavelet-based TVNGLV model. Furthermore, the hidden variable $w(t)$ in Equation (1) is rewritten as:

$$w(t) = \sum_m \sum_{k \in \Gamma_m} \alpha_{p,k}^{0,m} V_{p,k}^{0,m}(t)$$

$$+ \sum_{n=1}^{N_0} \sum_{j=1}^{L} \sum_m \sum_{k \in \Gamma_m} \alpha_{j,p,k}^{r_n,m} V_{j,p,k}^{r_n,m}(t)$$

$$+ \sum_{n=1}^{N_0} \sum_{j_1=1}^{L} \sum_{j_2=1}^{j_1} \sum_m \sum_{k \in \Gamma_m} \alpha_{j_1,j_2,p,k}^{r_n,m} V_{j_1,j_2,p,k}^{r_n,m}(t)$$

$$+ \sum_{n_1=1}^{N_0} \sum_{n_2=1}^{n_1-1} \sum_{j_1=1}^{L} \sum_{j_2=1}^{L} \sum_m \sum_{k \in \Gamma_m} \alpha_{j_1,j_2,p,k}^{r_{m_1},r_{m_2},m} V_{j_1,j_2,p,k}^{r_{m_1},r_{m_2},m}(t) \qquad (18)$$

$$+ \sum_{n=1}^{N_0} \sum_{j_1=1}^{L} \sum_{j_2=1}^{j_1} \sum_{j_3=1}^{j_2} \sum_m \sum_{k \in \Gamma_m} \alpha_{j_1,j_2,j_3,p,k}^{r_n,m} V_{j_1,j_2,j_3,p,k}^{r_n,m}(t)$$

$$+ \dots$$

$$+ \sum_{j=1}^{L} \sum_m \sum_{k \in \Gamma_m} \alpha_{j,p,k}^{h,m} V_{j,p,k}^{h,m}(t) + \varepsilon(t)$$

Equation (18) can thus be simplified as a linear-in-the-parameters regression model

$$w(t) = \sum_{i=1}^{M} \alpha_i x_i(t) + \varepsilon(t), \; t = 1, 2, \dots, T \qquad (19)$$

where α_i represents the unknown coefficients, $x_i(t)$ denotes the model terms in the combination of $\{V_{p,k}^{0,m}(t), V_{j,p,k}^{r_n,m}(t), V_{j_1,j_2,p,k}^{r_n,m}(t), V_{j_1,j_2,p,k}^{r_{m_1},r_{m_2},m}(t), V_{j_1,j_2,j_3,p,k}^{r_n,m}(t), V_{j,p,k}^{h,m}(t)\}$, and M represents the total number of candidate model terms in the multiwavelet-based TVNGLV model in (18). Therefore, the original TVNGLV model in E\equations (7) and (8) can now be converted to a time invariant regression one, which enables the parameter estimation problem to be solved using the maximum likelihood strategy.

2.5 Model Structure Selection Based on OMP Algorithm

Model structure selection is an important part in the process of modeling and identification. A reduced model is generated by selecting meaningful candidate model items or regression factors from an enormous number of redundant model items. For instance, the B-spline basis function expansion model described by Equation (18) may involve myriad candidate model terms or regression elements when the parameters W, L and N are large. However, many of these candidate model terms may be highly possible linear dependence among the multiwavelet-based basis function expansions, which may lead to overfitting of the identified models, and the reliable results cannot be obtained by using the direct maximum-likelihood method from these ill-posed problems. Therefore, it is important to decide which model items should be included in the final model [32].

In this study, the efficient OMP [5, 28] is employed to solve the difficult problem of model subset selection or model structure detection from the time-varying nonlinear system with spiking data, which has proved effective for select significate model terms in system identification [16].

OMP algorithm is an iterative greedy algorithm, which chooses the column most related to the current residual in each step [29]. This column can be then added to the collection of selected columns. The algorithm updates the residuals by means of projecting the observations into a linear subspace composed of selected columns and then the algorithm iterates.

In each iteration, a new signal approximation $\bar{Y}^n$ is calculated by OMP. The approximation error can thus be obtained with $r^n = Y - \bar{Y}^n$, which will be used to determine which new element should be selected in the next iteration. Specially, the selection depends on the inner products between the column vectors x_i of X and the current residual r^n. These inner products can be expressed as:

$$\alpha_i^n = x_i^T r^n \tag{20}$$

The new element with the largest magnitude of α_i^n is then selected, *i.e.*

$$i_{max}^n = ar\,g_i\,max\left|\alpha_i^n\right| \tag{21}$$

Therefore,

$$\Gamma^{n+1} = \Gamma^n \cup i_{max}^n \tag{22}$$

Considering the binary characteristics for the spike train data involved in this study, the model terms selection algorithm based on the OMP algorithm in the complete model in Equation (18) is described in detail as follows:

Step 1: Combine the complete model candidates $\{V_{p,k}^{0,m}(t), V_{j,p,k}^{r_n,m}(t), V_{j_1,j_2,p,k}^{r_n,m}(t), V_{j_1,j_2,p,k}^{r_{m_1},r_{m_2},m}(t), V_{j_1,j_2,j_3,p,k}^{r_n,m}(t), V_{j,p,k}^{h,m}(t)\}$ and the output spike train data y, we can obtain the maximum likelihood estimator $\widehat{\alpha}_{mle}$ based on a generalized linear models (GLM). The values for the synaptic potential and after-potential can be roughly calculated based on Equations (16) and (17). Furthermore, the hidden variable $w(t) = 1, ..., T$ can be derived which denotes the measured outputs;

Step 2: Initializes the residuals $r^0 = w$, initializes the selected variable index and the selected variable set as $\Gamma^0 = \varnothing, \Phi^0 = \varnothing$;

Step 3: The variables $X = x_i, i = 1, 2, 3, ..., M$ are the candidates for the selected important variables. The inner product can be calculated as $\alpha_i^1 = x_i^T r^0$. The first index for the selected model term is $i_{max}^1 = ar\,g_i\,max\left|\alpha_i^1\right|$, and $\Gamma^0 \cup i_{max}^1, \Phi^1 = X_{\Gamma^1}$. The approximation error can be given as $r^1 = w - \Phi^1 X$;

Step 4: Subsequent valid model terms or regressors can therefore be selected step by step in the same way until terminated at the M_s-th step;

Step 5: If the stop condition is reached, the algorithm is stopped. Otherwise, set and return to Step 2.

2.6 Likelihood and Parameter Estimation

Supposing that the selected important model terms $\phi = \{\varphi_1, ..., \varphi_p\}$ have been obtained using the OMP algorithm, the associated time-invariant expansion coefficients $\alpha = \{\alpha_1, ..., \alpha_p\}$ can thus be estimated based on the maximum-likelihood method. The log likelihood function can be described as follows:

$$L(y\,|\,x, k, h, \sigma, \theta) = ln\ \Pi_{t=0}^{T} P\,(y\,|\,x, k, h, \sigma, \theta) \qquad (23)$$

$$= \Sigma_{t=0}^{T}\ ln\ P\,(y\,|\,x, k, h, \sigma, \theta)$$

where

$$P\,(y\,|\,x, k, h, \sigma, \theta) = \begin{cases} Prob\,(w \geq \theta, x, k, h, \sigma, \theta),\ when\ y = 1 \\ Prob\,(w < \theta, x, k, h, \sigma, \theta),\ when\ y = 0 \end{cases} \qquad (24)$$

$P\,(y\,|\,x, k, h, \sigma, \theta)$ represents the estimated firing probability which can be calculated with error function, *i.e.* integral of Gaussian function.

$$P(t) = 0.5 - 0.5\ erf\left(\frac{\theta - w(t)}{\sqrt{2}\sigma}\right) \qquad (25)$$

The gradient of the log-likelihood function over the coefficients a is shown below:

$$G(i) = \frac{\partial L}{\partial \alpha_i} = \sum_{t=0}^{T} \frac{1}{p(t)} \times \frac{\partial P(t)}{\partial \alpha_i} \qquad (26)$$

$$= \sum_{t=0}^{T} \frac{1}{\sigma\sqrt{2\pi}p(t)} \times e^{-\frac{(\theta - w(t))^2}{2\sigma^2}} \times \frac{\partial w(t)}{\partial \alpha_i}$$

where

$$\frac{\partial w(t)}{\partial \alpha_{p,k}^{0,m}} = \psi_{p,k}^{m}\left(\frac{t}{T}\right)$$

$$\frac{\partial w(t)}{\partial \alpha_{j,p,k}^{r_n,m}} = \psi_{p,k}^{m}\left(\frac{t}{T}\right)v_{j_1}^{(r_n)}(t)$$

$$\frac{\partial w(t)}{\partial \alpha_{j_1,j_2,p,k}^{r_n,m}} = \psi_{p,k}^{m}\left(\frac{t}{T}\right)v_{j_1}^{(r_n)}(t)v_{j_2}^{(r_n)}$$

$$\frac{\partial w(t)}{\partial \alpha_{j_1,j_2,p,k}^{r_{n_1},r_{n_2},m}} = \psi_{p,k}^{m}\left(\frac{t}{T}\right)v_{j_1}^{(r_{n_1})}(t)v_{j_2}^{(r_{n_2})}(t)$$

$$\frac{\partial w(t)}{\partial \alpha_{j_1,j_2,j_3,p,k}^{r_n,m}} = \psi_{p,k}^{m}\left(\frac{t}{T}\right)v_{j_1}^{(r_n)}(t)v_{j_2}^{(r_n)}(t)v_{j_3}^{(r_n)}(t) \qquad (27)$$

$$\vdots$$

$$\frac{\partial w(t)}{\partial \alpha_{j,p,k}^{h,m}} = \psi_{p,k}^{m}\left(\frac{t}{T}\right)v_{j}^{h}(t)$$

A generalized linear model (GLM) fitting method based on maximizing the likelihood function can be used to estimate the unknown parameters $\{\tilde{\alpha}_1, ..., \tilde{\alpha}_p\}$. The connection function of GLM is chosen as the *probit* function, that is, the inverse cumulative distribution function of normal distribution [26].

A parsimonious model will be obtained, based on the selected significant model terms and the corresponding parameters estimated. The estimated parameters $\{\tilde{\alpha}_1, ..., \tilde{\alpha}_p\}$ can be used to calculate Laguerre expansion coefficients and the feedforward and feedback kernels can thus be reconstructed.

2.7 Kernel Reconstruction and Interpretation

Once the coefficients $\{\tilde{\alpha}_1, ..., \tilde{\alpha}_p\}$ were estimated, the Laguerre expansion parameters $\{c_0, c_1^{(r_n)}, c_{2s}^{(r_n)}, c_{2x}^{(r_{n_1}, r_{n_2})}, c_{3s}^{(r_n)}, ..., c_h\}$ can be calculated, using Equation (10). The variables $\tilde{c}_1^{(r_n)}, \tilde{c}_{2s}^{(r_n)}, \tilde{c}_{2x}^{(r_{n_1}, r_{n_2})}, \tilde{c}_{3s}^{(r_n)}, ..., \tilde{c}_h$ and $\theta - \tilde{c}_0$, which represents the distance between the baseline value of w and the threshold, and can be scaled arbitrarily without affecting the relationship between x and y. Further, $\tilde{c}_0$ and θ can be transferred together. Therefore, θ and σ can be both set as unit values generally. θ remains unit value, and σ will be restored later.

By taking $\hat{c}_1^{(r_n)} = \dfrac{\tilde{c}_1^{(r_n)}}{(\theta - \tilde{c}_0)}, \hat{c}_{2s}^{(r_n)} = \dfrac{\tilde{c}_{2s}^{(r_n)}}{(\theta - \tilde{c}_0)}, ..., \hat{c}_h = \dfrac{\tilde{c}_h}{(\theta - \tilde{c}_0)}$ and $\hat{\sigma} = \dfrac{\sigma}{(\theta - \tilde{c}_0)}$,

respectively, the final coefficients $\hat{c}$ and $\hat{\sigma}$ can be easily calculated with the estimated Laguerre expansion coefficients $\tilde{c}$. Furthermore, the feedforward and feedback kernels can be reconstructed as follows:

$$\hat{k}_0 = 0$$

$$\hat{k}_1^{(n)}(t, \tau) = \sum_{j=1}^{L} \hat{c}_1^{(r_n)}(t, j) b_j^{(r_n)}(\tau)$$

$$\hat{k}_{2s}^{(n)}(t, \tau_1, \tau_2) = \sum_{j_1=1}^{L} \sum_{j_2=1}^{j_1} \hat{c}_{2s}^{(r_n)}(t, j_1, j_2) b_{j_1}^{(r_n)}(\tau) b_{j_2}^{(r_n)}(\tau)$$

$$\hat{k}_{2x}^{(n_1, n_2)}(t, \tau_1, \tau_2) = \sum_{j_1=1}^{L} \sum_{j_2=1}^{L} c_{2x}^{(r_{n_1}, r_{n_2})}(t, j_1, j_2) b_{j_1}^{(r_{n_1})}(\tau) b_{j_2}^{(r_{n_2})}(\tau)$$

$$\hat{k}_{3s}^{(n)}(t, \tau_1, \tau_2, \tau_3) = \sum_{j_1=1}^{L} \sum_{j_2=1}^{j_1} \sum_{j_3=1}^{j_1} c_{3s}^{(r_n)}(t, j_1, j_2, j_3) b_{j_1}^{(r_n)}(\tau) b_{j_2}^{(r_n)}(\tau) b_{j_3}^{(r_n)}(\tau) \qquad (28)$$

$$\vdots$$

$$\hat{k}(t, \tau) = \sum_{j=1}^{L} \hat{c}_h(t, j) b_j^h(\tau)$$

The normalized Volterra kernel can directly describe the time-varying nonlinear input-output dynamics. $\hat{k}_1^{(n)}(t, \tau)$ denotes the first-order linear response in u elicited by the n-th input spike signal; $\hat{k}_{2s}^{(n)}$ and $\hat{k}_{3s}^{(n)}$ describe the second- and third-order nonlinear response in u elicited by the n-th input spike signal; $\hat{k}_{2x}^{(n_1, n_2)}$ represents the second-order response to u triggered by the singles from both the n_1-th and n_2-th inputs, and h is the output spike elicited after-potential.

3. Simulation Studies on Time-varying Nonlinear Dynamical Model

In order to reveal the significant advantages for the proposed sparse time-varying nonlinear dynamical modeling method based on the B-spline basis expansion with OMP, we tested the scheme intensively using synthetic input-output spiking system with different kinds of kernels. In all simulations, the system is designed to be 4-input and single-output system, with two kernel functions to be zeros (*i.e.* k_3, k_4) and two functions to be significant (*i.e.* k_1, k_2). Both the first- and second-order model terms for the significant input signals are considered. The input signals are four independent Poisson process with a 10 Hz mean frequency. The length of the simulation system is 2500s.

3.1 Simulated Kernels with Step Changes

Firstly, a four-input and single-output second order time-varying spiking dynamical system with step changes is simulated. At the simulation time 1250 s, the amplitude of first-order kernel of the first input k_{1-1} is reduced to a quarter with the same waveforms; the amplitude of first-order kernel of the second input $k_{1,2}$ is quadruple with the same waveforms; the amplitude of second-order kernel of the first input k_{2-1} is reduced to a quarter with the same waveforms; and the other kernels (k_{2-2} and h) remain constant during the whole simulation time. Based on the simulated input and the associated output spike train data, we apply the sparse time-varying nonlinear dynamical modeling method based on the B-spline basis expansion with the OMP algorithm to identify the changes of the linear and nonlinear kernels. In the proposed method, the B-spline basis functions with the order of 2~5 and a scale index of $j = 3$ were utilized to construct a multiwavelet-based TVNGLV model. The group-LASSO regularization method was used to achieve functional group sparsity, and the parsimonious model structure can be obtained with the OMP algorithm. Therefore, the generalized linear time-invariant parameters in the sparse model can be approximated, based on the maximum-likelihood method, which will further be utilized to finish the reconstruction of the first- and second-order feedforward kernels, as well as the feedback kernels from Equation (28).

Fig. 3 shows the tracking results for the significant variables using the B-spline expansion method with OMP for the first- and second-order feedforward kernels as well as the first-order feedback kernel with the 3-D plots, where the x-axis denotes the simulated time, the y-axis denotes the time lag between present time and previous input events, and the z-axis denotes the amplitude of kernel functions. It is obvious that the proposed modeling scheme based on B-splines basis expansion and OMP cannot just accurately approximate the kernel shapes, but also can quickly track the different changes of kernel functions.

In order to reveal the advantages of the proposed multiple B-spline basis functions expansion method with OMP, Fig. 4 further shows the tracking performance of different kernel changes, where the SSPPF and Chebyshev polynomials expansion algorithms are also employed for comparison. Every single subfigure in Fig. 4 illustrates the estimated and actual kernels during the time evolution. Similarly, the x-axis indicates the simulated time, the y-axis indicates the time lag between the present time and the previous input events, and the amplitudes for different kernels are color-coded. Results illustrate that the four different approaches can approximate various kinds of changes of nonstationary feedback and feedforwards kernels. In particular, though SSPPF can estimate the variation of steady time-varying parameters, it cannot approximate the transient characteristics of sharp changing

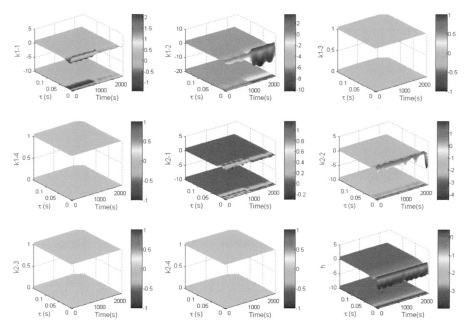

Fig. 3: The estimated first-order feedforward kernels for four inputs (k_{1-1}, k_{1-2}, k_{1-3}, k_{1-4}), second-order feedforward kernels for four inputs (k_{2-1}, k_{2-2}, k_{2-3}, k_{2-4}), and the feedback kernel (h) from a four-input, single-output time-varying nonlinear system with step changes based on the B-spline basis functions expansion method with OMP

nonstationary systems as a result of its poor convergence rate. Compared with SSPPF, the Chebyshev polynomials expansion algorithm can obtain relatively higher kernels estimation accuracy, while as a result of the global nature, satisfactory estimates cannot be achieved at the edge of the dataset. The performance of B-spline expansion algorithm with OMP is better than SSPPF and Chebyshev polynomial expansion method. The B-spline basis functions expansion method with OMP proposed in this paper can approximate the evolution of time-varying kernel smoothly and accurately and track the change of time-varying kernels quickly even at the starting point and end point for the spike trains.

The Mean Absolute Error (MAE) and the normalized Root Mean Square Error (RMSE) are utilized to quantitatively evaluate the estimation accuracy for the proposed method. Relative to the corresponding real values, the MAE and RMSE for the estimated kernels from the simulated spiking data are calculated, and are shown in Table 1. Compared with the SSPPF and Chebyshev polynomial expansion algorithms, the proposed B-spline basis functions expansion method with OMP has relatively smaller MAE and RMSE. The MAE and RMSE of the peak amplitude of the feedforward and feedback kernels can be given as:

$$\text{MAE} = \begin{cases} \dfrac{1}{T}\sum_{t=1}^{T}|\widehat{K(t)} - K(t)| \\[2mm] \dfrac{1}{T}\sum_{t=1}^{T}|\widehat{H(t)} - H(t)| \end{cases} \tag{29}$$

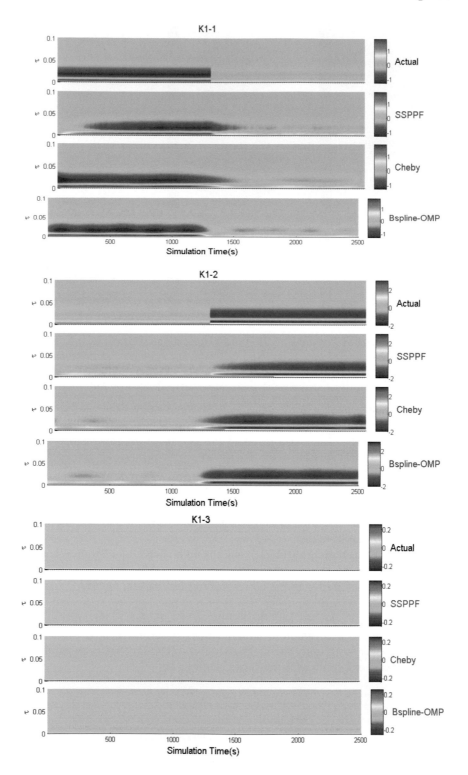

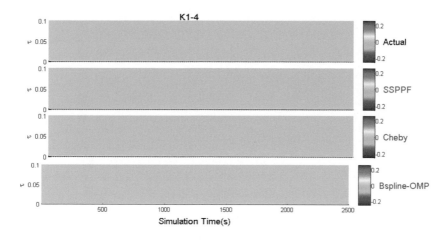

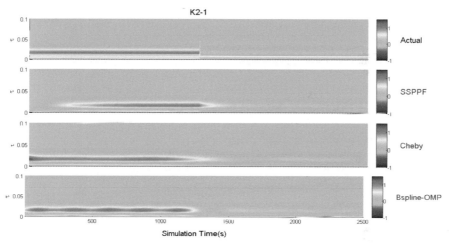

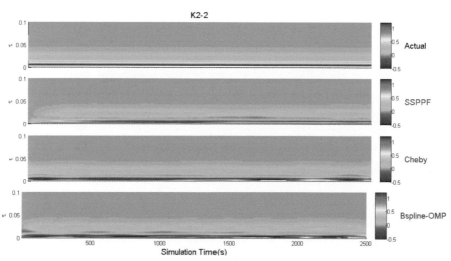

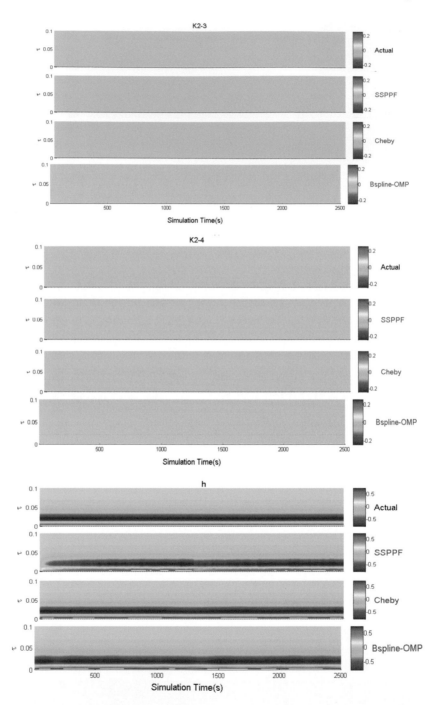

Fig. 4: Actual and estimated kernel shapes in 2-D for a second-order, four-input, single-output system with step changes based on three different approaches during the simulated time. The amplitudes of the kernel function are color-coded

Table 1: A performance comparison of nonstationary baseline estimations using different approaches

Performance Indicators	Estimated Kernels	Identification Methods		
		SSPPF	Cheby	B-spline + OMP
MAE	$\hat{k}_{1-1}$	0.1815	0.1163	0.0866
	$\hat{k}_{1-2}$	0.2222	0.2122	0.1088
	$\hat{k}_{1-3}$	0	0	0
	$\hat{k}_{1-4}$	0	0	0
	$\hat{k}_{2-1}$	0.1564	0.0691	0.0611
	$\hat{k}_{2-2}$	0.3007	0.1249	0.1454
	$\hat{k}_{2-3}$	0	0	0
	$\hat{k}_{2-4}$	0	0	0
	$\hat{h}$	0.0341	0.0112	0.0151
RMSE	$\hat{k}_{1-1}$	0.03657	0.2846	0.2321
	$\hat{k}_{1-2}$	0.2111	0.2152	0.1443
	$\hat{k}_{1-3}$	-		-
	$\hat{k}_{1-4}$	-		-
	$\hat{k}_{2-1}$	0.4987	0.2742	0.1886
	$\hat{k}_{2-2}$	0.2967	0.1117	0.1339
	$\hat{k}_{2-3}$	-		-
	$\hat{k}_{2-4}$	-		-
	$\hat{h}$	0.1498	0.0413	0.558

$$\text{RMSE} = \begin{cases} \sqrt{\dfrac{1}{T}\sum_{t=1}^{T} \dfrac{\left\|\widehat{K(t)} - K(t)\right\|^2}{\left\|K(t)\right\|^2}} \\[3ex] \sqrt{\dfrac{1}{T}\sum_{t=1}^{T} \dfrac{\left\|\widehat{H(t)} - H(t)\right\|^2}{\left\|H(t)\right\|^2}} \end{cases} \qquad (30)$$

where $K(t)$ and $H(t)$ represent the peak amplitudes for the actual feedforward and feedback kernels, and the peak amplitudes for the estimated feedforward and feedback kernels are given as $\widehat{K}(t)$ and $\widehat{H}(t)$, respectively. Table 1 illustrates the good performance of the proposed B-spline basis function expansion method with OMP. Simulation results show that compared with the traditional SSPPF, this proposed approach can achieve much higher estimation accuracy for different kernels.

The simulation results show that compared with the traditional adaptive filter estimation method (SSPPF), the B-spline basis functions expansion method with OMP can approximate the sharp and gradual changes in artificial spiking signal more accurately and clearly. The proposed multiwavelet expansion method makes the time-varying nonlinear models more suitable and flexible for capturing the transient changes of nonlinear and non-stationary signals only by observing the spiking input and output signals.

3.2   Simulated Kernels with Input-independent Random Changes

In the time-varying model, the zeroth-order time-varying kernel k_0 (*i.e.*, c_0) represents the input-independent nonstationarity, i.e. changes of the output firing probability provoked by potential factors other than the observed input spike trains. In order to test whether the algorithm can deal with this kind of nonstationarity well, we change the step-changes simulation of random noise sequence by changing the input independent baseline transmission probability (Fig. 5). The tracking results and the estimation error can be given respectively as Fig. 5 and Table 2. Results illustrate that the estimated baselines using the three methods (blue for SSPPF, green for Chebyshev polynomial expansion, and red for B-splines basis function expansion with OMP) can eventually converge to the actual baselines, while the proposed B-splines basis function expansion with OMP can obtain more fast tracking rate and a smaller estimation error.

Table 2: A Performance Comparison of Nonstationary Baseline Estimations Using Different Approaches

Identification Methods	MAE	RMSE
SSPPF	0.1044	0.004400
Cheby	0.0391	0.000272
B-spline	0.0277	0.000133

4.   Summary and Conclusion

This paper describes a novel sparse systems identification approach based on a multi-wavelet basis expansion scheme combined with an OMP algorithm for studying the time-varying nonlinear dynamical system from spiking activities, where the corresponding time-varying coefficients can be approximated by means of a set of multiple B-spline basis functions. The identification problem of the time-varying nonlinear systems is transformed into the linear system identification problem with time-invariant parameters, which can be solved by the OMP algorithm and

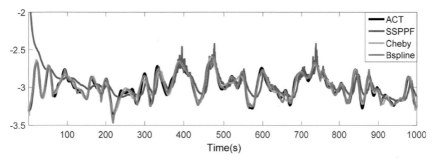

Fig. 5: Tracking the input-independent baseline k_0, which is a random noise sequence. The mean and low pass filtered of k_o are -3 and 0.001 Hz, respectively

generalized linear model fitting algorithm. The proposed computational scheme, based on multi-wavelet basis expansion, makes the TVGLV model more flexible and suitable for capturing the transient changes of non-stationary systems from spiking activities.

The results in the simulation show that the group-LASSO method can efficiently select the significant inputs and thus faithfully capture the sparsity of the system. In the next step, the multi-wavelet basis functions expansion method can be utilized to capture different changes for the feedforward and feedback kernels accurately, even for the nonlinear kernel terms. In addition, the OMP model terms selection method can select the important terms from a large number of candidate data while not remarkably increasing the computational complexity. The identification method proposed in this paper can, not only provide a calculation scheme for studying time-varying nonlinear characteristics and tracking more general forms of neuro-dynamic changes, but also reveal potential spiking transformations, which promote understanding of the transient information transmission across brain region evolution during behavior.

References

1. Abd-Elhameed, W.M. and Youssri, Y.H. (2018). Fifth-kind orthonormal Chebyshev polynomial solutions for fractional differential equations, *Computational and Applied Mathematics*, 37(3): 2897-2921.
2. Bloesch, M., Burri, M., Omari, S., Hutter, M. and Siegwart, R.Y. (2017). Iterated extended Kalman filter based visual-inertial odometry using direct photometric feedback, *The International Journal of Robotics Research*, 36(10): 1053-1072.
3. Cai, L., Zhang, Z., Yang, J., Yu, Y. and Qin, J. (2019). A noise-immune Kalman filter for short-term traffic flow forecasting, *Physica A: Statistical Mechanics and its Applications*, 536: 122601.
4. Chen, L., Huang, T., Machado, J., Lopes, A.M. and Wu, R. (2019). Delay-dependent criterion for asymptotic stability of a class of fractional-order memristive neural networks with time-varying delays, *Neural Networks*, 118: 289-299.
5. Cui, M. and Prasad, S. (2016). Sparse representation-based classification: Orthogonal least squares or orthogonal matching pursuit?, *Pattern Recognition Letters*, 84: 120-126.

6. Eden, U.T., Frank, L.M., Barbieri, R., Solo, V. and Brown, E.N. (2004). Dynamic analysis of neural encoding by point process adaptive filtering, *Neural Computation*, 16(5): 971-998.

7. Fan, J. and Li, R. (2001). Variable selection via nonconcave penalized likelihood and its oracle properties, *Journal of the American Statistical Association*, 96(456): 1348-1360.

8. Hong, Yiguang, Yuan, Deming, Daniel, W., *et al.* (2016). On convergence rate of distributed stochastic gradient algorithm for convex optimization with inequality constraints, *SIAM Journal on Control and Optimization*, 54(5): 2872-2892.

9. Hua, L.W., Billings, S.A. and Liu, J.J. (2010). Time-varying parametric modeling and time-dependent spectral characterization with application to EEG signal using multiwavelets, *International Journal of Modelling Identification and Control*, 9(3): 215-224.

10. Huang, J., Ma, S., Xie, H. and Cun-Hui, Z. (2009). A group bridge approach for variable selection, *Biometrika*, 96(2): 339-355.

11. Jian, P., Xiao, J. and Wan, X. (2017). A filtering based multi-innovation extended stochastic gradient algorithm for multivariable control systems, *International Journal of Control, Automation and Systems*, 15(3): 1189-1197.

12. Jiang, Z. Q. , Gao, X. L. , Zhou, W. X. and Stanley, H. E. (2017). Multifractal cross wavelet analysis, *Fractals*, 2017, 25(06): 1750054.

13. Kaltofen, E.L. and Yang, Z.H. (2020). Sparse Interpolation with Errors in Chebyshev Basis Beyond Redundant-Block Decoding, *IEEE Transactions on Information Theory*, 2021, 67(1): 232-243.

14. Kressner, D. and Roman, J.E. (2014). Memory-efficient Arnoldi algorithms for linearizations of matrix polynomials in Chebyshev basis, *Numerical Linear Algebra with Applications*, 21(4): 569-588.

15. Kuznetsov, D.F. (2018). Expansion of iterated Stratonovich stochastic integrals of multiplicity 2, based on double Fourier-Legendre series, summarized by Prinsheim method, arXiv preprint arXiv:180101962.

16. Lee, J., Gil, G.T. and Yong, H.L. (2016). Channel estimation via orthogonal matching pursuit for hybrid MIMO systems in millimeter wave communications, *IEEE Transactions on Communications*, 64(6): 2370-2386.

17. Li, Y., Liu, Q., Tan, S.R. and Chan, R. (2016). High-resolution time-frequency analysis of EEG signals using multiscale radial basis functions, *Neurocomputing*, 195: 96-103.

18. Li, Y., Lei, M.Y., Guo, Y.Z., Hu, Z.Y. and Wei, H.L. (2018). Time-varying nonlinear causality detection using regularized orthogonal least squares and multi-wavelets with applications to EEG, *IEEE Access*, 6: 17826-17840.

19. Ln, A., Dong, W.A., Vps, B., Jw, A., Yw, A. and Yt, A. (2020). Streamflow and rainfall forecasting by two long short-term memory-based models, *Journal of Hydrology*, 583: 124296.

20. Martel, F., Rancourt, D., Remond, D., Chochol, C., Chesne, S. and St-Amant, Y. (2015). Time-varying torsional stiffness identification on a vertical beam using Chebyshev polynomials, *Mechanical Systems and Signal Processing*, 54: 481-490.

21. Ming, Y. and Yi, L. (2006). Model selection and estimation in regression with grouped variables, *Journal of the Royal Statistical Society: Series B* (*Statistical Methodology*), 68(1): 49-67.

22. Park, D.S., Chan, W., Zhang, Y., Chiu, C.C., Zoph, B. and Cubuk, E.D. (2019). Specaugment: A simple data augmentation method for automatic speech recognition, arXiv preprint arXiv:190408779.

23. Phillips, R.C., Denise, G. and Mariapaz, E. (2018). Cryptocurrency price drivers: Wavelet coherence analysis revisited, *PloS One*, 13(4): e0195200.

24. Qi, S., Chen, G. and Chan, R. (2016). Evaluating the small-world-ness of a sampled network: Functional connectivity of entorhinal-hippocampal circuitry, *Scientific Reports*, 6: 21468.

25. Song, D., Chan, R., Robinson, B.S., Marmarelis, V.Z., Opris, I. and Ham (2015). Identification of functional synaptic plasticity from spiking activities using nonlinear dynamical modeling, *Journal of Neuroscience Methods*, 244: 123-135.

26. Stafford, B.K., Sher, A., Litke, A.M. and Feldheim, D.A. (2009). Spatial-temporal patterns of retinal waves underlying activity-dependent refinement of retinofugal projections, *Neuron*, 64(2): 200-212.

27. Tibshirani, R. (1996). Regression shrinkage and selection via the lasso, *Journal of the Royal Statistical Society Series B (Methodological)*, 58(1): 267-288.

28. Wang, D., Li, L., Yan, J. and Yan, Y. (2016). Model recovery for Hammerstein systems using the auxiliary model based orthogonal matching pursuit method, *Applied Mathematical Modelling*, 54: 537-550.

29. Wang, D., Yan, Y., Liu, Y. and Ding, J. (2019). Model recovery for Hammerstein systems using the hierarchical orthogonal matching pursuit method, *Journal of Computational and Applied Mathematics*, 345: 135-145.

30. Wei, Y., Kang, Y., Yin, W. and Wang, Y. (2020). Generalization of the gradient method with fractional order gradient direction, *Journal of the Franklin Institute*, 357(4): 2514-2532.

31. Wei, Y., Zhang, K., Zhang, H., Xu, X. and Hu, L. (2018). Laguerre-polynomial-weighted squeezed vacuum: Generation and its properties of entanglement, *Laser Physics Letters*, 15(2): 025204.

32. Wen, J., Wang, J. and Zhang, Q. (2017). Nearly optimal bounds for orthogonal least squares, *IEEE Transactions on Signal Processing*, 65(20): 5347-5356.

33. Windeatt, T. and Zor, C. (2013). Ensemble pruning using spectral coefficients, *IEEE Transactions on Neural Networks and Learning Systems*, 24(4): 673-678.

34. Xu, S., Li, Y. and Huang, T. (2017). A sparse multiwavelet-based generalized laguerre-volterra model for identifying time-varying neural dynamics from spiking activities, *Entropy*, 19(8): 425.

35. Xu, S., Li, Y. and Guo, Q. (2017). Identification of time-varying neural dynamics from spike train data using multiwavelet basis functions. *Journal of Neuroscience Methods*, 278(46-56).

36. Xu, S., Li, Y., Wang, X. and Chan, R. (2016). Identification of time-varying neural dynamics from spiking activities using Chebyshev polynomials. *Proceedings of the 38th Annual International Conference of the IEEE Engineering in Medicine and Biology Society (EMBC)*, IEEE.

37. Yang, Li, Mei-Lin, Luo, Ke, and Li. (2016). A multiwavelet-based time-varying model identification approach for time–frequency analysis of EEG signals, *Neurocomputing*, 2016, 193(106-14).

38. Yoo, J., Uh, Y., Chun, S., Kang, B. and Ha, J.W. (2019). Photorealistic style transfer via wavelet transforms, *Proceedings of the IEEE International Conference on Computer Vision*, 9035-9044, doi: 10.1109/ICCV.2019.00913.

39. Zhang, D. (2019). Wavelet transform, *Fundamentals of Image Data Mining*, 35-44. Springer.

40. Zümray Dokur and Lmez, T. (2020). Heartbeat classification by using a convolutional neural network trained with Walsh functions, *Neural Computing and Applications*, 1-20.

How Weather Conditions Affect the Spread of COVID-19: Findings of a Study Using Contrastive Learning and NARMAX Models

Yiming Sun[1] and Hua-Liang Wei[1,2]*

[1] Department of Automatic Control and System Engineering, University of Sheffield, Sheffield, S1 3JD, UK
[2] INSIGNEO Institute for in Silico Medicine, University of Sheffield, Sheffield, S1 3JD, UK
E-mail: w.hualiang@sheffield.ac.uk

1. Introduction

Machine learning (ML) has demonstrated its powerful ability in learning complex patterns or inherent dynamics from observed data and has drawn considerable attention for its ability to predict complex phenomenon [19]. Most machine learning models are black-box in which the internal behavior of the models is opaque and thus unknown to either the model builders or the end-users. However, in many applications, there is a high demand for the model's accountability or explanability. This means that in addition to accurate prediction, ML should also tell, from the observed information and knowledge, the domain relationships contained in the data. This is referred to as interpretability [8]. In many application domains, for example, health care and medicine [14, 30], policymaking [15], and material design [36], researchers are concerned with not only the forecast of machine learning systems but also the explanation of machine learning models, or the relationship between the system outputs and inputs so as to make reliable decisions and provide clear guidance according to the explanation [17]. In healthcare and medicine, the interpretability of models and the interactions of system variables, usually treated as a prerequisite of using ML models as the 'reason' behind the prediction by ML models, is usually most desirable and useful for making important decisions [38].

Coronavirus (COVID-19), the new global pandemic and the latest largest threat to global health, has been the focus of the past year [21]. There were totally over 192 million confirmed cases and over 4 million deaths according to the World Health Organization, while the new confirmed cases and deaths still keep rising up [22].

Thus, besides the prediction of new cases, it is vital for us to better understand the main factors that may influence the spread of the COVID-19 virus [26]. As at the first stage of the pandemic, it was widely hoped that higher temperature would slow the spread of the virus last summer [32], but the reality shows that not only the temperate but also other factors may lead to a significant impact on the spread of the pandemic [42].

Since the start of the pandemic, a huge amount of research related to the prediction of new cases and spread have been carried out. A variety of methods have been proposed for predicting the number of new cases. These methods include ML, such as deep learning [5], support vector machine [13], fuzzy system [31], neural networks [20], and dynamical Bayesian models [33]. However, as mentioned earlier, most ML models can only generate a prediction of new cases without providing any information on the inner relationship of relevant factors.

This study introduces a novel interpretable machine learning method based on Contrastive Learning and Non-linear AutoRegressive Moving Average with eXogenous inputs (CL-NARMAX) model for medical data analysis. Unlike most ML methods which are the black-box, the proposed method provides a glass-box model, where the input-output relationship and interactions between the input variables can be explained. This implies that the model cannot only be applied for predicting future new cases and deaths but also provide an explanation on how the spread of the pandemic is affected by different climatic and weather variables. Such a model may be used for better understanding of the pandemic dynamics and for carrying out further studies in future.

The remainder of the paper is as follows. In Section 2, the NARMAX model structure is described in detail. In Section 3, the novel CL-NARMAX method is presented. In Section 4, two case studies are provided with one focusing on modeling and analyzing weather conditions against the COVID-19 data in the UK and France and the other concerning the relationship between influenza-like illness (ILI) incidence rate and the relevant mortality. The work is briefly summarized in Section 5.

2. NARMAX Model

2.1 The Regression Model

For convenience of description, take NARX model (as a special case of NARMAX) as an example and consider a multivariate regression problem, with n predictor variables, $x_1, x_2,...,x_n$, and one response variable y. The modelling task is to investigate the quantitative dependent relationship of the response on the predictors. Mathematically, the objective is to establish a model that links the predictors to the response via a function f as follows:

$$y(t) = f(x_1(t), x_2(t),...,x_n(t)) + e(t) \qquad (1)$$

where $x_i(\cdot)(i = 1, 2,..., n)$ and $y(\cdot)$ represent the sequence of the observed predictor and response variables, respectively, $e(t)$ represents the model error; f represents some linear or non-linear functions. Usually, f is unknown but can be approximated from given observational data. There are a diversity of methods for building a function to approximate the true system, such as polynomials [12, 39], radial basis functions [2], and wavelet functions [7, 10, 18, 3]. In this study, a polynomial-based regression model is considered due to its superb properties [37]. By applying the polynomial form with the non-linear degree of up to ℓ, model (1) can be represented as:

$$y(t) = \theta_0 + \sum_{i_1=1}^{d} f_{i_1}(x_{i_1}(t)) + \sum_{i_1=1}^{d} \sum_{i_2=i_1+1}^{d} f_{i_1 i_2}(x_{i_1}(t), x_{i_2}(t))$$

$$+ \sum_{i_1=1}^{d} \cdots \sum_{i_m=i_{m-1}+1}^{d} f_{i_1 i_2 \ldots i_m}(x_{i_1}(t), x_{i_2}(t), \ldots, x_{i_m}(t)) + e(t) \quad (2)$$

where $\theta_{i_1 i_2 \ldots i_l}$ are parameters, $d = n_y + n_u$ and

$$f_{i_1 i_2 \ldots i_m}(x_{i_1}(t) x_{i_2}(t), \ldots, x_{i_m}(t)) = \theta_{i_1 i_2 \ldots i_m} \times \prod_{k=1}^{m} x_{i_k}(t), \quad 1 \le m \le \ell \quad (3)$$

The degree of the multivariate polynomial is defined as the highest order among the terms. For example, the non-linear degree of the polynomial term $x_1(t)$ is 1, $x_1(t) x_2(t)$ is 2 and $x_1^2(t) x_2^3(t)$ is 5. Therefore, the degree of any term in model (2) is not higher than ℓ. In most practical implementations, the non-linear mapping function f can be approximated by a linear combination of a predefined set of functions, $\phi_i(\varphi(k))$. Note that the polynomial NARX model described by Equation (2) can be written as the following linear-in-the-parameters form:

$$y(t) = \sum_{m=1}^{M} \theta_m \varphi_m(t) + e(t) \quad (4)$$

where $\varphi_m(t)$ are the model terms generated from the candidate variables, e.g. a model with the non-linear degree $\ell = 2$, involving only two variables, contains the following six terms: $\varphi_0(t) = 1$, $\varphi_2(t) = x_1(t)$, $\varphi_3(t) = x_2(t)$, $\varphi_4(t) = x_1^2(t)$, $\varphi_5(t) = x_1(t) x_2(t)$, $\varphi_6(t) = x_2^2(t)$.

2.2 NARMAX Model Structure

Taking the case of a single-input and single-output system as an example, the NARMAX model is written as [3]:

$$y(t) = f(y(t-1), \ldots, y(t-n_y), u(t-1), \ldots, u(t-n_u),$$

$$e(t-1), \ldots, e(t-n_e)) + e(t) \quad (5)$$

where $u(t)$, $y(t)$, and $e(t)$ are the measured system input, output, and noise signal respectively at time instant t; n_u, n_y, and n_e are the maximum lags for the system input, output, and noise; $f(\cdot)$ is some non-linear function to be identified. The NARMAX model (5) can be accommodated in the linear-in-the-parameters form (4) by defining $\varphi_m(t)$ as:

$$x_k(t) = \begin{cases} y(t-m) & 1 \le m \le n_y \\ u(t-m+n_y) & n_y+1 \le m \le n_y+n_u \\ e(t-m+n_y+n_u) & n_y+n_u+1 \le m \le n \end{cases} \tag{6}$$

where $n = n_y + n_u + n_e$. Normally the noise signal $e(t)$ in model (5) is unmeasurable. Therefore, $e(t)$ is often replaced by the model residual sequence in model identification procedure:

$$e(t) = \varepsilon(t) = y(t) - \hat{y}(t) \tag{7}$$

where $\hat{y}(t)$ is the predicted value at time instant t generated by an estimated model. Detailed information about how to calculate model parameters and update model errors can be found in [40].

2.3 NARMAX Model Term Selection and Estimation

The initial linear-in-the-parameters model (4) may involve a large number of candidate model terms. However, only a small number of significant model terms are necessary in the final model to represent the given observed data in most cases. Most candidate model terms are either redundant or make minimal contribution to the system output and can therefore be removed from the model. Thus, efficient model term selection and estimation method is needed.

The forward regression orthogonal least squares (FROLS) algorithms [20, 26] provide an efficient and powerful method for non-linear model term selection and model structure estimation. More detailed discussion of the FROLS algorithm and ERR index can be found in [11, 9]. In this paper, we only give a very brief summary of the algorithm. FROLS searches in a set consisting of all the specified possible candidate model terms or regressors to select the most significant model terms iteratively, in a stepwise manner, through an orthogonalization procedure, where the significance of model terms is measured by an index called the error reduction ratio (ERR) [37]. The FROLS algorithm will stop once the specified conditions are met, e.g. when the Error Signal Ratio (ESR) or index reaches its minimum [37]. Also, some statistical criteria, e.g. AIC [16], BIC [35], APRESS [34], can be used to monitor the model selection procedure and determine the model complexity.

3. Contrastive Learning Enhanced NARMAX Method

3.1 Contrastive Learning

Contrastive learning (CL), as a representative learning approach, has delivered impressive results on various scenarios [1]. This self-supervised training process can be understood as learning representation through contrastive positive data pairs against negative ones, where separate encoders could achieve the representation [23]. As the pre-definition or categorization for the candidate encoders is unnecessary, the process of CL is more effective and flexible.

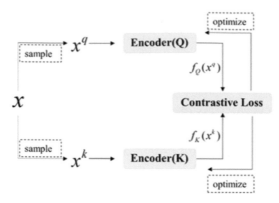

Fig. 1: Contrastive learning

The general training process of contrastive learning is shown in Fig. 1. For general CL process, the input samples will be re-sampled into two independent parts: positive samples and negative samples by the related labels [24]. Normally, positive and negative samples are vital to the process of CL, while it is also possible to use the entire data set rather than to split it [25]. Note that the encoder Q generates representation for positive samples x^q and encoder K generates representation for negative samples x^k. When the contrastive loss between outputs of two encoders converges, positive samples become closer, while negative samples get less similar.

Normally, contrastive learning focuses on comparing the embeddings with a Noise Contrastive Estimation (NCE) function that is defined as [1]:

$$L_{NCE} = -\log \frac{\exp(cos\_sim(q,k_+)/\tau)}{\exp(cos\_sim(q,k_+)/\tau) + \exp(cos\_sim(q,k_-)/\tau)} \tag{8}$$

$$cos\_sim(A,B) = \frac{A \cdot B}{\|A\|\|B\|} \tag{9}$$

where q is the original output, $k_+ = f_Q(x^q)$ represents the prediction by Encoder Q with the positive sample x^q, and $k_- = f_K(x^k)$ represents the prediction by Encoder K with the negative sample x^k, τ is a hyperparameter. More information of contrastive learning can be found in [27].

3.2 The CL-NARMAX Model

The scheme of the proposed framework is shown in Fig. 2. Unlike the general CL process, the proposed CL-NARMAX modeling shares the same dataset $X_{D \times n}$ by two modeling frameworks, the true system (considering the true system as a model), and NARMAX models. The outputs of the two are models: $y = F(\cdot)$ (observed system output) and $\tilde{y}_l = F_{NARX}^l (l = 1,2,...,L)$ (model output), respectively. The system output y is set as the positive pair, and the outputs by NARMAX models $\hat{y}_l(l = 1,2,...,L)$ are set to be the negative pairs.

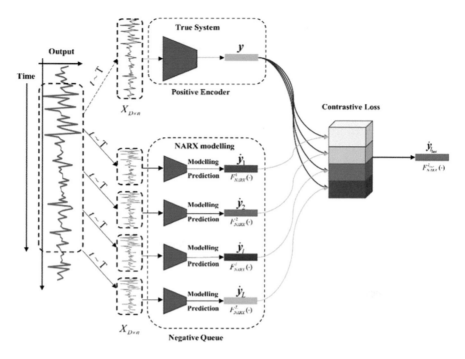

Fig. 2: The general process of CL-NARMAX

Notice that the identified models F_{NARX}^l and predicted values $\tilde{y}_l$ by NARMAX vary due to the different choices of the parameters. Thus, the group generated by the NARMAX should be a set of possible options, which could also be called the negative queue. Then, for each negative pair in the negative queue, there will be a related contrastive loss with the positive pair, which is defined as follows:

$$CL_{NCE}^l = -\log \frac{\exp(h(y, y))}{\sum_{l=1}^{L} \exp(h(y, y)) + \exp(h(y, \tilde{y}_l))} \tag{10}$$

where $h(u, v) = \exp(u^T v / \tau \cdot \|u\| \cdot \|v\|)$ is used to compute the similarity between u and v vectors with an adjustable parameter temperature, τ. Hence, the most similar negative pair compared with the positive pair is chosen as:

$$l_{best} = \arg \max_{1 \le l \le L} CL_{NCE}^l \tag{11}$$

Where $L = n_y \times n_u \times \ell$, n_u and n_y are the maximum input and output lags of the NARMAX models; ℓ is the maximum non-linear degree of the NARMAX models. In this study here, the most effective NARMAX model, $F_{Nl_{best}}$ that generates the closest output $\tilde{y}_{\ell, best}$ to the corresponding measurement is included in the negative queue.

4. Case Studies

4.1 Modeling and Analyzing Weather Conditions against Daily Confirmed COVID-19 Cases in the UK and France

4.1.1 Data

The daily confirmed cases for UK and France between 1st January 2020 and 27th July 2020 were acquired from the World Bank's World Development Indicators database [28] which conveys the first wave of the pandemic. The dataset contains a total of 213 daily confirmed cases. Though there were no confirmed cases on 1st January in UK and France, that day was still assumed to be the starting time for the pandemic and the raw data are plotted in Fig. 1. Meanwhile, the daily meteorological data and the daily mobility data were collected from the DELVE program [, 28] and Apple mobility reports data, respectively. The definition of the five climatic factors and two mobility variables, treated as input variables in this study, is summarized in Table 1.

A total of 213 data points for each country are split into two parts: the first 180 samples are used for model training and validation, while the remaining 33 points are used for model testing. Note that each input variable does not impact the spread of the coronavirus immediately. A previous study shows that their impacts become obvious after seven days until 14 days [29]. The variation of the confirmed cases (denoted by y) is treated to be a dynamic process in this study. Therefore, a number of delayed output variables (autoregressive variables), namely, $y(t-1)$, $y(t-2)$,…, $y(t-7)$, are also included in the model. Thus, to build an CL-NARMAX model, for inputs variables, the value of n_u varies from 7 to 14, while for output, n_y vary from 1 to 7. In addition, cross-product variables are also included in the model to reflect the impacts of the interactions between input and output variables on the spread of the virus. It argues that the impact of higher order interactions between input and output variables (e.g. higher than 3) on the pandemic becomes insignificant [41]. Thus, the degree ℓ of CL-NARMAX model is set to be 2.

Table 1: Climate and mobility variables

Variable	Symbol	Definition	Unit
Temperature	T	Average daily mean of temperatures	Celsius Degrees
Humidity	H	Average daily humidity of air	Kilograms of water vapor per kilogram
Wind speed	Ws	Average daily wind speed	Meters per second
Solar radiation	Sr	Average daily short-wave radiation	W/m² (Watts per square meter)
Precipitation	P	Average daily precipitation	mm/hr
Driving	Dr	Percentage of change in routing requests by driving based on 100	N/A
Walking	W	Percentage of change in routing requests by walking based on 100	N/A

4.1.2 Results

The proposed CL-NARMAX model is applied to the training data of the two counties, and the identified best CL-NARMAX model for the UK case is:

$$
\begin{aligned}
y(t) = {} & 0.9104\,y(t-1) + 0.0062 Dr(t-14)y(t-6) \\
& - 96.0698 H(t-7)Sr(t-12) - 0.0187 Dr(t-8) \\
& + 0.0126 Dr(t-6) + 0.0053 W(t-11)
\end{aligned}
\tag{12}
$$

and the best CL-NARMAX model for France's case is:

$$
\begin{aligned}
y(t) = {} & 0.1365 P(t-9) + 0.0017 Sr(t-10)y(t-2) \\
& + 0.0689 P(t-9)y(t-3) - 0.0089 Sr(t-13) \\
& + 1.8559\,y(t-6)
\end{aligned}
\tag{13}
$$

Note that all the model terms involving noise variables, such as $Sr(t-1)e(t-1)$ are omitted and not included in the final model, because all these noise terms are not useful for model prediction but are only used to reduce bias in model estimation. A comparison of the model predicted daily cases and the corresponding true values, on the training and test datasets, are shown in Figs. 4 and 5, respectively. Several statistical indicators are used for measuring the performance of the proposed method on training and test dataset, such as MSE (mean-squared-error), RMSE (root-mean-squared-error), MAE (mean absolute error), and R2 (coefficient of determination), where the values are listed in Table 2.

Model (12) (for the UK data) depends on the following four climatic factors: humidity, solar radiation, driving, and walking. More specifically, it shows that climate factors and mobility factors of one week ago have more influence on the present new cases. Similarly, model (13) shows that in France, the daily new cases is correlated to precipitation, solar radiation, and past cases, where only climate factors of one week ago make an impact on the present cases. From Figs. 4, and 5 and Table 2, it can be seen that the proposed CL-NARMAX model shows an excellent prediction performance.

Table 2: Statistical indicators of proposed method for case 1

	MSE		RMSE		MAE		R2	
	UK	FR	UK	FR	UK	FR	UK	FR
Train	0.016	0.117	0.127	0.342	0.080	0.188	0.984	0.883
Test	0.100	0.569	0.998	0.754	0.677	0.407	0.974	0.414

4.2 The Relationship between Influenza-like Illness (ILI) Incidence Rate and Relevant Mortality

4.2.1 Data

The weekly influenza-like illness (ILI) incidence rate and deaths data of England were obtained from the Office for National Statistics (ONS), The Royal College of General Practitioners Research and Surveillance Centre and Public Health, Wales.

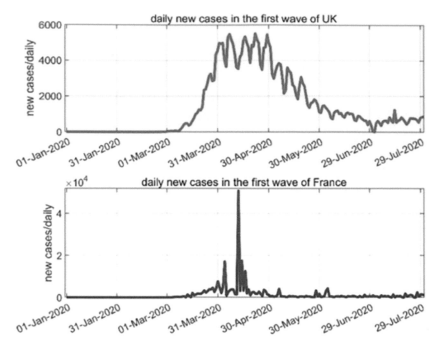

Fig. 3: Daily confirmed cases in the first wave in UK and France

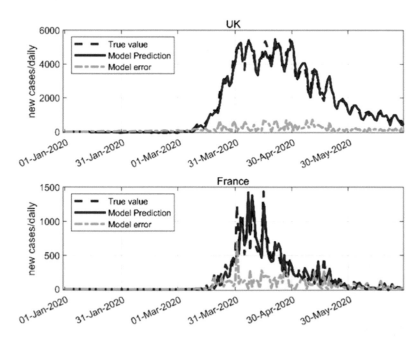

Fig. 4: The comparison of the model prediction with related true daily confirmed cases of UK and France on the training dataset

The dataset contains 835 weekly data points, starting in week 31 of 2004 and ending in week 30 of 2020. The raw data are plotted in Fig. 6.

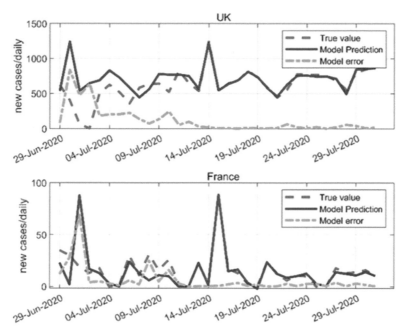

Fig. 5: The comparison of the model prediction with related true daily confirmed cases of UK and France on the test dataset

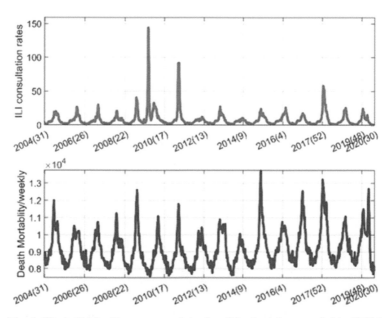

Fig. 6: Weely ILI incidence rate and deaths of England from week 31 of 2004 to week 30 of 2020

Following the previous studies [4, 6], the relationship between ILI incidence rate and weekly deaths is treated as a single-input single-output (SISO) system. Thus, in this problem, there is only one input variable, which is weekly ILI incidence rate. The number of weekly deaths is the system output. The initial dataset is split into two parts: the first 600 data points are for training and validation and the remaining 235 are for test.

4.2.2 Results

The identified predictive model by the proposed CL-NARMAX method is as follows:

$$y(t) = 1.0027y(t-1) - 0.0037u(t-3)y(t-1)$$
$$+28.0887u(t-3) + 0.0005u(t-1)y(t-1) \tag{14}$$

where $u(t)$ represents the weekly ILI incidence rate and $y(t)$ represents the number of weekly deaths. Similarly, there is no noise variables in model (14). A comparison between the model prediction and the corresponding observations on the training and test dataset are shown in Figs. 7 and 8. Similarly, several statistical indicators, such as MSE, RMSE, MAE, and R2, are used on training and test dataset, where the values are listed in Table 3.

From Figs. 7 and 8 and Table 3, it can be seen that the identified CL-NARMAX model (14) shows an excellent prediction performance. More importantly, model (14) shows that the number of weekly deaths in the present week is closely related to the weekly ILI incidence rate one week and three weeks ago.

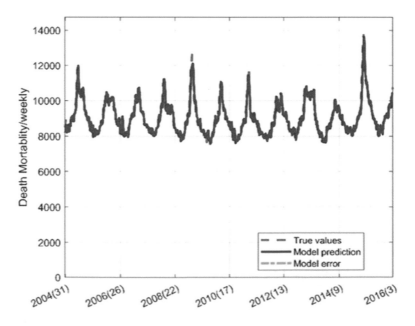

Fig. 7: A comparison between the model prediction and the corresponding observations on the training dataset

Table 3: Statistical indicators of proposed method for case 2

	MSE	RMSE	MAE	R2
Train	0.0936	0.3059	0.2318	0.9062
Test	0.0914	0.3023	0.2303	0.9082

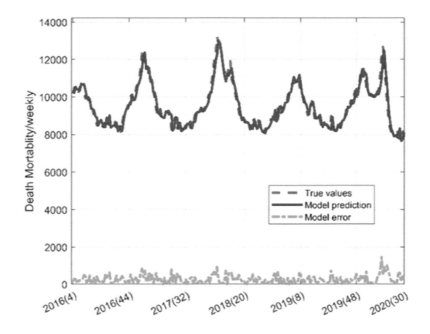

Fig. 8: A comparison between the model prediction and the corresponding observations on the training dataset

5. Conclusion

This paper focuses on presenting a novel interpretable machine learning method based on contrastive learning and NARMAX model (CL-NARMAX) for solving healthcare data modeling problems. The proposed method can, not only provide excellent predictions, but also possesses the ability to explain how the output variable (response) is linked to the most important input variables (explanatory variables) and their interactions. The main contributions of the work are as follows: firstly, the proposed CL-NARMAX method takes advantage of contrastive learning and the NARMAX method, significantly improving the prediction ability of NARMAX models and meanwhile maintaining the distinct and attractive properties of NARMAX model, e.g. the 'SIT' (sparse, interpretable, and transparent) and 'SMART' (simple and simulatable, meaningful, accountable, reproducible, and transparent) properties [4, 6], which are significantly important and highly desirable in many real applications. Secondly, based on the data of the UK and France daily confirmation cases (both for the first wave), two models were provided, demonstrating how the climatic and weather conditions affected the spread of the coronavirus. These models can be useful for understanding the spread of the coronavirus.

This paper does not analyze the model uncertainty and its effect on model generalization performance. Meanwhile, along with new variants and the development of vaccine since the start of second wave of the pandemic, more possible factors affecting the spread of the virus shall be considered. Therefore, in future, the model uncertainty of the proposed model will be further studied to improve the performance and robustness of this approach. Also, more candidate input variables will be investigated to develop a more comprehensible model for the spread of the virus.

Acknowledgements

This work was supported in part by the Natural Environment Research Council (NERC) under the Grant NE/V001787 and Grant NE/V002511, the Engineering and Physical Sciences Research Council (EPSRC) under Grant EP/I011056/1 and the Platform Grant EP/H00453X/1.

References

1. Aguirre, L.A. and Billings, S.A. (1995). Dynamical effects of overparametrization in nonlinear models, *Physica D: Nonlinear Phenomena*, 80(1-2): 26-40.
2. Ahmad, M.A., Eckert, C. and Teredesai, A. (2018). Interpretable Machine Learning in Healthcare. *In: Proceedings of the 2018 ACM International Conference on Bioinformatics, Computational Biology, and Health Informatics*, pp. 559-560. Association for Computing Machinery: Washington, DC, USA.
3. Akaike, H. (1987). *Factor Analysis and AIC, in Selected papers of Hirotugu Akaike*, pp. 371-386. Springer.
4. Alazab, M., Awajan, A., Mesleh, A., Abraham, A., Jatana, V. and Alhyari, S. (2020). COVID-19 prediction and detection using deep learning, *International Journal of Computer Information Systems and Industrial Management Applications*, 12: 168-181.
5. Alimadadi, A., Aryal, S., Manandhar, I.., Munroe, P.B., Joe, B. and Cheng, X. (2020). Artificial intelligence and machine learning to fight COVID-19, *Physiol Genomics*, 52(4): 200-202.
6. Ashish, J., Ashwin Ramesh, B., Mohammad Zaki, Z., Debapriya, B. and Fillia, M. (2021). A survey on contrastive self-supervised learning, *Technologies* (Basel), 9(1): 2.
7. Babukarthik, R.G., Adiga, V.a.K., Sambasivam, G., Chandramohan, D. and Amudhavel, J. (2020). Prediction of COVID-19 Using Genetic Deep Learning Convolutional Neural Network (GDCNN), *IEEE Access*, 8: 177647-177666.
8. Bhoopchand, A., Paleyes, A., Donkers, K., Tomasev, N. and Paquet, U. (2020). *DELVE Global COVID-19 Dataset*, The Royal Society.
9. Billings, S. and Wei, H.L. (2019). NARMAX model as a sparse, interpretable and transparent machine learning approach for big medical and healthcare data analysis. *In: 2019 IEEE 21st International Conference on High Performance Computing and Communications*; IEEE 17th International Conference on Smart City; IEEE 5th International Conference on Data Science and Systems (HPCC/SmartCity/DSS), IEEE.

10. Billings, S.A., Wei, H.-L. and Balikhin, M.A. (2007). Generalized multiscale radial basis function networks, *Neural Netw.*, 20(10): 1081-1094.

11. Billings, S.A. and Wei, H.L. (2005). The wavelet-NARMAX representation: A hybrid model structure combining polynomial models with multiresolution wavelet decompositions, *International Journal of Systems Science*, 36(3): 137-152.

12. Billings, S.A. (2013). *Nonlinear System Identification: NARMAX Methods in the Time, Frequency, and Spatio-temporal Domains*, Chichester, England: Wiley.

13. Billings, S.A. and Wei, H.-L. (2008). An adaptive orthogonal search algorithm for model subset selection and non-linear system identification, *International Journal of Control*, 81(5): 714-724.

14. Brennan, T. and Oliver, W.L. (2013). The emergence of machine learning techniques in criminology, *Criminology and Public Policy*, 12(3): 551-562.

15. Bukhari, Q., Massaro, J.M., D'agostino, S.R.B. and Khan, S. (2020). Effects of Weather on Coronavirus Pandemic, *Int. J. Environ. Res. Public Health*, 17(15): 5399.

16. Burnham, K.P. and Anderson, D.R. (2004). Multimodel inference: Understanding AIC and BIC in model selection, *Sociological Methods & Research*, 33(2): 261-304.

17. Chen, T., Kornblith, S., Norouzi, M. and Hinton, G. (2020). A Simple Framework for Contrastive Learning of Visual Representations. *In:* D. Hal, III and S. Aarti (Eds.). *PMLR: Proceedings of 37th International Conference on Machine Learning Research*, pp. 1597-1607.

18. Chen, S., Billings, S.A. and Luo, W. (1989). Orthogonal least squares methods and their application to non-linear system identification, *International Journal of Control*, 50(5): 1873-1896.

19. Cruz-Cano, R., Ma, T., Yu, Y., Lee, M. and Liu, H. (2021). Forecasting COVID-19 cases based on social distancing in Maryland, U.S.A.: A time-series approach, *Disaster Med. Public Health Prep.*, pp. 1-4.

20. Damette, O., Mathonnat, C. and Goutte, S. (2021). Meteorological factors against COVID-19 and the role of human mobility, *PloS One*, 16(6): e0252405-e0252405.

21. Du, M., Liu, N. and Hu, X. (2019). Techniques for interpretable machine learning, *Communications of the ACM*, 63(1): 68-77.

22. Guidotti, R., Monreale, A., Ruggieri, S., Turini, F., Giannotti, F. and Pedreschi, D. (2019). A survey of methods for explaining black-box models, *ACM Computing Surveys*, 51(5): 1-42.

23. Hassan, S.A., Sheikh, F.N., Jamal, S., Ezeh, J.K. and Akhtar, A. (2020). Coronavirus (COVID-19): A review of clinical features, diagnosis, and treatment, *Cureus*, 12(3): e7355-e7355.

24. Jha, P.K., Cao, L. and Oden, J.T. (2020). Bayesian-based predictions of COVID-19 evolution in Texas using multispecies mixture-theoretic continuum models, *Computational Mechanics*, 66(5): 1055-1068.

25. Litjens, G., Kooi, T., Bejnordi, B.E., Setio, A.a.A., Ciompi, F., Ghafoorian, M., Van Der Laak, J.a.W.M., Van Ginneken, B. and Sánchez, C.I. (2017). A survey on deep learning in medical image analysis, *Med. Image Anal.*, 42: 60-88.

26. Liu, Y., Li, Z., Pan, S., Gong, C., Zhou, C. and Karypis, G. (2021). Anomaly detection on attributed networks via contrastive self-supervised learning, *IEEE Trans Neural Netw. Learn Syst.*, pp. 1-15.

27. Lolli, S., Chen, Y.-C., Wang, S.-H. and Vivone, G. (2020). Impact of meteorological conditions and air pollution on COVID-19 pandemic transmission in Italy, *Sci. Rep.*, 10(1): 16213.

28. Mandal, M., Jana, S., Nandi, S.K., Khatua, A., Adak, S. and Kar, T.K. (2020). A model based study on the dynamics of COVID-19: Prediction and control, *Chaos Solitons Fractals*, 136: 109889-109889.

29. Hannah, R., Edouard, M., Lucas, R.-G., Cameron, A., Charlie, G., Esteban, O., Joe, H., Bobbie, M., Diana, B. and Max, R. (2020). Coronavirus Pandemic (COVID-19). Published online at OurWorldInData.org. Retrieved from: https://ourworldindata.org/coronavirus [Online Resource].

30. Pinter, G., Felde, I., Mosavi, A., Ghamisi, P. and Gloaguen, R. (2020). COVID-19 Pandemic prediction for Hungary: A hybrid machine learning approach, *Mathematics*, 8(6): 890.

31. Qin, S., Mudur, N. and Pehlevan, C. (2021). Contrastive similarity matching for supervised learning, *Neural Computation*, 33(5): 1300-1328.

32. Rao, H., Xu, S., Hu, X., Cheng, J. and Hu, B. (2021). Augmented skeleton based contrastive action learning with momentum LSTM for unsupervised action recognition, *Information Sciences*, 569: 90-109.

33. Vu, M.a.T., Adalı, T., Ba, D., Buzsáki, G., Carlson, D., Heller, K., Liston, C., Rudin, C., Sohal, V.S., Widge, A.S., Mayberg, H.S., Sapiro, G. and Dzirasa, K. (2018). A shared vision for machine learning in neuroscience, *J. Neurosci.*, 38(7): 1601-1607.

34. Wei, H.L. and Billings, S.A. (2004). A unified wavelet-based modelling framework for non-linear system identification: The WANARX model structure, *International Journal of Control*, 77(4): 351-366.

35. Wei, H.L. and Billings, S.A. (2006). Long-term prediction of non-linear time series using multiresolution wavelet models, *International Journal of Control*, 79(6): 569-580.

36. Wei, H.L., Billings, S.A., Yifan, Z. and Lingzhong, G. (2009). Lattice dynamical wavelet neural networks implemented using particle swarm optimization for spatio-temporal system identification, *IEEE Trans Neural Netw.*, 20(1): 181-185.

37. Wei, H.L., Billings, S.A., Zhao, Y.F. and Guo, L.Z. (2010). An adaptive wavelet neural network for spatio-temporal system identification, *Neural Netw.*, 23(10): 1286-1299.

38. Wei, H.L., Billings, S.A. and Liu, J. (2004). Term and variable selection for non-linear system identification, *International Journal of Control,* 77(1): 86-110.

39. Wei, H.L. and Billings, S. (2008). Model structure selection using an integrated forward orthogonal search algorithm assisted by squared correlation and mutual information, *Int. J. Model. Identif. Control.*, 3: 341-356.

40. Wei, H.L. (2019). *Sparse, Interpretable and Transparent Predictive Model Identification for Healthcare Data Analysis*, pp. 103-114. Springer International Publishing: Cham.

41. Wu, H., Ruan, W., Wang, J., Zheng, D., Liu, B., Geng, Y., Chai, X., Chen, J., Li, K., Li, S. and Helal, S. (2021). Interpretable machine learning for COVID-19: An empirical study on severity prediction task, *IEEE Transactions on Artificial Intelligence*, pp. 1-1.

42. Yadav, U., Pathrudkar, S. and Ghosh, S. (2021). Interpretable machine learning model for the deformation of multiwalled carbon nanotubes, *Physical Review B*, 103(3): 035407.

New Measurement of the Body Mass Index with Bioimpedance Using a Novel Interpretable Takagi-Sugeno Fuzzy NARX Predictive Model

Changjiang He[1], Yuanlin Gu[2]*, Hua-Liang Wei[3] and Qinggang Meng[4]

[1] Department of Mathematics and Statistics, Lancaster University, UK
[2] Department of Computer Science, Roehampton University, UK
[3] Department of Automatic Control and Systems Engineering, University of Sheffield, UK
[4] Department of Computer Science, Loughborough University, UK
Email: guyuanlin@hotmail.com

1. Introduction

Body Mass Index (BMI) describes the relationship between the mass and height of human. It is a critical indicator for the medical research and diagnosis, especially in the field of obesity, maturity and heritability [14, 22, 23]. The conventional measurement of human BMI is derived from the weight and the square of body height. Such a direct measurement can be severely interfered by many factors, such as race, gender, and abdominal body structure. These individual discrepancies can increase the uncertainty and error in decision making, hence, the conventional BMI measurement is not recommended for clinical judgment [29].

Bioelectrical impedance measures the body impedance via a weak electric current. According to the design, it fundamentally depends on the body water measurement within the tissue. Therefore, conventional body composition analysis rarely employs bioelectrical impedance for its high variability. However, with current technological developments, bioelectrical impedance shows a great potential in BMI analysis [6].

A recent research has achieved a significant improvement in BMI measurement via integrating multi-frequency bioelectrical impedance into the conventional method [7]. However, similar to all other biomarkers, bioelectrical impedance is under the regulation of multiple bioprocesses and its correlation with the BMI is dependent on various conditions [16, 17]. Meanwhile, it is also worth noting that the most frequency bands of bioelectrical impedance share a strong collinearity and may not be directly associated with the BMI.

In order to establish effective and efficient models with bioelectrical impedance, it is important to explore the frequency bands that are consistently correlated with human BMI. Such a filtering process can significantly limit the interference of the collinearity within bioelectrical impedance and the noise from irrelevant psychophysiological activities. In addition, it can also identify the key frequency bands, which can be used for further signal analysis to increase the BMI measurement reliability.

Meanwhile, the distribution and quantity of body water in humans is subject to psychophysiological state and individual conditions, and it is a dynamic process under the influence of external factors. As shown in the previous research, the direct correlations between some frequency bands of bioelectrical impedance and human BMI are quite low [7]. Therefore, it is necessary to apply feature extraction methods before integrating the bioelectrical impedance signals into the model.

Based on the above considerations, this paper proposes a novel Takagi-Sugeno Fuzzy NARX model (TSF-NARX) for prediction of BMI. The main contributions of this work are:

- Establish a hybrid model with interpretable structure for nonlinear BMI modelling.
- Identify the key frequency bands of bioimpedance and anthropometric factors that are correlated with BMI.
- Explore inter- and intra-uncertainty within the BMI measurement using fuzzy logic inference.

The remainder of this paper is as follows. Section 2 presents related works. The proposed TSF-NARX model is introduced in Section 3. The experimental results are presented in Section 4. Finally, the paper is concluded in Section 5.

2. Related Work

The interpretability of machine learning models has become an important topic in recent years [2]. For many real applications, the predictive model should not only achieve good prediction accuracy but also be able to reveal the insights, e.g. how the predictions are produced and what are the effective and most important features. For example, in BMI prediction, it is crucial to identify the frequency bands and anthropometric factors that are highly correlated with BMI. The transparency of these features and model structures are important for obtaining an insightful understanding of human BMI. Roughly speaking, there are two common types of interpretable models, that is, post-explainable models [2] (e.g. most advanced machine learning models) and intrinsically-explainable models [3] (e.g. nonlinear parametric polynomial regression models). Post-explainability techniques aim to interpret an established machine learning model after the completion of the training process. For example, SHapley Additive exPlanations (SHAP) [18] is used to measure the influence and importance of each feature for the prediction purpose. However, the model itself remains to be opaque and it cannot reveal the process of how the prediction was generated. On the other hand, a regression-based model employs fully interpretable structures and features to generate predictions [3]. A limitation of traditional regression-based models may be that they lack prediction accuracy when the required information is incomplete. The Nonlinear Auto Regressive Moving Average with Exogenous Input (NARMAX) model was developed for data modeling and system identification in the time, frequency, and spatiotemporal domains [3]. The NARMAX model can derive nonlinear terms from data of original variables and identify the most effective variables and their interactions (i.e. the product-terms) to build the model. These terms can describe the nonlinear dynamics of complex systems and can usually be linked to the original physical system or process. More importantly, the selected terms and model structure are fully interpretable. The Nonlinear Auto Regressive with Exogenous Input (NARX) model is a special case of the NARMAX model, and has been successfully applied to solve real data modeling problems in many areas [1, 3, 12, 24, 27]. In this study, we take advantage of the NARX model, especially its TIPS (transparent, interpretable, parsimonious, and simple/sparse/simulatable) properties [5, 27] and integrate it into our proposed model framework for BMI prediction.

Fuzzy logic combines objective knowledge and subjective knowledge via computing on 'degree of truth' rather than the traditional Boolean logic of 'true or false'. Fuzzy logic handles uncertainty, small size data, and data sparsity better than other machine learning paradigms. Compared with the conventional modeling approaches, such as neural network, models based on fuzzy logic share significant advantages, such as being flexible, simple, and intuitive [19]. Fuzzy logic-based models have been wildly applied in biological research for the high complexity and uncertainty within these systems, especially in the area related to medical diagnosis [15, 20, 21]. These fuzzy logic-based systems have proved efficient and effective tools to support health condition analyses and clinical decisions. However, current existing fuzzy logic models developed for BMI prediction are limited to a few specific groups, such as athlete and obesity. The findings from the bioimpedance may allow for new and ground-breaking path-opening to the prediction of general human BMI with fuzzy logic-based model.

3. The Proposed TSF-NARX Model

The TSF-NARX model combines the type-1 Sugeno fuzzy inference system with the NARX representation. The NARX models are built, based on multiple subsets resampled from the original dataset. The means and standard deviations of each identified terms are utilized to create the Gaussian distribution membership functions of each fuzzy rule correspondingly. Given an output variable $y(t)$ and a number of R input variables $u_1(t)$, $u_2(t)$,…,$u_R(t)$, the general NARX model can be represented as [3]:

$$y(t) = F[u_1(t-1), u_1(t-2),…, u_1(t-n_x), u_2(t-1),$$
$$u_2(t-2)], …, u_2(t-n_u), …, u_R(t-1), u_R(t-2),$$
$$…, u_R(t-n_u), y(t-1), y(t-2), …, y(t-n_y),] + e(t) \quad (1)$$

where t indicates the index of the data samples, $F[\cdot]$ is some nonlinear function, n_u and n_y are the maximum time lags of the input and output variables, $e(t)$ is the noise sequence. Note that equation (1) is the general NARX model for time series prediction with a delay of 1. In some situations, the system is static and there is no delay or time dependencies between the input and output variables. In these cases, the NARX model can be simplified as follows:

$$y(t) = F[u_1(t), u_2(t), \ldots, u_R(t)] + e(t) \tag{2}$$

where t indicates the index of the data samples. Usually, the R input variables $u_1(t)$, $u_2(t), \ldots, u_R(t)$, and their interactions, such as $u_1(t)\, u_2(t)$ (the nonlinear degree of the term is 2) and $[u_2(t)]^2\, u_3(t)$ (the nonlinear degree of the term is 3) are used to build a NARX model. A number of model terms $\varphi_1, \varphi_2, \ldots, \varphi_M$ can be derived from the regression vector $[u_1(t), u_2(t), \ldots, u_R(t)]$ via some functions, for example, linear term $\varphi = u_i$ and nonlinear term $\varphi = u_i(t) \times u_j(t)$, $i = 1, 2, \ldots, R, j = i, i + 1, \ldots, R$. The initial NARX model that contains all the specified model terms can be written in a vector form:

$$y = \theta_1\varphi_1 + \theta_2\varphi_2 + \ldots + \theta_M\varphi_M + e \tag{3}$$

where $\varphi_1, \varphi_2, \ldots, \varphi_M$ are a number of M terms derived from the original inputs u_1, $u_2, \ldots, u_R$, and $\theta_1, \theta_2, \ldots, \theta_M$ are the estimated parameters of these terms. The nonlinear terms are derived using the original inputs through some nonlinear conversions. These derived terms together can approximate the nonlinear relationship between the input and output. Details of the term generation procedures can be found in [3, 25, 26]. Note that usually the initial full model contains a huge number of candidate model terms to guarantee that the true or exact nonlinearities of the system to be studied can be sufficiently approximated using the specified model terms. However, more than often, not all the candidate model terms are equally important and useful. In practice, the final identified model only contains a relatively small number of the most significant model terms, meaning that the final model is sparse. This is important to avoid overfitting; otherwise, the model may show poor generalization ability and more importantly fail to reveal the key drivers. For example, in the case of BMI prediction, it is crucial to identify the key bioimpedance frequencies that are consistently correlated with BMI. The NARX model employs an orthogonal forward regression (OFR) algorithm to select the most important model terms from all the candidate model terms. The OFR algorithm chooses the model terms in a stepwise manner, by ranking the contributions of the candidate model terms to explaining the output variable. The Error Reduction Ratio (ERR) is employed to measure the contribution of each term, which is defined as follows [3, 8, 25]:

$$ERR_i = \frac{\left[y^T q_i^{(s)} \right]^2}{\left(y^T y \right)\left(q_i^{(s)T} q_i^{(s)} \right)} \tag{4}$$

where $q_i^{(s)}$ is the basis of the i-th model term at s-th selection iteration. At the first iteration when $S = 1$, the initial basis is selected as $q_i^{(1)} = \varphi_i, i = 1, 2, \ldots M$. At each step, the significant term can be selected as $q_s = q_{l_s}^{(s)}$, where $l_s = arg \max\limits_{1 \le j \le M, j \notin l} \left\{ ERR_j^{(s)} \right\}$

Then, $q_i^{(s+1)}$ is re-calculated and updated during by an orthogonalization procedure,

as $q_i^{(s+1)} = q_i^{(s)} - \dfrac{q_i^{(s)^T} q_s}{\left(q_i^{(s)^T} q_s\right)} q_s$. The selection procedure stops when an optimal number

of terms is reached, which is indicated by adjustable prediction error sum of squares (APRESS) [4]. The details of the OFR algorithm and term selection procedure can be found in [3, 25, 26]. When we assume that a number of n model terms are selected, the final NARX model can be represented as:

$$y = \theta_{l_1} \varphi_{l_1} + \theta_{l_2} \varphi_{l_2} + \ldots + \theta_{l_n} \varphi_{l_n} + e \tag{5}$$

where $\varphi_{l_1}, \varphi_{l_2}, \cdots, \varphi_{l_n}$ are the selected terms and $\theta_{l_1}, \theta_{l_2}, \ldots, \theta_{l_n}$ are the parameters which can be calculated by least square estimation.

In this study, the size of original datasets is limited (there are only a total of 135 samples), and the target system is highly nonlinear. Thus, a single NARX model may not sufficiently summarise the underlying uncertainty, e.g. the different scales of bioimpedance will determine different correlation degree with BMI. Therefore, following the method in [13, 28] the original dataset is resampled several times; in this way, more possible patterns within the system can be explored. The inputs and output of the k-th resampled sub-dataset can be represented as $y^{(k)}$ and $u_1^{(k)}, u_2^{(k)}, \ldots, u_R^{(k)}$ Following the procedures of OFR algorithm, the NARX model for the k-th sub-dataset can be built as:

$$y^{(k)} = \theta_{l_1}^{(k)} \varphi_{l_1}^{(k)} + \theta_{l_2}^{(k)} \varphi_{l_2}^{(k)} + \cdots + \theta_{l_n}^{(k)} \varphi_{l_n}^{(k)} + e \tag{6}$$

where $\theta_{l_1}^{(k)} \ldots \theta_{l_n}^{(k)}$ are the estimated weights of the selected terms $\varphi_{l_1}^{(k)}, \ldots, \varphi_{l_n}^{(k)}$ for the k-th sub-dataset.

These multiple NARX models are combined as a whole by designing and implementing the Takagi-Sugeno fuzzy inference. Compared to the other fuzzy inferences, Takagi-Sugeno fuzzy logic systems, due to their good inference properties, are able to integrate the polynomial NARX models. Specifically, the linguistic form of a typical Takagi-Sugeno fuzzy rule reads as follows:

IF x_1 is A_1, x_2 is A_2, … and x_n is A_n, **THEN** $y = f(x_1, x_2, \ldots, x_n)$,

where x_m represents the m-th (m = 1, 2, …, n) input variable for the antecedence, and as a consequence, the output variable y is obtained from a linear function based on the n variables. $A_1, A_2, \ldots, A_n$ are the fuzzy labels with membership functions for calculating the intensity of the rule for final output. Meanwhile, the fuzzification of the inputs also expend the feasible range of each NARX model. For example, a Gaussian membership function can provide a membership degree over all feasible ranges rather than a few limited points. This results in a soft partition between NARX models and a finer tuned surface for prediction.

In this study, the fuzzy rules are generated from the pattern extraction based on the NARX regression. For each fuzzy rule, an NARX model defines the output function as the consequent, and the corresponding training data ranges of each input features define the fuzzy labels and the correlated membership functions in the antecedence. The boundaries of linguistic labels are determined on the means μ and variances σ of parameters in the training data, i.e., $\mu \pm 2\sigma$.

Table 1 summarizes the linguistic labels used for the model. The proposed fuzzy model uses the Gaussian basis function for representing the membership functions within the fuzzy rules.

Table 1: Linguistic labels of selected features for TSF-NARX model

Linguistic Labels	φl_1	φl_2	φl_3	φl_4	φl_5	φl_6
Small	<145	<850	<900	<40.0	<200	<2500
Medium	145~165	850~1050	900~1100	40.0~60.0	200~300	2500~3500
Large	>165	>1050	>1100	>60.0	>300	>3500

A typical rule (rule no.5) implemented in the TSF-NARX reads as following:

IF φ_{l_1} is medium and φ_{l_2} is medium and φ_{l_3} is medium and φ_{l_4} is medium and φ_{l_5} is medium and φ_{l_6} is medium, **THEN** BMI $y^{(5)} = \theta_{l_1}^{(5)}\varphi_{l_1}^{(5)} + \theta_{l_2}^{(5)}\varphi_{l_2}^{(5)} + \ldots + \theta_{l_n}^{(5)}\varphi_{l_n}^{(5)}$, and its corresponding membership functions are shown in the Fig. 1.

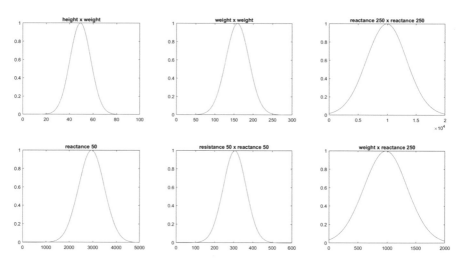

Fig. 1: Membership functions for Rule no. 5

4. Experimental Results

The data used in this study is described in Section 4.1. Section 4.2 presents the term generation and selection results. Section 4.3 shows the TSF-NARX model and performance comparison. Discussions are in Section 4.4.

4.1 Data Description

We applied the proposed model on a BMI dataset with bioimpedance. The dataset was collected at the Food Science and Human Nutrition Research Unit of the Department of Experimental Medicine of Sapienza, Rome University. A number of 135 data samples were obtained from overweight and obese women via dual-energy X-ray absorptiometry (DXA) examination (Hologic 4500 RDR) [9, 10]. The BMI index was considered as model output, and the bioimpedance signals, along with some other external variables, were used as model inputs. Detailed descriptions of the data and variables are presented in Table 2. Bioimpedance data contains resistance and reactance signals, which were measured at frequencies 5, 10, 50, 100, 250 kHz, respectively. Height, weight, and age were used as external anthropometric variables for model construction [7].

Table 2: Data description

	Variable	Description	Mean	Min	Max
y	BMI	Body Mass Index [kg]	55.1473	36.5900	74.9500
u1	height	height [m]	1.6216	1.4500	1.8000
u2	weight	weight [kg]	97.7400	56.2000	136.8000
u3	age	age [year]	44.8593	18.0000	69.0000
u4	R5	logarithm of resistance at 5 kHz	6.3090	5.9440	6.6840
u5	R10	logarithm of resistance at 10 kHz	6.2842	5.9220	6.6540
u6	R50	logarithm of resistance at 50 kHz	6.1761	5.8310	6.5390
u7	R100	logarithm of resistance at 100 kHz	6.1253	5.7900	6.4910
u8	R250	logarithm of resistance at 250 kHz	6.0516	5.7250	6.4170
u9	X5	reactance at 5 kHz	25.7164	9.9300	41.9200
u10	X10	reactance at 10 kHz	35.7914	18.8400	59.6900
u11	X50	reactance at 50 kHz	49.3887	29.8100	74.4100
u12	X100	reactance at 100 kHz	44.0784	26.1400	62.0200
u13	X250	reactance at 250 kHz	30.6726	17.1000	44.9700

As can be seen from Table 2, the variation ranges of these variables are as follows. The age of participants ranges from 18 to 69 with an average of 45. The mean, minimum, and maximum height is 1.62 m, 1.45 m, and 1.80 m, respectively. The cases within the dataset are comprehensive, covering a wide range of anthropometric conditions. A graphical illustration of seven variables is given in Fig. 2, which presents large variations in each variable. No missing value or outlier was detected in the dataset. Around 80% of the samples were used for model training and the remaining 20% samples were used for testing the model performance.

4.2 The Identified NARX Model

Previous studies have investigated the effectiveness of linear models [7]. However, the nonlinear effect among the input variables has not been analyzed. To explore the impact of nonlinear interactions, we performed a number of trial experiments and compared the performances of linear models and nonlinear NARX models. First,

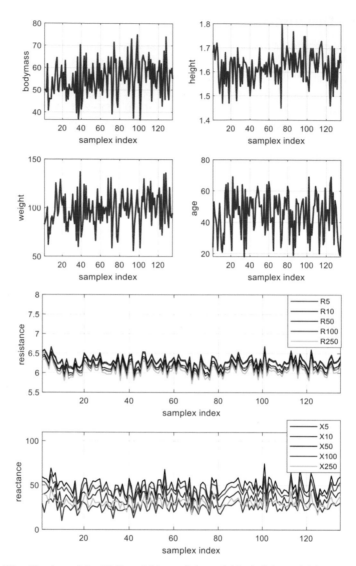

Fig. 2: Visualization of the BMI variable, weight variable, height variable, age variable, logarithm of resistance at 5, 10, 50, 100, 250 kHz, and reactance at 5, 10, 50, 100, 250 kHz

we applied the OFR (orthogonal forward regression) algorithm to generate a number of nonlinear NARX models with different number of model terms. Second, we built the associated linear models using exactly the same input variables. The only difference is that the nonlinear NARX models contain nonlinear model terms, but the linear models only use original linear inputs. The performances of the models were evaluated with the correlation coefficient (CC) and the prediction efficiency (PE) as presented in Fig. 3. For the definition of PE, interested readers can refer to [11]. From the results, the linear models have better performances when the model contains up to four model terms. However, as the number of terms increases, the nonlinear

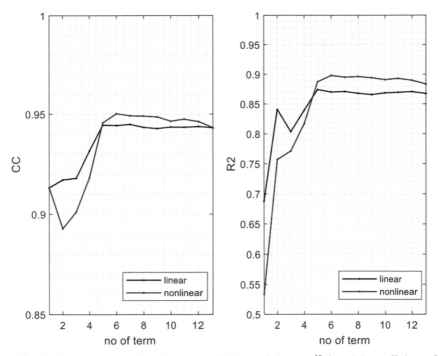

Fig. 3: The comparison of performances (CC: correlation coefficient, R2: coefficient of determination) of linear models and nonlinear NARX models versus number of terms

NARX models achieve significantly better performances than linear models. It may be understood that there are indeed some nonlinear model terms (interaction product model terms) that can better explain the variation of BMI. In other words, the value of BMI nonlinearly depends on these specified variables.

It can be observed that the performances vary with the number of terms. A too simple model (e.g. with a too small number of terms) cannot sufficiently represent the relation between the inputs and output, whereas the inclusion of excessive terms increases the model complexity, leading to overfitting and deteriorating the model's generalization ability. To identify the optimal number of terms, we applied the term length selection criteria, called adjustable prediction error sum of squares (APRESS) [4]. According to the APRESS, the optimal number of the model terms is six. We applied the OFR algorithm and identified six most effective model terms; *see* Table 3 for details.

The associate NARX model in Table 3 reads as:

$$y = 0.4143 \times u1 \times u2 - 0.0015 \times u2 \times u2 + 0.0023 \times u13 \times u13$$
$$+ 2.8189 \times u11 - 0.4219 \times u6 \times u11 - 0.0029 \times u2 \times u13 \qquad (7)$$

The ERR values indicate the importance of the selected terms. Note that the ERR value of the first term is much higher than those of the following terms. This indicates that the combined effect of the two factors, height and weight, plays a dominant role in determining a person's BMI value. The finding here accords closely

Table 3: The identified terms, the associated ERR values, estimated weights and t values

Terms	Description	Weights	ERR	T Value
'$u1 \times u2$'	Height × weight	0.4143	99.3188	15.0239
'$u2 \times u2$'	Weight × weight	−0.0015	0.2939	4.6767
'$u13 \times u13$'	Reactance 250 kHz × reactance 250 kHz	0.0023	0.0199	0.9794
'$u11$'	Reactance 50 kHz	2.8189	0.0275	7.7502
'$u6 \times u11$'	Resistance 50 kHz × reactance 50 kHz	−0.4219	0.1080	7.3365
'$u2 \times u13$'	Weight × reactance 250 kHz	−0.0029	0.0102	2.1716

with that currently used in hospitals, i.e. BMI = weight/(height square), where the units of weight and height are kg and m, respectively.

The significance of the selected terms was validated with the student t-test, which is used to determine if there is a significant difference between two groups [25]. The confidence interval is chosen to be 95%. If the t value is larger than 1, it means that this term contributes significantly to the regression model. From the results, all but one term are significant, and the t value of the term $u13 \times u13$ is slightly smaller than (but very close to) 1; it is reasonable to keep the term in the model.

From the results, some insights can be revealed from the identified model. First, the selected terms indicate the most impactful factors on BMI. For example, $u1 \times u2$ and $u2 \times u2$ indicates that the interactions between height and weight and square of weigh significantly affect BMI.

This finding is in line with previous study which explored the effectiveness of anthropometric variables [7]. In addition, the model reveals the key bioimpedance frequency bands that impact BMI. For example, the selected terms $u11$, $u6 \times u11$ and $u2 \times u13$ are derived from the variable reactance at 50 kHz, resistance at 50 kHz, and reactance at 250 kHz, indicating that these are the most important frequency bands.

4.3 TSF-NARX Results

Fig. 4 shows the surface plots for TSF-NARX model membership functions. The BMI roughly increases with the values of height, weight, and the reactance at 50 kHz and decreases with the reactance at 250 kHz. It can be observed that the surface is not flat. Therefore, these selected features are correlated with each other to a certain degree, and the relationships between them and BMI are nonlinear.

Table 4 summarizes the prediction performances of all models measured, using the following three metrics: correlation coefficient (CC), coefficient of determination (R2), and root mean square error (RMSE). CC measures the correlation between the target observations and the model predictions, R2 describes the match between the predictions and the observations, whereas RMSE measures the overall residuals between the estimations and the observations.

For comparison purposes, we tested the linear model, the NARX model, a neural network modelm and adaptive neuro-fuzzy inference system (ANFIS) model on the same test dataset. The linear model was constructed by using all the original input variables listed in Table 2. The NARX model was constructed by the six selected important terms. The ANFIS model was generated from Matlab built-in function

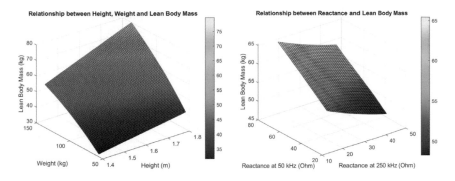

Fig. 4: Membership function for height & weight, reactance at 50kHz and reactance at 250 kHz of fuzzy-NARX model

with default settings. As this is a static system which does not contain time series data, the tested neural network was constructed by one input layer, one output layer, and several fully connected layers. ANFIS combines the fuzzy logic interference with neural network, and the fuzzy rules of the system are extracted from the data using the conventional neural network structure. The fuzzy rule-base of the ANIFS in this study was generated with subtractive clustering (range of influence 0.5, squash factor 1.25, accept ratio 0.5 and reject ratio 0.15), and then the fuzzy inference system was trained with hybrid optimization (error tolerance 0 and epochs 3).

As shown in Table 4, the nonlinear NARX and the TSF-NARX outperform the other models in terms of all the three metrics. It suggests that data relationship of such a biological dataset is severely nonlinear; thus, its modeling requires nonlinear elements. Meanwhile, compared to the NN and ANFIS, the NARX and the TSF-NARX models can deal with the small sample size problem very well. NARX model has achieved the highest CC and R2 while TSF-NARX has achieved the lowest root mean square error. This indicates that NARX is more specialized in the pattern extraction with existing data, whereas the TSF-NARX focuses on exploring the boundary of feasible data range.

Table 4: The summary of model performances

Model	CC	R2	RMSE
Linear model	0.9417	0.8624	2.8013
NARX model	0.9503	0.8981	2.4653
TSF-NARX model	0.9484	0.8937	2.4620
Neural network	0.9417	0.0000	57.4197
ANFIS	0.9153	0.7191	4.0012

Figure 5 summarizes the residuals between the predictions and observations for the ANFIS and the TSF-NARX model. Compared to the ANFIS, the residuals of the TSF-NARX model are closer to the desired value 0 and are evenly distributed on both negative and positive side. It suggests that the TSF-NARX model has relatively high accuracy and high precision. Figure 6 shows the relations between

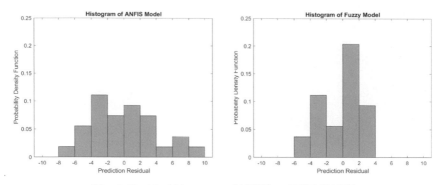

Fig. 5: Residual histogram of ANFIS and TSF-NARX

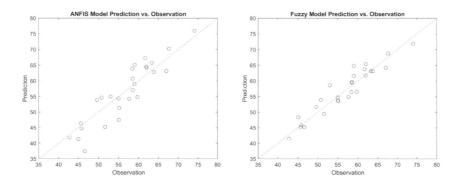

Fig. 6: Scatter plot of ANFIS and TSF-NARX

the predictions and observations for the ANFIS and TSF-NARX model. It can be observed that compared to the ANFIS, the TSF-NARX model can provide more accurate prediction with less bias in general. This indicates that the pattern extraction based on the NARX model is more efficient and robust than the subtractive clustering method in this small sample data case.

The experimental results show that the TSF-NARX provides excellent generalization properties. The use of the Takagi-Sugeno fuzzy rule-base interference created a simple and intuitive linguistic-based model that can be easily interpreted by end-users of the model. The relationships within the features can be directly displayed in the surface plots as shown earlier and be implemented into quick estimation. Meanwhile, the fuzzy inference expands the prediction range with nonlinear membership functions of the rules. In addition, it ensures that the model is robust to the intra and extra uncertainty within the system. Therefore, fuzzy logic modeling approaches can normally achieve excellent performance even with small sample size. However, such models are heavily dependent on the quality of rule-base. Thus, the pattern extraction or the clustering method is important for fuzzy logic-based models and is of great importance in the current research, especially for developing the associated fuzzy logic. Compared to the ANFIS, the pattern extraction of NARX achieves better results than the neural network model and explores the full potential of fuzzy inference. Therefore, the combination of NARX and fuzzy logic has proved to be the most efficient and effective approach to model the clinical data.

5. Conclusion

This work proposed a novel hybrid TSF-NARX model for BMI prediction. It identified the key factors for BMI measurement, i.e. reactance at 50 kHz, resistance at 50 kHz, reactance at 250 kHz, weight and height. The prediction accuracy of the proposed model outperformed other existing state of the art models. The integration of fuzzy logic inference provides a model structure that can be fully interpreted by humans, and this is important for further applications. Furthermore, such a paradigm creates the foundation for more advanced predictive approach for medical research and clinical diagnosis. The proposed method can be applied to the exploration of underlying mechanism of complicated data modeling problems in other areas, such as environment and space weather.

Acknowledgement

This work was partially supported by Innovate UK under grant reference 26526.

References

1. Aguirre, L.A. (2019). *A Bird's Eye-view of Nonlinear System Identification*, arXiv preprint arXiv: 1907.06803.
2. Arrieta, A.B., Díaz-Rodríguez, N., Del Ser, J., Bennetot, A., Tabik, S., Barbado, A., García, S., Gil-López, S., Molina, D., Benjamins, R. and Chatila, R. (2020). Explainable Artificial Intelligence (XAI): Concepts, taxonomies, opportunities and challenges toward responsible AI, *Information Fusion*, 58: 82-115.
3. Billings, S.A. (2013). *Nonlinear System Identification: NARMAX Methods in the Time, Frequency, and Spatio-temporal Domains*, John Wiley & Sons.
4. Billings, S.A. and Wei, H.L. (2008). An adaptive orthogonal search algorithm for model subset selection and non-linear system identification, *International Journal of Control*, 81(5): 714-724.
5. Billings, S. and Wei, H.L. (2019, August). NARMAX model as a sparse, interpretable and transparent machine learning approach for big medical and healthcare data analysis. *In: 2019 IEEE 21st International Conference on High Performance Computing and Communications; IEEE 17th International Conference on Smart City; IEEE 5th International Conference on Data Science and Systems (HPCC/SmartCity/DSS)*, pp. 2743-2750, IEEE.
6. Borga, M., West, J., Bell, J.D., Harvey, N.C., Romu, T., Heymsfield, S.B. and Leinhard, O.D. (2018). Advanced body composition assessment: From body mass index to body composition profiling, *Journal of Investigative Medicine*, 66(5): 1-9.
7. Cammarota, C. and Pinto, A. (2021). Variable selection and importance in presence of high collinearity: An application to the prediction of lean body mass from multi-frequency bioelectrical impedance, *Journal of Applied Statistics*, 48(9): 1644-1658.

8. Chen, S., Billings, S.A. and Luo, W. (1989). Orthogonal least squares methods and their application to non-linear system identification, *International Journal of Control*, 50(5): 1873-1896.

9. Earthman, C.P. (2015). Body composition tools for assessment of adult malnutrition at the bedside: A tutorial on research considerations and clinical applications, *Journal of Parenteral and Enteral Nutrition*, 39(7): 787-822.

10. Ellis, K.J. (2000). Human body composition: *In vivo* methods, *Physiological Reviews*, 80(2): 649-680.

11. Gu, Y., Wei, H.L., Boynton, R.J., Walker, S.N. and Balikhin, M.A. (2019). System identification and data-driven forecasting of AE index and prediction uncertainty analysis using a new cloud-NARX model, *Journal of Geophysical Research: Space Physics*, 124(1): 248-263.

12. Gu, Y., Yang, Y., Dewald, J.P., Van der Helm, F.C., Schouten, A.C. and Wei, H.L. (2020). Nonlinear modeling of cortical responses to mechanical wrist perturbations using the NARMAX method, *IEEE Transactions on Biomedical Engineering*, 68(3): 948-958.

13. Gu, Y. and Wei, H.L. (2018). A robust model structure selection method for small sample size and multiple datasets problems, *Information Sciences*, 451: 195-209.

14. Jaquish, C.E., Dyer, T., Williams-Blangero, S., Dyke, B., Leland, M. and Blangero, J. (1997). Genetics of adult body mass and maintenance of adult body mass in captive baboons (*Papio hamadryas* subspecies), *American Journal of Primatology*, 42(4): 281-288.

15. Keivanian, F., Chiong, R. and Hu, Z. (2019, April). A fuzzy adaptive binary global learning colonization-MLP model for body fat prediction. *In: 2019 3rd International Conference on Bio-engineering for Smart Technologies (BioSMART)*, pp. 1-4, IEEE.

16. Kyle, U.G., Bosaeus, I., De Lorenzo, A.D., Deurenberg, P., Elia, M., Gómez, J.M., Heitmann, B.L., Kent-Smith, L., Melchior, J.C., Pirlich, M. and Scharfetter, H. (2004). Bioelectrical impedance analysis – Part I: Review of principles and methods, *Clinical Nutrition*, 23(5): 1226-1243.

17. Kyle, U.G., Bosaeus, I., De Lorenzo, A.D., Deurenberg, P., Elia, M., Gómez, J.M., Heitmann, B.L., Kent-Smith, L., Melchior, J.C., Pirlich, M. and Scharfetter, H. (2004). Bioelectrical impedance analysis – Part II: Utilization in clinical practice, *Clinical Nutrition*, 23(6): 1430-1453.

18. Lundberg, S.M. and Lee, S.I. (2017, December). A unified approach to interpreting model predictions. *In: Proceedings of the 31st International Conference on Neural Information Processing Systems*, pp. 4768-4777.

19. Mendel, J.M. (2017). *Uncertain Rule-based Fuzzy Systems: Introduction and New Directions.* pp. 684. Springer.

20. Rathore, H., Mohamed, A., Guizani, M. and Rathore, S. (2021). Neuro-fuzzy analytics in athlete development (NueroFATH): A machine learning approach, *Neural Computing and Applications*, 1-14.

21. Ribeiro, A.C., Silva, D.P. and Araujo, E. (2014, July). Fuzzy breast cancer risk assessment. *In: 2014 IEEE International Conference on Fuzzy Systems (FUZZ-IEEE)*, pp. 1083-1087, IEEE.

22. Schultz, S.R. and Johnson, M.K. (1995). Effects of birth date and body mass at birth on adult body mass of male white-tailed deer, *Journal of Mammalogy*, 76(2): 575-579.

23. Shea, J.R., Henshaw, M.H., Carter, J. and Chowdhury, S.M. (2020). Lean body mass is the strongest anthropometric predictor of left ventricular mass in the obese paediatric population, *Cardiology in the Young*, 30(4): 476-481.

24. Spinelli, W., Piroddi, L. and Li, K. (2004, December). Nonlinear modeling of NO/sub x/emission in a coal-fired power generation plant. *In: 2004 43rd IEEE Conference on Decision and Control (CDC)* (IEEE Cat. No. 04CH37601), vol. 4, pp. 3850-3855, IEEE.

25. Wei, H.L. and Billings, S.A. (2008). Model structure selection using an integrated forward orthogonal search algorithm assisted by squared correlation and mutual information, *International Journal of Modeling, Identification and Control*, 3(4): 341-356.

26. Wei, H.L., Billings, S.A. and Liu, J. (2004). Term and variable selection for non-linear system identification, *International Journal of Control*, 77(1): 86-110.

27. Wei, H.L. (2019 June). Sparse, interpretable and transparent predictive model identification for healthcare data analysis. *In: International Work-Conference on Artificial Neural Networks*, pp. 103-114. Springer, Cham.

28. Wei, H.L. and Billings, S.A. (2009). Improved model identification for non-linear systems using a random subsampling and multifold modelling (RSMM) approach. *International Journal of Control*, 82(1): 27-42.

29. Zuercher, G.L., Roby, D.D. and Rexstad, E.A. (1999). Seasonal changes in body mass, composition, and organs of northern red-backed voles in interior Alaska, *Journal of Mammalogy*, 80(2): 443-459.

Training Therapy with BCI-based Neurofeedback Systems for Motor Rehabilitation

Jingjing Luo[1,2,3], **Qiying Cheng**[1], **Hongbo Wang**[1,2,3], **Youhao Wang**[1] and **Qiang Du**[1]

[1] Institute of AI and Robotics, Academy for Engineering and Technology, Fudan University, Shanghai, China
[2] Shanghai Engineering Research Center of AI Robotics, Shanghai, China
[3] Engineering Research Center of AI Robotics, Ministry of Education, Shanghai, China

1. Introduction

Stroke, as an acute cerebrovascular disease, is a leading cause of brain damage and has a very high mortality and disability rate worldwide [6, 60]. Patients with hemiplegia caused by stroke account for 3‰ to 5‰ of the total population [42]. Studies show that the healing effect of upper limb function after stroke is poor [13], and about 70% of patients will have upper limb dysfunction, such as hemiplegia and spasticity, especially the distal fine function. The problem of upper limb and hand dysfunction caused by nerve damage in stroke needs to be solved urgently [64, 48]. Muscles and nerves slowly degenerate, and the longer after the nerve is injured, the more difficult it is to restore motor function [17, 67]. The long-term lack of hand function also brings heavy spiritual and economic burdens to the family and society [16].

Closed-loop motor-sensory therapies can promote nerve repair through brain reorganization or compensation [26]. Brain plasticity is the basis of neurorehabilitation [15]. The key to recovery training of motor function is to increase the direct training intensity as much as possible and repeat the training, based on the required tasks [27]. Brain-Computer-Interface (BCI) technology uses brain signal acquisition devices to detect the electrical signals of the human brain's proactive activities in real time, and uses feedback to artificially stimulate or guide the corresponding movement to strengthen nerve-muscle connectivity. This motor neurorehabilitation is referred to as BCI rehabilitation and has been widely studied for stroke treatment. However, most of the BCI system training aimed at the recovery of hand motor function has unstable intervention outcome, and it is in the critical stage of translational and application research [9].

Study of the training mode of closed-loop neurofeedback has very important theoretical significance and application value for improving the degree of recovery and achieving precise rehabilitation. This study conducts an in-depth investigation on the robot-assisted hand motor neurorehabilitation system and explores hand rehabilitation by introducing paradigms inducing hand motor neural activities as well as neural decoding methodologies. Finally, we discuss future directions to investigate BCI-based motor rehabilitation concerning the efficacy of the training system in both paradigms designing and neural decoding.

2. Overview of Closed-loop Neurofeedback Systems

BCI-based motor neurofeedback system is composed of signal acquisition and processing module, control system and feedback executing [61], and rehabilitation evaluation. The signal processing module is to train the classifier that decodes neuronal signal. The use of artificial intelligence algorithms, such as machine learning and deep learning, allow the BCI system to accurately recognize brain activity. The signal processing module sends the recognized brain-activity pattern to the control system in the form of instructions, and the latter controls the mechanical feedback mechanism through the control algorithm to perform exercise executions for the affected hand. Feedback process can provide feedback through visual, auditory, tactile, and somatosensory ways, combined with display screens, speakers, electrodes, or robotic mechanisms. In this way, BCI combines the top-down method of neural activity intervention to change peripheral behaviors and the bottom-up method of using peripheral nerves to induce changes in nerve levels to achieve closed-loop training of the nerve-muscle circuit [58]. At present, rehabilitation evaluation can be investigated from the aspects of clinical function evaluation, motor parameter evaluation, and brain function evaluation, with clinical outcome as the main purpose and supplementing analysis of brain functional changes through data or image calculation.

3. Paradigms for Inducing Hand Motor Nerve Activity

3.1 Motor Imagery Task

Motor imagery (MI) task is one of the continuous, stable, and controllable mental activities, which is also the most widely used paradigm for induction of motor rehabilitation training. MI is an endogenous EEG signal, that is, the system must collect the subject's own mental activity, or brain activity, without external stimuli. Quantitative calculations have shown that motor imagination and motor execution (ME) tasks can activate the same area network in the human brain, including the auxiliary motor area (SMA), the contralateral posterior central gyrus, the contralateral superior lobule, and the ipsilateral prefrontal cortex [21]. MI technology mainly extracts the characteristics of brain activity through the corresponding electroencephalography (EEG) response in the specific frequency band of the brain's sensory motor area (SM) when a person is performing motor imagination, which is the so-called sensorimotor rhythm (SMR). It is distinct that the μ (8~12 Hz)

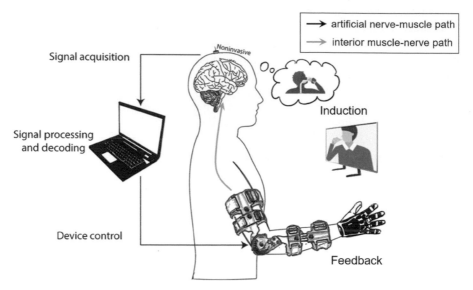

Fig. 1: BCI-based closed loop rehabilitation system composition

rhythm and the β (12~30 Hz) rhythm will have a significant attenuation during MI. The phenomenon of increase or attenuation of the electrical signal before and after the appearance of the imaging task is called event-related synchronization (ERS) or event-related desynchronization (ERD). This feature makes the MI-EEG signal comparably recognizable. However, the MI-EEG signal also has some shortcomings. Some experiments have shown that a long-term motion imaging task will gradually fatigue people, thereby reducing the classification accuracy [3]. Besides, there exist inherent differences in abilities among different individuals about pure motion imagination (do not watch movement videos and rely entirely on imagination); thus the difficulty of completing rehabilitation training. Since most stroke patients are elderly people with cognitive decline, it will greatly limit their rehabilitation efficacy [56].

3.2 Motor Attempt Task

Motor attempt (MA) is the brain activity when the patient wants to move the affected hand but cannot achieve it. It can induce the brain to produce the same ERD/ERS phenomenon as MI and ME. When the patient performs the MA task, it does not involve the effort of deliberately inhibiting the physical activity and thus is not prone to fatigue. Research has shown that such mental activity tasks have more information than MI [7]. Neuroimaging studies have shown that the cortical activity of motor intention closely follows the organization of specific areas of the body during motor execution after spinal cord injury [49]. MA can induce the Hebbian neural plasticity mechanism, that is, the continuous and repeated stimulation of presynaptic neurons to postsynaptic neurons, which can lead to an increase in synaptic transmission performance. Although stroke patients have lost the ability to proactively exercise, they retain the ability of exercise planning [22]. The MA-based BCI system can promote the patient's exercise preparation, execution, and normal recovery sequence of peripheral muscle effects [37].

3.3 Action Observation Task

Action observation (AO) originated from experiments related to 'mirror neurons' in non-human primates [14]. In these experiments, it was found that the brain regions activated while performing actions are the same as those activated in action observation [47]. It indicates that we can use motion observation as an alternative to MI. Through experiments, Murata Hospital in Osaka, Japan, in cooperation with Meiji University [55], proved that AO can promote the motor cortex activity of hemiplegic stroke patients, thereby improving the classification accuracy of rehabilitation equipment. Robert M. Hardwick *et al.* [19] compared the activated brain areas of motor imagination, action observation, and motor execution from the perspective of sports science, and found that the three tasks' brain networks overlap in the bilateral premotor area, parietal lobe, and motor sensory area, proving the feasibility of MI and AO for neurorehabilitation. Moreover, the combination of MI and AO can compensate for each other's shortcomings and improve classification accuracy while activating more brain regions.

3.4 Hybrid Paradigm

MI is the most widely used induction paradigm, but the patient's training experience is affected by personal imagery abilities. MA is the brain activity that occurs spontaneously before exercise. Action observation can cause brain activities similar to exercise execution. All of these can circumvent the shortcomings of motor imagination and are a new trend in current research. At present, the decoding system of most BCI motor neurofeedback training systems performs binary classification between bilateral hand movements or ipsilateral hand movements. Under this training paradigm, different activation patterns of the brain are significantly different, and the recognition task is comparably simple. The signals of fine finger activities are difficult to distinguish, such as making a fist and stretching the palm, dorsiflexion of the wrist joint, palm flexion, radial deviation, and ulnar deviation of the radioulnar joint, finger movements between the thumb, etc. This article believes that the study of the combination of MI and AO or MA and AO can help enhance the characteristics of brain activity, not only improving the accuracy of decoding, but also improving the patients' training experience and increased participation, thereby hoping to improve the rehabilitation effect.

4. Neural Decoding Methodologies

SMR is one of the most widely used paradigms in sports imagination. Specifically, when the patient performs MI or MA tasks, ERD/ERS modulation will be produced in his/her sensorimotor cortex. Recognizing SMR energy changes in the μ and β frequency bands can manipulate the robot exoskeleton to guide the patient to exercise the affected hand and achieve effective feedback. Norman *et al.* [40] used the FINGER robot exoskeleton to assist stroke patients with finger extension

Table 1: Comparisons of three induction paradigms for hand motor nerve activities

Paradigms	Activated Brain Areas	Characteristics	Advantages	Disadvantages	Comparisons
Motor Imagery (MI) Mattia, 2020 [35] Chci, 2016 [11] Kaiser, 2011 [25]	A network of bilateral anterior motor cortex, lower and upper middle parietal rostral, basal ganglia, and cerebellum, involving the dorsolateral prefrontal cortex [35]	Activating more brain areas involved in actual exercise	Brain activity characteristics of different actions are distinguishable.	Certain requirements of patient's cognitive ability. Long-term training is prone to fatigue.	MI, MA and AO can activate the motor sensory area similarly to the actual movement, and use the same BCI configuration and decoding features. MI is most close to exercise execution state, MA is rich in more activation information, and AO involves a wider range of brain areas. Considering their advantages and disadvantages, we recommend using hybrid induction paradigms to help patients improve their attention, reduce fatigue, and increase the discrimination of brain signals during training, thereby improving system performance and improving training effects.
Motor Attempt (MA) Li, 2019 [30] Rathee, 2019 [44] Shu, 2018 [50]	Two banks of the contralateral central sulcus (primary motor and sensory cortex) extending towards the central anterior and posterior sulcus (premotor and parietal lobe) [28]	No need to deliberately suppress the willingness to move.	Less fatigue and rich in more brain activity information [27]	Brain activity caused by the intention is short. Difficult for real-time decoding algorithms.	
Action Observation (AO) Chci, 2019 [10] Tani, 2018 [55]	Bilateral network of the premotor cortex and parietal lobe including a larger parietal volume and more bilateral areas than MI[35].	Reactive brain activity patterns triggered by mirror neurons.	The requirement for cognitive ability is low, and less easy to cause fatigue.	Stimulated brain area not quite overlap with motor execution. Large but not specific areas of brain activity.	
Hybrid Aoyama, 2020 [4] Nagai, 2019 [39] Ono, 2018 [43]	Involving brain regions activated in multiple tasks.	Simulate real-world multiple brain activities.	Patients are more concentrated, and tasks are easier to complete.	-	

training, and practiced the regulation of brain motor sensory rhythm (SMR) under multimodal feedback. P300 shows that 220 to 500 milliseconds after an event, a positive peak of 5 to 10 microvolts appears in the EEG signal. The visual P300 potential is one of the most studied event-evoked potentials, which is induced by events that occur infrequently. Most subjects can use the Vision P300 efficiently and complete the calibration within a few minutes. An overview of neural decoding algorithms is summarized in Figs. 2 and 3, consisting of multi-step algorithms, end-to-end algorithms, and brain network analysis algorithms.

4.1 Machine Learning Algorithms with Multiple Steps

Machine learning algorithms perform pattern classification after extracting signal features. EEG signals often use time-frequency domain and spatial domain

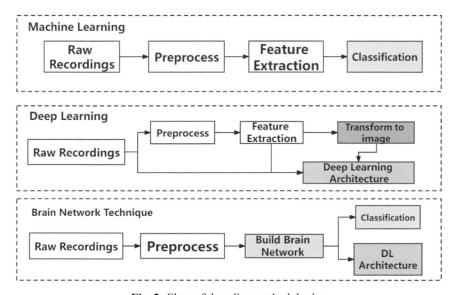

Fig. 2: Flow of decoding methodologies

	Machine Learning	Deep Learning	Brain Network Techniques
Characteristic	✓ mature technology ✓ diverse combinations ✓ short training time ✓ stable performance	✓ end-to-end structure ✓ automatically learn features ✓ high accuracy ✓ promising	✓ calculating the spatial position relationship between brain regions (electrodes) ✓ extracting neurobiological features.
Difficulties	● difficult to extract features of high-dimensional biological signals ● limited accuracy of classifier	● complicated model design & parameter adjustment ● requiring large amounts of data for training ● long training time ● poor interpretability	● still need to design machine learning or deep learning models for decoding

Fig. 3: Comparisons of neural decoding methodologies

feature-extraction methods. Time-frequency analysis can extract energy change characteristics in the time-frequency domain, which is suitable for SMR paradigm. Typical time-frequency algorithms include Short-time Fourier Transform (STFT), Wavelet Transform (WT), and Discrete Wavelet Transform (DWT). In particular, the latter two methods can decompose EEG signals in multi-resolution and multi-scale. Usually, independent component analysis (ICA) and principal component analysis (PCA) in the space domain are used to extract the characteristics of the P300 signal. Widely used classification methods include Support Vector Machines (SVMs) and Linear Discriminative Analysis (LDA). Many experiments have shown that under the same conditions, the classification performance of Gaussian kernel SVM is better than other classifiers, such as LDA, K-nearest neighbor (k-NN), naïve Bayes, and autoregression tree (Regression Tree, RT), etc. The overfitting problem can be solved by using Regularization LDA or by adjusting the training program of SVM.

4.2 End-to-end Deep Learning Algorithms

With the increasing amount of data and the development of computing power, deep learning technology is receiving more and more attention. The excellent deep network performance in the EEG field is more robust than machine learning algorithms [31]. There are two main problems in building a deep learning architecture to decode EEG signals – the data form of the network input and the design of the network structure. Studies have shown that compared to the calculated features (such as CSP) and the calculated image form (such as continuous wavelet time-frequency image), the network performance of directly inputting the original data is better [1]. Another problem is the design of the network structure, among which exceeding 70% of the research paper use Convolutional Neural Network (CNN) [2, 32, 34, 54]; others also have Recurrent Neural Network (RNN) [33, 62] and Hybrid Neural Network (h-CNN) [65, 68], a small number of which use stacked autoencoder (SAE) [20] and other structures. The comparison shows that the deep learning network based on CNN and h-CNN performs better. Zhang *et al.* [65] arranged the signal at each sampling moment in the time window as a picture according to the projection of the electrode distribution system on the two-dimensional plane, and input a three-layer CNN and LSTM cascade or parallel hybrid convolutional neural network. The spatio-temporal information of the signal is retained, and the accuracy rate in the multi-class MI classification of 108 healthy people in the PyhsioNet data set is over 98%. Compared with traditional machine learning and simple networks, the decoding performance improved by 10% to 30%. However, the use of deep networks requires a lot of training data, many parameters need to be adjusted, and the network structure design is also more complicated. Therefore, many scholars focus on research of lightweight network structure to solve the EEG decoding problem. Lawhern *et al.* [28] designed a compact convolutional neural network architecture EEGNet for EEG signals. It uses time convolution and deep convolution operations successively, besides utilizing fewer network parameters to extract spatiotemporal features. It has advantages in the general decoding performance of four paradigms, including P300, ERN, and MRCP, and SMR.

4.3 Brain Network Analysis

Brain network analysis is an intuitive expression of the interaction between different dynamic activities among neurons, neuron clusters or cerebral cortex regions [67]. It can characterize the topology and dynamic characteristics of the brain network. Based on the different brain function network connections of different brain activities, we can identify the corresponding brain state by calculating the functional brain network (FBN) of the patient under different training tasks. FBN is widely used in the recognition of emotional and cognitive states [24, 36], and is gradually applied to the study of decoding the rehabilitation status of upper limbs after stroke [66]. EEG electrode sensors are regarded as the nodes of the brain network diagram, and the relationship between the electrodes is regarded as the edges. Directed Transfer Function (DTF) and Partial Directed Coherence (PDC) are proposed to calculate vertex values in EEG graphs, including coherence (Coh), Pearson correlation coefficient (PCC), phase lock value (PLV), and Phase Lag Index (PLI), etc. When the brain function network is decoded by traditional machine learning algorithms, feature selection algorithms, such as sequential forward selection (SFS), can be used to reduce feature redundancy. In terms of data structure characteristics of the brain network graph, graph convolution can be used for network learning. Ghazani *et al.* [52] proposed an adaptive spatio-temporal graph convolutional network (ASTGCN) that simultaneously explores the characteristics of MI-EEG signals from the spatial relationship between time and channels, and shows more efficient and stable performance than CNN-SAE, EEGNet, and STGCN on 25 test data.

5. Existing Problems

Although the post-stroke BCI motor neurofeedback training system can promote the functional and structural reorganization of the brain and reflect improvement in motor planning and execution ability, the experimental results of some systems have not reached the measurable level or the smallest clinically significant change value (Minimal Clinically Important Difference, MCID), and is still in a sub-clinical effect stage [13]. The reasons for this problem may be the following:

(1) The real-time classification performance of the current BCI decoding system still needs to be improved. The training time of most machine learning algorithms is short, but the selection of features has a great impact on classification accuracy, and the real-time window data it uses can only reflect short-term brain activity. When using machine learning algorithms, it is necessary to deal with the noise of the original signal (such as eye movements and electromyography, etc.). So it is difficult to extract the features of high-dimensional biological signals, and the accuracy of the classifier is limited. Deep learning algorithms can take advantage of end-to-end's superiority to avoid feature selection, but their model training process takes a long time and requires a large number of training datasets to increase generalization capabilities. It takes much effort to collect data and train the model before formal feedback training. The brain function network method is based on the brain function reorganization mechanism, which can

extract neurobiological features, combined with deep learning algorithms, and is expected to improve real-time decoding performance.

(2) The experimental samples of existing research are limited and most lack control variables. Due to the mobility of stroke patients undergoing rehabilitation training in hospitals, many patients cannot complete the entire experiments for various reasons. It takes a long time to collect a large amount of patient data. At present, most studies have not accumulated such a large amount of patient's data. The number of patients in most studies ranges from four to 30 [8, 12, 40, 44, 53, 66]. Many academic platforms provide data from healthy people, but healthy people's brains have functional and structural differences. Thus they are only for reference purposes. Some experiments have carried out a randomized controlled trial (RCT) design, but due to the small sample size, there are many variables (such as the patient's lesion area, disease severity, age, disease time, disease type, gender, etc.). Thus, the reference value is yet to be verified.

(3) The BCI motor neurorehabilitation training system is affected by the individual differences of patients. The lesion location, severity of disease, type of disease, and other conditions of each stroke patient are not the same. Therefore, their neuroelectric signals are different. At present, the BCI system decoder needs to train targeted classification for each test subject; the neurofeedback mechanisms they apply are different. There is need to assign the best feedback method or a combination of multiple modes of feedback according to the specific conditions of the patient.

(4) There is lack of research on optimal training paradigm and training parameters. Some BCI motor neurofeedback training systems take a long time to train, and the patients are prone to fatigue. So the participation of the brain is reduced, and the effect of rehabilitation training is greatly limited. An experiment with a shorter training time may end without activating a desired pattern of the brain, and the rehabilitation effect will be poor. The duration of a single training session in existing research ranges from 30 minutes [57] to two hours [46]. In addition, more research is needed on the frequency of BCI treatment and the number of treatment courses to determine the best dose to promote upper limb exercise rehabilitation.

Acknowledgements

This research is supported by National Natural Science Foundation of China Regional Joint Project (No.U1913216); Guangdong Province Key Field Research Program, Jihua Laboratory Project (No.X190051TB190); Shanghai 'Science and Technology Action Innovation Plan' funded project (No.19441908200).

References

1. Al-Saegh A., Dawwd, S.A. and Abdul-Jabbar, J.M. (2021). Deep learning for motor imagery EEG-based classification: A review, *Biomedical Signal Processing and Control*, 63: 102172.

2. Amin, S.U., Alsulaiman, M., Muhammad, G., *et al.* (2019). Deep learning for EEG motor imagery classification based on multi-layer CNNs feature fusion, *Future Generation Computer Systems*, 101: 542-554.

3. An, X., Kuang, D., Guo, X., *et al.* (2014). A deep learning method for classification of EEG data based on motor imagery, *International Conference on Intelligent Computing*, pp. 203-210. Springer, Cham.

4. Aoyama, T., Kaneko, F. and Kohno, Y. (2020). Motor imagery combined with action observation training optimized for individual motor skills further improves motor skills close to a plateau, *Human Movement Science*, 73: 102683.

5. Bajaj, S., Butler, A.J., Drake, D., *et al.* (2015). Brain effective connectivity during motor-imagery and execution following stroke and rehabilitation, *NeuroImage: Clinical*, 8: 572-582.

6. Bhat, S.S., Fernandes, T.T., Poojar, P., *et al.* (2021). Low-field MRI of stroke: Challenges and opportunities, *Journal of Magnetic Resonance Imaging*, **54**(2): 372-390.

7. Blokland, Y., Vlek, R., Karaman, B., *et al.* (2012). Detection of event-related desynchronization during attempted and imagined movements in tetraplegics for brain switch control, *Annual International Conference of the IEEE Engineering in Medicine and Biology Society*, pp. 3967-3969. IEEE.

8. Carino-Escobar, R.I., Carrillo-Mora, P., Valdés-Cristerna, R., *et al.* (2019). Longitudinal analysis of stroke patients' brain rhythms during an intervention with a brain-computer interface, *Neural Plasticity,* 64.

9. Cervera, M.A., Soekadar, S.R., Ushiba, J., *et al.* (2018). Brain-computer interfaces for post-stroke motor rehabilitation: A meta-analysis, *Annals of Clinical and Translational Neurology*, 5(5): 651-663.

10. Choi, H., Lim, H., Kim, J.W., *et al.* (2019). Brain computer interface-based action observation game enhances mu suppression in patients with stroke, *Electronics*, **8**(12): 1466.

11. Choi, I., Bond, K. and Nam, C.S. (2016). A hybrid BCI-controlled FES system for hand-wrist motor function, *IEEE International Conference on Systems, Man, and Cybernetics (SMC)*, pp. 002324-002328. IEEE.

12. Chowdhury, A., Meena, Y.K., Raza, H., *et al.* (2018). Active physical practice followed by mental practice using BCI-driven hand exoskeleton: A pilot trial for clinical effectiveness and usability, *IEEE Journal of Biomedical and Health Informatics*, **22**(6): 1786-1795.

13. Desrosiers, J., Malouin, F., Richards, C., *et al.* (2003). Comparison of changes in upper and lower extremity impairments and disabilities after stroke, *International Journal of Rehabilitation Research*, **26**(2): 109-116.

14. Di Pellegrino, G., Fadiga, L., Fogassi, L., *et al.* (1992). Understanding motor events: A neurophysiological study, *Experimental Brain Research*, **91**(1): 176-180.

15. Eliassen, J.C., Boespflug, E.L., Lamy, M., *et al.* (2008). Brain-mapping techniques for evaluating poststroke recovery and rehabilitation: A review, *Topics in Stroke Rehabilitation*, **15**(5): 427-450.

16. GBD (2016). Neurology Collaborators. Global, regional, and national burden of neurological disorders, 1990-2016: A systematic analysis for the Global Burden of Disease Study 2016, *Lancet Neurol.*, **18**(5): 459-480.

17. Granger, C.V., Hamilton, B.B., Gresham, G.E., *et al.* (1989). The stroke rehabilitation outcome study: Part II. Relative merits of the total Barthel index score and a four-item subscore in predicting patient outcomes, *Archives of Physical Medicine and Rehabilitation*, **70**(2): 100-103.

18. Guggisberg, A.G., Koch, P.J., Hummel, F.C., *et al.* (2019). Brain networks and their relevance for stroke rehabilitation, *Clinical Neurophysiology*, **130**(7): 1098-1124.

19. Hardwick, R.M., Caspers, S., Eickhoff, S.B., *et al.* (2018). Neural correlates of action: Comparing metaanalyses of imagery, observation, and execution, *Neuroscience Biobehavioral Reviews*, 94: 31-44.
20. Hassanpour, A., Moradikia, M., Adeli, H., *et al.* (2019). A novel end-to-end deep learning scheme for classifying multi-class motor imagery electroencephalography signals, *Expert Systems*, 36(6): e12494.
21. Hétu, S., Grégoire, M., Saimpont, A., *et al.* (2013). The neural network of motor imagery: An ALE meta-analysis, *Neuroscience Biobehavioral Reviews*, 37(5): 930-949.
22. Jeon, Y., Nam, C.S., Kim, Y.J., *et al.* (2011). Event-related (De) synchronization (ERD/ERS) during motor imagery tasks: Implications for brain–computer interfaces, *International Journal of Industrial Ergonomics*, 41(5): 428-436.
23. JIA, Jie (2018). Implication of Left-Right Coordination and Counterbalance for Rehabilitation of Upper Limbs and Hand Function post Stroke (review), *Chinese Journal of Rehabilitation Theory and Practice*, 24(12) : 1365-1370.
24. Jiao, Z., Xia, Z., Ming, X., *et al.* (2019). Multi-scale feature combination of brain functional network for eMCI classification, *IEEE Access*, 7: 74263-74273.
25. Kaiser, V., Kreilinger, A., Müller-Putz, G.R., *et al.* (2011). First steps toward a motor imagery based stroke BCI: New strategy to set up a classifier, *Frontiers in Neuroscience*, 5: 86.
26. Krakauer, J.W. (2006). Motor learning: Its relevance to stroke recovery and neurorehabilitation, *Current Opinion in Neurology*, 19(1): 84-90.
27. Langhorne, P., Coupar, F. and Pollock, A. (2009). Motor recovery after stroke: A systematic review, *The Lancet Neurology*, 8(8): 741-754.
28. Lawhern, V.J., Solon, A.J., Waytowich, N.R., *et al.* (2018). EEGNet: A compact convolutional neural network for EEG-based brain–computer interfaces, *Journal of Neural Engineering*, 15(5): 056013.
29. Li, C., Jia, T., Xu, Q., *et al.* (2019). Brain-computer interface channel-selection strategy based on analysis of event-related desynchronization topography in stroke patients, *Journal of Healthcare Engineering*.
30. Li, G., Lee, C.H., Jung, J.J., *et al.* (2020). Deep learning for EEG data analytics: A survey, *Concurrency and Computation: Practice and Experience*, 32(18): e5199.
31. Li, Y., Zhang, X.R., Zhang, B., *et al.* (2019). A channel-projection mixed-scale convolutional neural network for motor imagery EEG decoding, *IEEE Transactions on Neural Systems and Rehabilitation Engineering*, 27(6): 1170-1180.
32. Luo, T. and Chao, F. (2018). Exploring spatial-frequency-sequential relationships for motor imagery classification with recurrent neural network, *BMC Bioinformatics*, 19(1): 1-18.
33. Mammone, N., Ieracitano, C. and Morabito, F.C. (2020). A deep CNN approach to decode motor preparation of upper limbs from time-frequency maps of EEG signals at source level, *Neural Networks*, 124: 357-372.
34. Mattia, D., Pichiorri, F., Colamarino, E., *et al.* (2020). The Promotoer, a brain-computer interface-assisted intervention to promote upper limb functional motor recovery after stroke: A study protocol for a randomized controlled trial to test early and long-term efficacy and to identify determinants of response, *BMC Neurology*, 20(1): 1-13.
35. Moon, S.E., Chen, C.J., Hsieh, C.J., *et al.* (2020). Emotional EEG classification using connectivity features and convolutional neural networks, *Neural Networks*, 132: 96-107.
36. Muralidharan, A., Chae, J. and Taylor, D. (2011). Extracting attempted hand movements from EEGs in people with complete hand paralysis following stroke, *Frontiers in Neuroscience*, 5: 39.
37. Murphy, T.H. and Corbett, D. (2009). Plasticity during stroke recovery: From synapse to behavior, *Nature Reviews Neuroscience*, 10(12): 861-872.

38. Nagai, H. and Tanaka, T. (2019). Action observation of own hand movement enhances event-related desynchronization, *IEEE Transactions on Neural Systems and Rehabilitation Engineering*, **27**(7): 1407-1415.

39. Norman S L, McFarland D.J., Miner, A., *et al*. (2018). Controlling pre-movement sensorimotor rhythm can improve finger extension after stroke, *Journal of Neural Engineering*, **15**(5): 056026.

40. Nudo, R. (2003). Adaptive plasticity in motor cortex: Implications for rehabilitation after brain injury, *Journal of Rehabilitation Medicine-Supplements*, 41: 7-10.

41. O'Brien, A.T., Bertolucci, F., 30. Torrealba-Acosta, G., *et al*. (2018). Non-invasive brain stimulation for fine motor improvement after stroke: A meta analysis, *Eur. J. Neurol.*, **25**(8): 1017-1026

42. Ono, Y., Wada, K., Kurata, M., *et al*. (2018). nhancement of motor-imagery ability via combined action observation and motor-imagery training with proprioceptive neurofeedback, *Neuropsychologia*, 114: 134-142.

43. Rathee, D., Chowdhury, A., Meena, Y.K., *et al*. (2019). Brain–machine interface-driven post-stroke upper-limb functional recovery correlates with beta-band mediated cortical networks, *IEEE Transactions on Neural Systems and Rehabilitation Engineering*, **27**(5): 1020-1031.

44. Remsik, A.B., Dodd, K., Williams, Jr. L., *et al*. (2018). Behavioral outcomes following brain–computer interface intervention for upper extremity rehabilitation in stroke: A randomized controlled trial, *Frontiers in Neuroscience*, 12: 752.

45. Rizzolatti, G., Cattaneo, L., Fabbri-Destro, M., *et al*. (2014). Cortical mechanisms underlying the organization of goal-directed actions and mirror neuron-based action understanding, *Physiological Reviews*, **94**(2): 655-706.

46. Schaechter, J.D. (2004). Motor rehabilitation and brain plasticity after hemiparetic stroke, *Progress in Neurobiology*, **73**(1): 61-72.

47. Shoham, S., Halgren, E., Maynard, E.M., *et al*. (2001). Motor-cortical activity in tetraplegics, *Nature*, **413**(6858): 793-793.

48. Shu, X., Chen, S., Meng, J., *et al*. (2018). Tactile stimulation improves sensorimotor rhythm-based bci performance in stroke patients, *IEEE Transactions on Biomedical Engineering*, **66**(7): 1987-1995.

49. Subramanian, S.K., Massie, C.L,, Malcolm, M.P., *et al*. (2010). Does provision of extrinsic feedback result in improved motor learning in the upper limb poststroke? A systematic review of the evidence, *Neurorehabilitation and Neural Repair*, **24**(2): 113-124.

50. Sun, B., Zhang, H., Wu, Z., *et al*. (2021). Adaptive Spatiotemporal Graph Convolutional Networks for Motor Imagery Classification, *IEEE Signal Processing Letters*, 28: 219-223.

51. Tabernig, C.B., Lopez, C.A., Carrere, L.C., *et al*. (2018). Neurorehabilitation therapy of patients with severe stroke based on functional electrical stimulation commanded by a brain computer interface, *Journal of Rehabilitation and Assistive Technologies Engineering*, 5: 2055668318789280.

52. Taheri, S., Ezoji,, M., Sakhaei, S.M. (2020). Convolutional neural network based features for motor imagery EEG signals classification in brain–computer interface system, *SN Applied Sciences*, **2**(4): 1-12.

53. Tani, M., Ono, Y., Matsubara, M., *et al*. (2018). Action observation facilitates motor cortical activity in patients with stroke and hemiplegia, *Neuroscience Research*, 133: 7-14.

54. Teo, W.P. and Chew, E. (2014). Is motor-imagery brain-computer interface feasible in stroke rehabilitation?, *PMR*, **6**(8): 723-728.

55. Tung, S.W., Guan, C., Ang, K.K., *et al.* (2013). Motor imagery BCI for upper limb stroke rehabilitation: An evaluation of the EEG recordings using coherence analysis, *35th Annual International Conference of the IEEE Engineering in Medicine and Biology Society (EMBC)*, IEEE, 261-264.

56. Van Dokkum, L.E.H., Ward, T. and Laffont, I. (2015). Brain computer interfaces for neurorehabilitation – its current status as a rehabilitation strategy post-stroke, *Annals of Physical and Rehabilitation Medicine*, **58**(1): 3-8.

57. von Bernhardi, R., Eugenín-von, Bernhardi, L. and Eugenín, J. (2017). What is neural plasticity?, *The Plastic Brain*, 1-15.

58. Wang Longde, Liu Jianmin, Yang Yi, *et al.* (2017). Essentials of report on the prevention for Chinese stroke 2017, *Chinese Journal of Cerebrovascular Diseases*, **15**(11): 611-617.

59. Wang, Mengya, Wang, Zhongpeng, Chen, Long, *et al.* (2019). Research Progress and Prospects of Motor Neurofeedback Rehabilitation Training after Stroke, *Chinese Journal of Biomedical Engineering*, **38**(6): 742-752.

60. Wang, P., Jiang, A., Liu, X., *et al.* (2018). LSTM-based EEG classification in motor imagery tasks, *IEEE Transactions on Neural Systems and Rehabilitation Engineering*, **26**(11): 2086-2095.

61. WHO. *World Health Report* (2010). Geneva: WHO.

62. Wilkins, Lippincott Williams (2008). *Occupational Therapy for Physical Dysfunction* [M].

63. Zhang, D., Yao, L., Chen, K., *et al.* (2019). Making sense of spatio-temporal preserving representations for EEG-based human intention recognition, *IEEE Transactions on Cybernetics*, **50**(7): 3033-3044.

64. Zhang, J., Wang, B., Li, T., *et al.* (2018). Non-invasive decoding of hand movements from electroencephalography based on a hierarchical linear regression model, *Review of Scientific Instruments*, **89**(8): 084303.

65. Zhang, L., Qiu, T., Lin, Z. *et al.* (2020). Construction and Application of Functional Brain Network Based on Entropy, *Entropy*, **22**(11): 1234.

66. Zhang, R., Zong, Q., Dou, L., *et al.* (2020). Hybrid deep neural network using transfer learning for EEG motor imagery decoding, *Biomedical Signal Processing and Control*, 63: 102144.

67. Zhang, Yan, Zhou, Huijun, Ni, Miaomiao, *et al.* (2020). Effect of stage rehabilitation nursing on upper limb and hand dysfunction in elderly stroke patients with hemiplegia, *Geriatrics Health Care*, **26**(04): 667-669.

A Modified Dynamic Time Warping (MDTW) and Innovative Average Non-self Match Distance (ANSD) Method for Anomaly Detection in ECG Recordings

Xinxin Yao[1] and Hua-Liang Wei[1,2]*

[1] Department of Automatic Control and Systems Engineering, University of Sheffield, Sheffield, S1 3JD, UK
[2] INSIGNEO Institute for in Silico Medicine, University of Sheffield, Sheffield, S1 3JD, UK
* Corresponding Author: w.hualiang@sheffield.ac.uk

1. Introduction

Cardiovascular diseases (CVDs) are the number one cause of death globally. According to the statistics of World Health Organization, in 2016, almost 17.5 million people died from CVDs, representing 31% of all global deaths [7]. In recent decades, diagnosis and prevention of CVDs have become the leading concern in clinical medicine. As the most commonly used biological signals in medical field, in the last few decades, ECGs are easy to collect via modern medical facilities. In clinical application, ECGs are used to determine the cardiac structure and function of patients. For CVDs, such as arrhythmia, myocardial and ischemia, they occur over a certain period when the heart of a cardiac patient does not work normally; in other words, corresponding ECG is an anomalous segment.

Anomalies are patterns in data that do not conform to a well-defined notion of normal behavior [22]. In daily life, anomalous signals appear in various fields because of different reasons, for instance, bank card fraud in economic field [14, 17] and cyber-intrusion in networking field [1, 3]. Similarly, ECG of cardiac patient is an anomaly at some periods. For example, ECG collected during abnormal work period of cardiac patient is different from normal period series [18]. Figure 1 gives a visual intuition of an anomalous segment which is highlighted by an ellipse.

Anomalies detection in ECG aims to detect the unexpected patterns from large volumes of non-anomalous data and such unexpected patterns are defined as anomalies. Nowadays, with an increasing aging population and the development of medical infrastructures, there are increasingly huge numbers of cardiac patients going to hospital for examination every day. Manual analysis (e.g. analysis by hand) of such mega data will consume a huge amount of manpower and resources. Moreover, errors may occur in the detection, resulting due to high-intensity handwork. Recently, applying machine learning and data analysis techniques to anomaly detection of ECG signals has become increasingly important in clinical medicine [23, 10, 9]. In [12], an anomaly detection method was introduced to find the most unusual time series subsequence. Later, in [16, 19, 4], a series of related works were proposed to improve the performance of the method in [12]. However, because these extended methods use the normalized Euclidean distance and non-self-match distance to find anomaly or anomalies, they cannot work well when there are two or more similar anomalies.

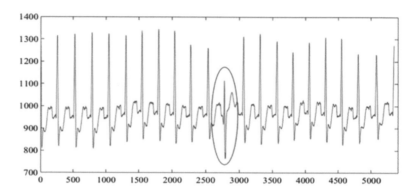

Fig. 1: Anomaly detection and labelled in an ECG (collected from MIT-BIH database)

In this study, in order to correctly detect all the anomalous segments via time series mining techniques, a modified dynamic time warping (MDTW) is used to calculate the distance between candidates, and a new notion, called the average non-self match distance (ANSD), is used to detect all the anomalous segments from raw data. To demonstrate the performance of the proposed ECG anomalies detection method, both the proposed method and previous published methods are applied to 30 ECGs from MIT-BIH arrhythmia database [13]. The experimental results show that the proposed method obviously outperforms previous published methods, which is evidenced by the following anomaly detection accuracy: for non-anomalous ECGs and ECGs only contain one anomaly, where all the three methods (BFDD, AWDD and the proposed method) work extremely well; for ECGs with two or more anomalies which are significantly different from each other, both AWDD and the proposed method work well while the detection accuracy of BFDD is 40%, for ECGs with more than one anomalies which are similar to each other, BFDD and AWDD do not work (the accuracy is zero), whereas the accuracy of the proposed method is 100%.

It is worth stressing that when the proposed method is applied to ECG data, it does not just simply use RQS complex but treats the ECG data of interest as a whole to reveal any anomaly through data segmentation and the analysis of the associated dynamic time warping and the average non-self match distance using the newly proposed algorithms.

The remainder of this paper is organized as follows: in Section 2, related works are described; in Section 3, the proposed time series similarity measure method and newly defined anomalies detection notion are presented; in Section 4, the proposed ECGs anomalies detection method and existing time series analysis-based approaches are applied to a series of ECGs, and some comparative analysis results are reported; and finally, this work is briefly summarized in Section 5.

2. Related Work

In the past few decades, anomalies detection has been used in diverse fields, such as cyber-intrusion detection [2], fraud detection [8], medical anomaly detection [11, 6], and so on. This study focuses on ECG anomaly detection via time series analysis, where ECG is regarded as a periodical time series reflecting the periodical change of heartbeat [21]. In this section, the basic notion (non-self match) of time series analysis-based anomaly detection is described firstly, then two popular ECG anomaly detection methods are reviewed, namely, brute force discord discovery (BFDD), and adaptive window-based discord discovery (AWDD).

2.1 Nearest Non-self Match

Given one time series A and one of its subsequence B beginning at position P, in general, in time series A, the beginning points of the best matches to B (apart from itself) should be $P \pm 1P \pm 1$ or $P \pm 2P \pm 2$. Therefore, excluding unnecessary matches is an important step prior to detecting anomalies, otherwise, distance between candidate subsequence and its corresponding best match subsequence will be lower than a pre-obtained threshold and it is impossible to find anomalies. In [12], one matching notion, called as *non-self match*, was introduced to remove trivial matches in the process of anomaly detection.

Given one time series T and its two subsequences $T1$ and $T2$, the beginning points of $T1$ and $T2$ are P and Q, and the length values of both $T1$ and $T2$ are set as n, $T2$ can be defined as a non-self match to $T1$ if the position distance greater or equal to n $(|P - Q|) \geq n)$. As an example, this is shown as below:

$$T = \overline{\fbox{a b c}} \, a b c a b c a b c a b c a b c \, \fbox{a b c} \, a b c a b c a b c$$

Note that $T1$ and $T2$ are labelled by dot box and solid box respectively, the length values of them are set as 3, the beginning point of $T1$ is 1 and the beginning point of $T2$ is 19. In this case, $T2$ can be defined as a non-self match to $T1$ because $|19 - 1|$ ≥ 3. On the contrary, another example shown in Fig. 3 describes that $T2$ cannot be defined as a non-self match to $T1$.

Below is another case, where the settings are the same as in the above illustration, the only difference being that the beginning points of $T1$ and $T2$ are 17 and 19 respectively. It can be noticed that $T1$ and $T2$ are partly folded together. For the position distance, it is $|19 - 17| = 2$ and hence lower than the length of sliding window is shown as follows:

$$T = a b c a b c a b c a b c a b c a \overline{\fbox{b c \fbox{a b c}}} a b c a b c a b c$$

Non-self match subsequences of one segment cannot be directly used to define whether the corresponding candidate is anomalous or not. For previous published anomaly detection methods, nearest non-self match of the candidate is used to detect the anomalous segments. Given one time series T and one candidate A, the distances between A and all its non-self matches are calculated and recorded as $D = [d1, d2, ..., dn]$, with the nearest non-self match distance being the minimum value in D and the corresponding segment is the nearest non-self match of A.

2.2 Brute Force Discord Discovery

Brute force discord discovery (BFDD) algorithm was initially proposed in [12, 15] and the advantage of this algorithm is that it is easy to understand and implement. Based on BFDD, the implemented procedure of time series anomaly detection is as follows: (1) the first segment to be tested is the subsequence whose length is equal to that of sliding window and its first point is the same with that of time series, as shown in Figs. 2(a) and 2(b); (2) after the definition of testing subsequence, sliding down the defined window on sample at a time and calculating the distance between the testing subsequence and the subsequence in the sliding window, this process can help us to find the nearest non-self match of the testing subsequence, as shown in Fig. 2(c); (3) once the first nearest non-self match distance and its corresponding subsequence are recorded, the first point of the testing subsequence will move from the first point of the time series to the second point. Meanwhile, the length of testing subsequence is still equal to that of sliding window, as shown in Fig. 2(d); (4) for every testing subsequence with length m, and an original time series with length n, in order to find the nearest non-self match, it needs to calculate at least $(n - 3m + 1)$ times, and sometimes $(n - 2m + 1)$ times, as shown in Fig. 2(e). (5) According to the obtained nearest non-self match distance values, the anomalous segment can be detected via the comparison between the recorded values and a pre-obtained threshold, which is calculated by applying BFDD to training time series. Fig. 2 illustrates the mechanism of BFDD.

The pseudo-code and procedure of BFDD is described in Algorithm 1.

Algorithm 1: Brute Force Discord Discovery

```
            The length of time series: n
            The length of sliding window: m
best_so_far_dist ← 0
for i = 1 to n − m + 1 do
        nearest_neighbour_dist = infinity
    for j = 1 to n − m + 1 do
        if |i − j| ≥ n do
            if Dist (Tp,…,T(p + n − 1),Tq,…,T(q + n −1))
                < nearest_neighbour_dist do
            nearest_neighbour_dist = Dist(Tp,…,T(p + n − 1), Tq,…,T(q + n − 1))
            end if
        end if
    end for
        if nearest_neighbour_dist>best_so_far_dist do
```

```
                best_so_far_dist = nearest_neighbour_dist
                best_so_far_loc = p
        end if
end for
```

From the above pseudo-code, it can be noticed that this method is achieved with nested calculations, where the outer calculation considers each possible candidate subsequence, and the inner calculation performs a linear scan to identify the nearest non-self match of corresponding candidate [28].

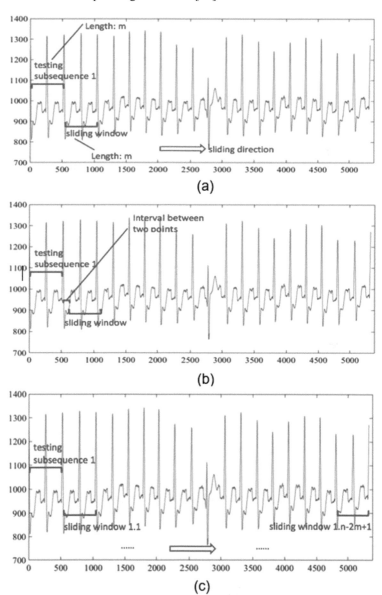

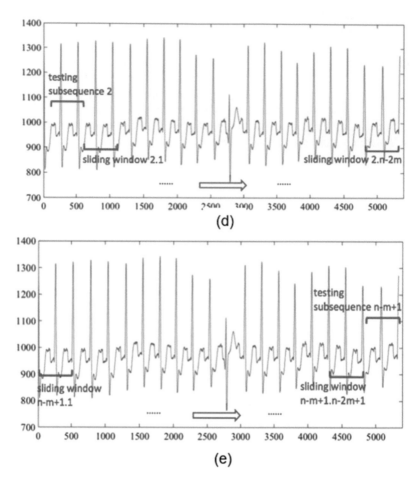

Fig. 2: Mechanism of BFDD-based anomaly detection. (a) Defining the first testing subsequence and sliding window; (b) sliding down the window on sample at a time; (c) every testing subsequence has at least $n - 3m + 1$ non-self match distances; (d) moving 1 unit backward of the testing subsequence; (e) the testing subsequence keeps moving to the end of the time series

As an example, the length of sliding window is 300, BFDD is applied to the ECG signal shown in Fig. 1, the associated nearest neighbor distances are shown in Fig. 3.

Once the nearest non-self match distances are computed, all these distances will be compared with a pre-obtained threshold (computed through applying BFDD to training dataset) and the corresponding segments are going to be identified. But it is worth mentioning the low computational efficiency of BFDD-based anomaly detection. As it can be observed that there is a nested calculation during the whole process, every calculation has to compute the distances between testing subsequence and its non-self subsequence at least $(n - 3m + 1)$ times (n is the length of time series and m is the length of sliding window). This will take a huge amount of time for even

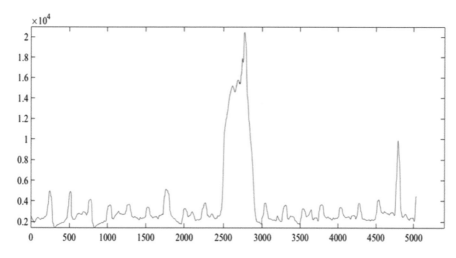

Fig. 3: Nearest neighbor distances

moderately large datasets. Assuming that the length values of sliding window and testing subsequence are set as 300, as the ECG signal shown in Fig. 1 is 15 seconds record and contains 5,400 data points, the whole calculation process has to take over 20 million times. For a normal computer, it has to take at least 50 seconds to finish the calculation. Although BFDD-based time series anomaly detection is time consuming, there is one advantage; it is its universality, which means this method can be used to detect anomalies for various kinds of time series.

2.3 Adaptive Window Discord Discovery

For some special types of time series, the application of universal anomaly detection methods may complicate the operation process and distort the calculation accuracy. In terms of ECG data, in order to overcome the heavy computational load involved in BFDD-based anomaly detection, adaptive window discord discovery (AWDD) was proposed in [16]. AWDD separates ECG into a number of segments based on the peak points; then it measures the distances between each segment and determines which subsequence can be treated as anomaly. It should be noted that the segments separated by peak points do not have the same length while the distance measure in AWDD is Euclidean distance, which requires that the length values of two candidates should be the same. To solve this problem, if the length values of two candidates are different, the longer one is compressed firstly so that its length is equal to that of the shorter one. In comparison with BFDD, the calculation time is significantly reduced through the application of AWDD without losing detection accuracy. The whole process of AWDD-based anomaly detection is described by Algorithm 2.

Algorithm 2: Adaptive Window Discord Discovery

Requirements: A time series: T

Location of peak points: P

$n \leftarrow$ length of input time series
$m \leftarrow$ number of peak points
for i = 1: $n - 1$ **do**
    outlength = P(i + 1) − P(i);
    for j = 1: $n - 1$ **do**
        innerlength = P (j +1) − P(j);
        if outlength > innerlength **do**
            B = imresize(A(P(i):P(i + 1)), [1, innerlength]);
            C = A(P(j):P(j + 1) − 1);
        else
            B = imresize(A(P(j):P(j + 1)), [1, outlength]);
            C = A(P(i):P(i + 1) − 1);
        end if
        ddd(j) = dist(B,C)
    end for
    nearest_neighbour_dist(i) = min(ddd);
    if nearest_neighbour_dist (i) > threshold **do**
        best_so_far_loc = i
    end if
end for

The inputs of Algorithm 2 include one ECG time series and the locations of peak points. The output of this Algorithm is the location or locations of anomaly or anomalies.

AWDD compresses the longer subsequence so that its length is equal to that of the shorter one. This enables the use of Euclidean distance to measure the distance between two candidates. As an illustration, Fig. 4(a) provides a simple example of two time-series with different length values. In Fig. 4(b), the longer time series is compressed.

AWDD is also applied to the ECG data in Fig. 1. As there are 21 normal peak points, prior to distance measure, the ECG is separated nto 20 segments. Then every segment is regarded as one testing subsequence and the corresponding nearest non-self match distances of these 20 segments are recorded, as shown in Fig. 5.

As the result of BFDD-based anomaly detection shown in Fig. 5, thenearest non-self match distances are anomalous between 2500 to 3000, the location of abnormal point of nearest non-self match distances in Fig. 5 is 11 and as the interval between two normal points is almost 260, it is obvious that AWDD-based nearest non-self match distances reserve most information of BFDD-based nearest non-self match distances. For the improvement of AWDD, the distances are calculated between every segment instead of sliding down the testing subsequence on one sample time. The whole procedure of AWDD-based anomaly detection is only $(20 - 1)^2$ times; for a normal computer, it only takes about 0.36 seconds.

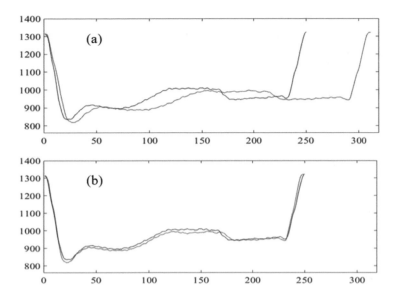

Fig. 4: Two time series. (a) before compression; (b) after compression.

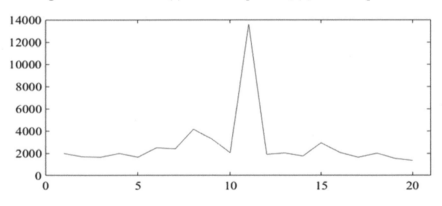

Fig. 5: Nearest neighbor distance based on AWDD

3. Proposed Method

If there is only one disordered segment or several significantly different disordered segments in an ECG signal, both BFDD and AEDD can correctly detect the anomalous segment(s), while AWDD outperforms BFDD in term of computational efficiency. But, both methods have two common drawbacks: (i) Euclidean distance measure method may influence the accuracy of anomaly detection if timeline drift exits during the process of calculating the non-self match distances; (ii) they cannot correctly detect the anomalies when there are two or more anomalous segments and the distance between anomalies is lower than the threshold. In this section, in order to improve the accuracy and efficiency of ECG anomalies detection, one modified distance measure method and one new notion are applied to ECG anomaly detection, and are introduced as follows: (1) modified DTW is presented to improve the accuracy of time series distance measure; (2) average non-self match distance is proposed to replace the nearest non-self match distance; (3) the analysis procedure using the proposed method for ECG anomaly detection is described.

3.1 Distance Measure

Unlike traditional DTW-based distance measure (which is detailed in [5, 20]), where DTW is used to directly calculate the distances between corresponding points and sum them as final distance. In this paper, the distance between two candidates is defined according to the DTW distance and the optimal align. Given two time series, A and B, the distance between them is defined as follows:

$$\text{Dist}(A, B) = d + \left(\frac{1-1a}{1a}\right) \times (\text{sum}(A\text{new}) - \text{sun}(A))$$

$$+ \left(\frac{1-1b}{1b}\right) \times (\text{sum}(B\text{new}) - \text{sun}(B)) \tag{1}$$

where Dist(A, B) represents the distance between A and B, d is the DTW distance between A and B, 1 is the length of the optimal align path; Anew and Bnew are two new time series segments which are constructed according to A, B and the optimal align path; 1a and 1b are the length values of Anew and Bnew; the function sum (…) returns the sum of elements of the input segment. The whole process of this distance measure method is summarized by Algorithm 3.

Algorithm 3: Distance Calculation

Requirements: Two time series A and B
1a ← length of A
1b ← length of B
distance ← DTWdistance (A,B)
optimal path ← DTWdrift (A,B)
wa ← first column of optimal path
wb ← second column of optimal path
1 ← length of optimal path
for i = 1 to 1 **do**
    A new(i) = A(wa(i))
    B new(i) = B(wb(i))
end for
final distance = distance + ((1 – 1a)/1a) × (sum(Anew)
                – sum(A)) + ((1–1b)/1b) × (sum(Bnew) -sun(B))

The inputs of this algorithm are two time series and the output is the distance between them.

To demonstrate the performance of the proposed method for time series similarity measure, two time series A and B are shown in Fig. 6. The proposed method, together with traditional dynamic time warping and Euclidean distance, are applied to calculate the distance between A and B.

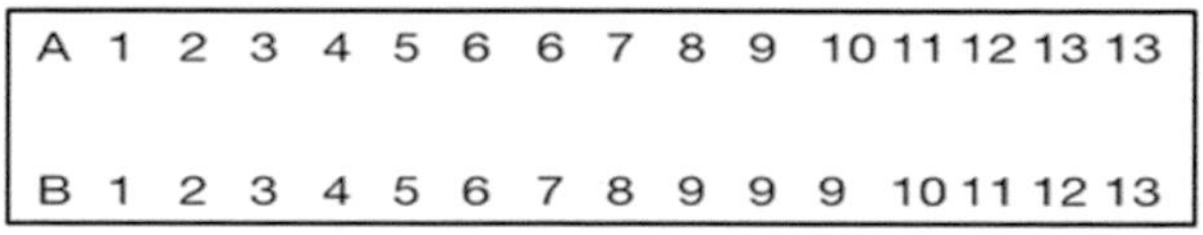

A 1 2 3 4 5 6 6 7 8 9 10 11 12 13 13
B 1 2 3 4 5 6 7 8 9 9 9 10 11 12 13

Fig. 6: Template time series

The calculation matching image of Euclidean distance is shown in Fig. 7 and the distance between *A* and *B* is 7. However, we can find that the subsequence from 7th point to 10th point in time series A is similar to the subsequence from 6th point to 9th point in time series B, but Euclidean distance directly calculates the distances between elements at same time point and sum them up as the distance between these two time series.

A 1 2 3 4 5 6 6 7 8 9 10 11 12 13 13
B 1 2 3 4 5 6 7 8 9 9 9 10 11 12 13

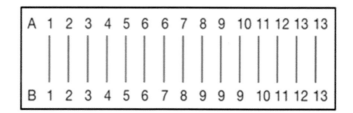

Fig. 7: Matching image of distance calculation based on Euclidean distance

The matching image of using traditional DTW to measure the distance between *A* and *B* is shown in Fig. 8. It can be noticed that the timeline has been warped and the most similar elements have aligned with each other. However, the final distance between these two time series is 0 although they are not identical.

A 1 2 3 4 5 6 6 7 8 9 10 11 12 13 13
B 1 2 3 4 5 6 7 8 9 9 9 10 11 12 13

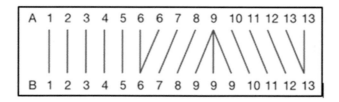

Fig. 8: Matching image of distance calculation based on DTW

Traditional distance measures, such as Euclidean distance and DTW do not work well for the above time series *A* and *B*. That is the motivation to propose the new method (Algorithm 3) to overcome the disadvantage of traditional distance measures. In Algorithm 3, in order to eliminate the impact of the neglect of timeline drift, the distance between two candidates is defined according to three variables: DTW distance, optimal align path between two time series, and the sum of distances between the extended new points and base point (these new points exist when there is timeline drift between two candidates, and vice versa). Compared with the results obtained based on Euclidean distance and traditional DTW, the distance between *A* and *B* computed, based on Algorithm 3 is 4.9333, not only eliminates the error caused by existence of timeline drift, but also removes the error caused by neglect of timeline drift.

3.2 Average Non-self Match Distance

To overcome the drawback of BFDD and AWDD that they can only work well for anomaly detection when all the anomalies in time series of interest are significantly different from each other, this subsection proposes a new notion, called as average non-self match distance.

Given one time series T, one of its segment is A, non-self matches of A in T are stored in $A_m = [A_1, A_2, ..., A_n]$, and the distances between A and all its non-self matches are obtained through the application of the proposed distance measure method and recorded in $D = [d_1, d_2, ..., d_n]$. In terms of anomaly detection based on BFDD and AWDD, the minimum value in D is recorded as nearest non-self match and used to identify whether A is anomaly or not. Different from nearest non-self match, average value of D is recorded and used in our proposed method to identify whether A is an anomaly or not.

In order to clearly state the advantage of average non-self match distance in anomalies detection of multi-anomalous time series, a time series contains two same anomalies as defined in Fig. 9, which is constructed through repeatedly using a time series (e.g. ECG signal in Fig. 1).

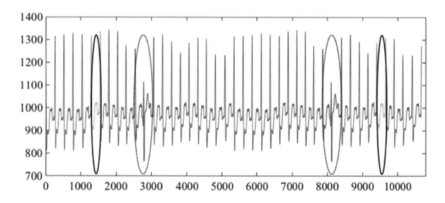

Fig. 9: Two-anomaly time series

Two normal segments and two anomalous segments (highlighted by solid and dot boxes in Fig. 9) are extracted as testing segments; both nearest non-self match and average non-self match distance are applied to these four segments. Table 1 illustrates the values that are used to identify whether the input segments are anomalous or not.

In Table 1, the second row states the distances between the extracted segments and their corresponding nearest non-self match segments, the third row illustrates the average values of distances between extracted segments and all their corresponding non-self match segments. The values in Table 1 show that anomaly detection methods based on nearest non-self match cannot correctly detect anomalies in some special conditions, while our proposed method correctly detects all the anomalous segments.

Table 1: Values used for anomalies identification

Distance	Normal 1	Normal 2	Anomaly 1	Anomaly 2
Nearest Non-self Match	0	0	0	0
Average Non-self Match Distance	1.9730×10^5	1.8508×10^5	6.6383×10^5	6.6383×10^5

3.3 Anomalies Detection in ECG Data

The analysis procedure using the proposed method for ECG anomaly detection is as follows: (1) separate the input ECG into several segments based on peak points; and (2) define the anomalies using the average non-self match distance of every segment.

3.3.1 Peak Points Collection

It is known that ECG can be defined as periodical time series because ECG derives from regularly heart muscle beat. Therefore, peak points-based ECG segmentation is applied prior to distance measure. Algorithm 4 briefly describes the procedure of peak points' collection.

Algorithm 4: Peak Points Collection
Requirements: An ECG signal: T
A defined value: h
$n \leftarrow$ length of input ECG signal
$m \leftarrow 1$
for i = 1 to n **do**
    if T(i) $\geq$ h **do**
        location(m) = i
        $m \leftarrow m + 1$
    end if
end for

The inputs of Algorithm 4 include one ECG signal and one threshold. This threshold is obtained through training the available ECG data. The output is a vector containing locations of peak points.

As an example, Algorithm 4 is applied to the ECG signal shown in Fig. 1, and the detected peak points are highlighted by red dots and shown in Fig. 10.

3.3.2 Anomaly Detection

As mentioned in Introduction, anomalies are patterns in data that do not conform to a well-defined notion of normal behavior. Therefore, at the beginning of anomaly detection, a criterion has to be defined and it can be obtained through application of the proposed anomaly detection method to the available training data. For example, average non-self match distances of all normal segments are stored in D_n, average non-self match distances of all anomalous segments are stored in D_a, the threshold is always defined as the average value of the maximum value in D_n and the minimum value in D_a.

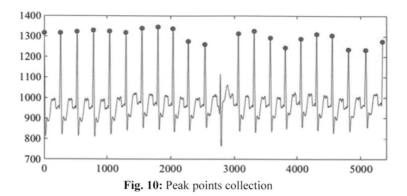

Fig. 10: Peak points collection

Once the threshold is obtained, the average non-self match distances of all the segments in testing signal have to be computed and compared with the threshold. If the value is greater than the threshold, the corresponding segment is defined as anomalous; on the contrary, if the value is lower than the threshold, the corresponding segment is defined as normal. Fig. 11 shows the average non-self distances of all the segments in ECG signal in Fig. 1, as the threshold is 1.8×10^4 (which is obtained by applying the proposed method to 10 training ECG signals). The average non-self match distance of the 11st segment is greater than the threshold, hence the 11st segment is anomaly and the others are normal segments.

It can be noticed that the average non-self match value of the anomalous segment in Fig. 11 is significantly greater than the values of normal segments, and the corresponding segment can be defined as anomalous directly without the comparison with the threshold. But it should be noted this is a special condition; in some common cases, comparing with threshold is the best way to correctly detect the anomalous segments.

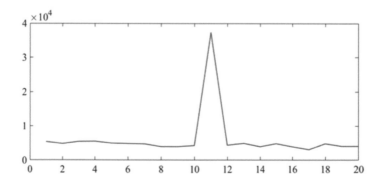

Fig. 11: Average non-self match distances of time series in Fig. 1

4. Case Study

The resource of ECGs (30 ECGs in total) included in the MIT-BIH database is a set of over 4,000 long-term Holter recordings that were obtained by Beth Israel Hospital

Arrhythmia Laboratory from 1975 to 1979. Approximately 60% of the recordings were obtained from inpatients [34]. Because of these, testing performance of ECG anomaly detection algorithms based on this database has strong conviction. The experimental database used in this study contains 30 ECGs, wherein 10 of them are used as training database to obtain the threshold and the rest 20 ECGs are used as testing database.

The 30 ECGs used in the study are listed in Table 2, where the 1st column is an index of the 10 training ECGs and 20 testing ECGs, the 2nd column illustrates the starting and ending time of corresponding ECG signal; the 3rd column is a location index to show where the anomaly occurs; and the 4th column is an indication of

Table 2: ECG excerpts from MIT-BIH record 109

No. ECG Datasets for Training	Start-end Points	Anomaly Location	Anomaly Identification
1	140s-180s	NA	NO
2	440s-480s	NA	NO
3	560s-600s	NA	NO
4	700s-740s	NA	NO
5	740s-780s	NA	NO
6	20s-60s	6758	YES
7	80s-120s	4474	YES
8	200s-240s	12890	YES
9	260s-300s	5690	YES
10	520s-560s	3205	YES
11	860s-900s	NA	NO
12	940s-980s	NA	NO
13	980s-1020s	NA	NO
14	1180s-1220s	NA	NO
15	1220s-1260s	NA	NO
16	620s-660s	11630	YES
17	660s-700s	7928	YES
18	820s-860s	9410	YES
19	1300s-1340s	6100	YES
20	1400s-1440s	7170	YES
21	900s-940s	6270,11590	YES
22	1060s-1100s	10320,12920	YES
23	1100s-1140s	2412,11970	YES
24	1140s-1180s	966,9957	YES
25	500s-540s	3655,10410	YES
26	NA	11630, 26030	YES
27	NA	7928, 22328	YES
28	NA	9410, 23810	YES
29	NA	6100, 20500	YES
30	NA	7170, 21570	YES

whether there is an anomaly or anomalies in the corresponding ECG. Each of the first five training ECGs does not contain anomalous segment whereas each of the last five training ECGs contains one anomalous segment. In testing dataset, in order to demonstrate the ability of the proposed method for detecting multiple anomalous ECGs, the 20 test ECGs were chosen as follows: No. 11-15 contain no anomaly, No. 16-20 contain one anomaly each, No. 21-25 contain two significantly different anomalies, and No. 26-30 are constructed through repeating one-anomalous (No. 16-20) segment and each of them contains two same anomalies.

4.1 BFDD-based Anomaly Detection

BFDD is applied to training ECGs (Nos. 1-10 in Table 2) to calculate the threshold. The nearest non-self match distances of normal segments and anomalous segments are shown in Table 3.

Table 3: Threshold calculation based on BFDD

No. ECG Datasets	Maximum Nearest Non-self Match Distances	Anomaly
1	5102	NO
2	4886	NO
3	9206	NO
4	5582	NO
5	6056	NO
6	21171	YES
7	22469	YES
8	16947	YES
9	9996	YES
10	22162	YES

On the basis of the obtained values, the threshold has to clearly tell whether the segment is an anomaly or not. Table 3 shows that the maximum value in the second column in relation to the normal segments is 9,206, and the minimum value in the second column in relation to the anomalous segments is 9,996. Therefore, the threshold is defined as the average value of 9,996 and 9,206, that is, 9,601.

With the threshold, BFDD-based anomaly detection is applied to testing ECGs. The first step is to calculate and record the nearest non-self match distance of every sliding window, and then compare the recorded values with the threshold to identify the corresponding segment. Table 4 shows the anomaly detection results of the 20 testing ECGs.

Table 4: Anomaly detection based on BFDD

No. ECG Datasets	Detected Location	Maximum Nearest Non-self Match Distance	Anomaly Identification	Operation Time (seconds)
11	NA	6093	NO	461.8
12	NA	7299	NO	458.6
13	NA	5351	NO	461.9
14	NA	5634	NO	459.2
15	NA	3831	NO	456.3
16	11673	24023	YES	458.8
17	7942	22513	YES	461.8
18	9421	17365	YES	455.1
19	6106	23793	YES	459.6
20	7178	20911	YES	460.7
21	6286, 11611	21261, 19876	YES	459.1
22	6750, 7659	8201, 6239	YES	457.8
23	4679, 7730	7299, 9206	YES	461.9
24	971, 9960	18481, 17365	YES	457.8
25	6089, 9409	6285, 6250	YES	460.7
26	NA	0	NO	1719.8
27	NA	0	NO	1843.5
28	NA	0	NO	1843.8
29	NA	0	NO	1855.1
30	NA	0	NO	1863.9

In Table 4, the 2nd column shows the location or locations of the detected anomaly or anomalies; the 3rd column illustrates the maximum nearest non-self match distance or the nearest non-self match distances that are greater than the threshold; the 4th column describes the results of anomaly detection, and the last column lists the calculation time used by BFDD. It can be seen that BFDD can correctly define that the ECG is normal or anomalous when there is no anomaly or only one anomaly in one ECG. When there are two significantly different anomalous segments in testing ECG, BFDD can also detect the presence of anomalies, but the accuracy is only 40%. What is worse is when there are two same or similar anomalies, BFDD cannot detect any of them. In terms of computation complexity, BFDD-based anomaly detection is not acceptable, as the length of testing ECG only contains the records in 40 seconds when the calculation time is over 450 seconds.

4.2 AWDD-based Anomaly Detection

AWDD-based anomaly detection is applied to the same ECG data. Based on Algorithm 2, the first step is the same as that of BFDD-based anomaly detection, which is to define the threshold through applying AWDD to training dataset. In this subsection, the threshold for ECG anomaly detection is 8,325. The results of applying AWDD to training ECGs are shown in Table 5.

Table 5: Threshold calculation based on AWDD

No. ECG Datasets	Maximum Nearest Non-self Match Distances	Anomaly
1	2326	NO
2	1960	NO
3	2636	NO
4	2133	NO
5	3068	NO
6	1433	YES
7	15990	YES
8	13583	YES
9	13830	YES
10	15164	YES

Once the threshold is known, AWDD is then applied to the 20 testing ECGs. Table 6 shows the results of the AWDD-based anomaly detection.

Table 6 shows that AWDD can correctly tell the normal ECGs (No. 11-15). For testing ECGs (No. 16-20), AWDD can also correctly detect anomalous segments and identify their corresponding locations. When there are two different anomalies in testing ECGs (No. 21-25), AWDD can detect the existences of all the anomalies, which outperform the results of BFDD-based anomaly detection, while for the remaining five testing ECGs (No. 26-30), the results are same with BFDD-based anomaly detection and no anomaly is detected. One improvement to be mentioned is that the whole process of anomaly detection for every ECG only takes about 1.5 seconds. To sum up, AWDD is more trustworthy and efficient when compared with BFDD in terms of ECG anomaly detection.

4.3 Proposed Method, based Anomaly Detection

The whole process of the proposed method-based ECG anomaly detection is similar with that of BFDD and AWDD. Specifically, the first step is to compute a threshold, which is obtained through training the available ECGs. The second step is to calculate the average non-self match distances of testing ECGs and identify the testing ECGs through comparing the distances with the threshold. The results generated by applying the proposed method to train ECGs are shown in Table 7.

Table 6: Anomaly detection based on AWDD

No. ECG Datasets	Detected Location	Maximum Nearest Non-self Match Distance	Anomaly Identification	Operation Time
11	NA	2224	NO	1.9752
12	NA	2006	NO	1.8634
13	NA	3670	NO	1.7463
14	NA	1873	NO	1.8055
15	NA	1630	NO	1.9323
16	11440	15650	YES	2.1261
17	7800	15346	YES	1.9347
18	9360	13206	YES	1.9914
19	5880	15263	YES	1.7558
20	7020	14696	YES	1.9283
21	6240, 11440	13008, 12578	YES	1.9469
22	10140, 12740	15788, 15678	YES	1.9222
23	2340, 11960	11520, 13250	YES	2.2854
24	780, 9800	14967, 12191	YES	1.8531
25	3640, 10400	8542, 10254	YES	1.8206
26	NA	0	NO	6.9522
27	NA	0	NO	7.0700
28	NA	0	NO	7.3153
29	NA	0	NO	7.8347
30	NA	0	NO	7.3965

Table 7: Threshold calculation based on proposed method

No. ECG Datasets	Maximum Nearest Non-self Match Distances	Anomaly
1	5308	NO
2	5531	NO
3	5832	NO
4	7012	NO
5	5246	NO
6	32031	YES
7	30694	YES
8	30120	YES
9	35691	YES
10	29098	YES

As shown in Table 7, a threshold can be computed to allow us to define whether the testing segment is anomaly or not. Here the threshold is 18,055. With this threshold, the new method is applied to testing ECGs. The results are shown in Table 8.

Table 8: Anomaly detection based on proposed method

No. ECG Datasets	Detected Location	Maximum Nearest Non-self Match Distance	Anomaly Identification	Operation Time
11	NA	4143	NO	3.6727
12	NA	7199	NO	3.3580
13	NA	4865	NO	3.1531
14	NA	4040	NO	3.3303
15	NA	5149	NO	3.1878
16	11440	34477	YES	3.3558
17	7800	33981	YES	3.1755
18	9360	32119	YES	3.3403
19	5880	36618	YES	3.4315
20	7020	41718	YES	3.3230
21	6240, 11440	27768, 26318	YES	3.3121
22	10140, 12740	32760, 31071	YES	3.2435
23	2340, 11960	38546, 34453	YES	3.3711
24	780, 9800	35035, 28146	YES	3.1587
25	3640, 10400	37826, 36183	YES	3.2899
26	10400, 26000	34527, 34527	YES	12.5846
27	7800, 22100	33916, 33916	YES	12.3937
28	9360, 23660	31946, 31946	YES	12.8550
29	5880, 20280	36730, 36730	YES	12.5591
30	7020, 21580	41759, 41759	YES	12.8025

From Table 8, it is clear that the new method has the ability to detect all the anomalies when there are more than one similar or same anomalies in an ECG. For the five ECGs that each of them contains two different anomalies and the five ECGs contain one anomalous segment, this new method can correctly detect the existence or existences of anomaly or anomalies. For the rest non-anomalous ECGs, this new method can also correctly identify that they are normal.

As shown in Table 4, Table 6 and Table 8, BFDD and AWDD cannot detect anomalies when there are two or more similar or same anomalies in one ECG, while the proposed method can correctly detect all the anomalies. For ECG containing two or more significantly different anomalies, BFDD-based anomaly detection has the accuracy of 40%, while the proposed method and AWDD has the accuracy of 100%. For the ECG which only contains one anomalous segment and non-anomalous ECG, these three methods work well, with an accuracy of 100% for all of them. In terms of computation complexity, BFDD anomaly has to take over 460 seconds while AWDD and the proposed method only take 1.9 seconds and 3.5 seconds respectively. To sum up, the proposed methods provide a promising improvement in terms of detecting anomalies from ECG signals. The overall performances of the three methods are briefly summarized in Table 9.

Table 9: Anomaly detection accuracy comparison

	2 or more Anomalies Similar or Same with Each Other	2 or more Anomalies Significantly Different from Each Other	1 Anomaly ECG	Non-anomalous ECG
New Method	100%	100%	100%	100%
BFDD [12]	0	40%	100%	100%
AWDD [16]	0	100%	100%	100%

5. Conclusion

Given the fact that cardiovascular disease has been a focus in society and clinical fields for ages, we believe that the application of data mining method to ECG anomaly detection will make a great contribution to heart disease detection. With the nature of fast calculation and high accuracy, data mining methods are helpful for patients to get fast and accurate treatment. In this paper, we proposed an ECG anomaly detection method based on a new distance measure method and a new anomaly detection notion. For the proposed distance measure method, with the purpose of eliminating the error caused by existence of timeline drift and remove the error caused by neglect of timeline drift, the distance between two candidates is calculated according to their DTW distance and the optimal aligning path between them. For the new anomaly detection notion, in order to correctly detect all the anomalies in one time series, the average value of non-self match distance is used to replace the minimum value of non-self match distances. By applying the proposed method and the other two famous anomaly detection methods to 30 real ECGs, experimental results show that the proposed method is promising in terms of calculation complexity and outperforms the two compared methods with regard to the accuracy of anomaly detection.

Acknowledgements

This work was supported in part by the Natural Environment Research Council (NERC) under the Grant NE/V001787 and Grant NE/V002511, the Engineering and Physical Sciences Research Council (EPSRC) under Grant EP/I011056/1, and the EPSRC Platform Grant EP/H00453X/1.

References

1. Benson, Edwin Raj, S. and Annie Portia, A. (2011). Analysis on credit card fraud detection methods. *In: 2011 International Conference on Computer, Communication and Electrical Technology (ICCCET)*, pp. 152-156.
2. Berndt, D.J. and Clifford, J. (1994). Using dynamic time warping to find patterns in time series, *KDD Workshop*, 10(16): 359-370.
3. Bu, Y., Leung, T.W., Fu, A.W.C., Keogh, E., Pei, J. and Meshkin, S. (2007). WAT: Finding Top-K Discords in Time Series Database. *In: Proceedings of the 2007 SIAM International Conference on Data Mining*, pp. 449-454.
4. Cardiovascular diseases, *World Health Organization*, 2016; available at: http://www.who.int/cardiovascular_diseases/en/
5. Chauhan, S. and Vig, L. (2015). Anomaly detection in ECG time signals via deep long short-term memory network. *In: IEEE International Conference on Data Science and Advanced Analytics (DSAA)*, pp. 1-7.
6. Chandola, V., Banerjee, A. and Kumar, V. (2009). Anomaly detection: A survey, *ACM Computing Surveys (CSUR)*, 41(3).
7. Chuah, M.C. and Fu, F. (2007). ECG anomaly detection via time series analysis. *In: Proceedings of the 2007 international Conference on Frontiers of High Performance Computing and Networking-ISPA 2007 Workshops*, pp. 123-135.
8. Goldberger, A.L., Amaral, L.A.N., Glass, L., Hausdorff, J.M., Ivanov. P.C., Mark, R.G., Mietus, J.E., Moody, G.B., Peng, C.K., Stanley, H.E. (2000). PhysioBank, PhysioToolkit, and PhysioNet: Components of a New Research Resource for Complex Physiologic Signals. Circulation, 101: e215–e220; available at: https://physionet.org/physiobank/database/mitdb/.
9. Guo, T. and Li, G.Y. (2008). Neural data mining for credit card fraud detection. *In: International Conference on Machine Learning and Cybernetics, 2008*, pp. 3630-3634.
10. Huang, Z., Dong, W., Ji, L., Yin, L. and Duan, H. (2015). On local anomaly detection and analysis for clinical pathways, *Artificial Intelligence in Medicine*, 65(3): 167-177.
11. Kaur, R. and Singh, S. (2015). A survey of data mining and social network analysis based anomaly detection techniques, *Egyptian Informatics Journal*, 17(2): 119-216.
12. Keogh, E., Lin, J. and Fu, A. (2005). HOT SAX: Efficiently finding the most unusual time series subsequence. *In: Data Mining, Fifth IEEE International Conference*, pp. 226-233.
13. Keogh, E., Lin, J., Lee, S.H. and Van Herle, H. (2007). Finding the most unusual time series subsequence: Algorithms and applications, *Knowledge and Information Systems*, 11(1): 1-27.
14. Keogh, E., Lin, J., Fu, A.W. and Van Herle, H. (2006). Finding unusual medical time-series subsequencies: Algorithms and applications, *IEEE Transactions on Information Technology in Biomedicine*, 10(3): 429-439.
15. Khanh, N.D.K. and Anh, D.T. (2012). Time series discord discovery using WAT algorithm and iSAX representation. *In: Proceedings of the Third Symposium on Information and Communication Technology*, pp. 207-213.
16. Lemos, A.P., Tierra-Criollo, C.J. and Caminhas, W.M. (2007). ECG Anomalies Identification Using a Time Series Novelty Detection Technique. *In: IV Latin American Congress on Biomedical Engineering 2007, Bioengineering Solutions for Latin America Health*, Springer, Berlin, Heidelberg, pp. 65-88.
17. Lin, J., Keogh, E., Fu, A. and Van Herle, H. (2005). Approximations to magic: finding unusual medical time series. *In: Proceeding 18th IEEE Symposium on Computer-Based Medical Systems, 2005*, pp. 329-334.

18. Monowar, H.B., Dhruha, K.B. and Jugal, K.K. (2013). Network anomaly detection: Methods, systems and tools, *IEEE Communications Surveys and Tutorials*, 16(1): 303-336.

19. Müller, M. (2007). Dynamic time warping, *Information Retrieval for Music and Motion*, pp. 69-84.

20. Ogwueleka, F.N. (2011). Data mining application in credit card fraud detection system, *Journal of Engineering Science and Technology*, 6(3): 311-322.

21. Oresko, J.J., Jin, Z., Cheng, J., Huang, S., Sun, Y., Duschl, H. and Cheng, A.C. (2010). A wearable smartphone-based platform for real-time cardiovascular disease detection via electrocardiogram processing, *IEEE Transactions on Information Technology in Biomedicine*, 14(3): 734-740.

22. Sahoo, P.K., Thakkar, H.K. and Lee, M.Y. (2017). A cardiac early warning system with multi-channel SCG and ECG monitoring for mobile health, *Sensors*, 17(4): 711.

23. Ten, C.W., Hong, J. and Liu, C.C. (2011). Anomaly detection for cyber security of the substations, *IEEE Transactions on Smart Grid*, 2(4): 865-873.

An Investigation on ECG-based Cardiological Diagnosis via Deep Learning Models

Alex Meehan, Zhaonian Zhang, Bryan Williams and Richard Jiang

LIRA Center, Lancaster University, Lancaster LA1 4YW, UK

1. Introduction

Acute Coronary Syndromes (ACS) are caused by an imbalance between the demand for oxygen and the blood flow. It may be caused by either an acute reduction of blood supply or an increase in demand that cannot be matched by the blood flow [4]. The main ACS are ST-elevation myocardial infarction (STEMI), non-ST-elevation myocardial infarction (NSTEMI), and unstable angina. These ACS are an important cause of morbidity and mortality in the UK and worldwide [6], with the World Health Organization estimating that 17.9 million people died from cardiovascular diseases in 2016 worldwide, representing 31% of all global deaths. Rapid and accurate diagnosis can dramatically improve patient outcomes, with treatments ranging from surgical interventions for the more acute cases to medications for less severe cases.

ACS can be diagnosed using an electrocardiogram (ECG), a focused history, and physical examination. The ECG checks the heart's rhythm and electrical activity, using sensors attached to the skin to detect the electrical signals produced by the heart, each time it beats. ACS could affect any component of the heart's electrical activation, including the *P* wave, *PR* interval, *QRS* complex, ST segment, and the *T* and *U* waves. One of the dramatic ECG manifestations is ST-segment elevation in the leads facing the ischemic zone and ST-segment depression in the leads facing the anatomically opposite myocardial segments [4].

The electrocardiogram dates back to Willem Einthoven receiving the Nobel Prize in Physiology or Medicine, in 1924. Originally it was visually examined by a qualified professional, who tried to identify certain types of arrhythmias. However, many arrhythmias are very hard to detect due to inherent ECG noise and because there can be subtle changes in frequency over time. ECGs were identified as one of the earliest applications for neural networks (NN), in the 1980s, to identify complex patterns within a one-dimensional signal [17]. As computational power has grown and neural networks have become more sophisticated, automatic ECG diagnosis has continued to be an active area in academic research, with 4,910 papers published in the last three years (according to Google Scholar).

This paper aims to compare three of the most popular and suitable types of neural networks for identifying abnormalities in ECG data: a Multi-Layer Perceptron, a Convolution Neural Network, and a Long Short Term Memory network.

The dataset contains 4,998 heartbeat samples, with 140 readings for each data sample and classification as to whether an expert has identified an abnormality or not. The dataset originates from the BIDMC Congestive Heart Failure Database on physionet.org and it was heavily pre-processed to extract each heartbeat, make each heartbeat equal length using interpolation and then standardized. The dataset was fairly well balanced, with normal/abnormal split of 2,919 samples (58%)/2,079 (42%).

Multi-Layer Perceptron (MLP) and Convolutional Neural Network (CNN) were considered in this paper. A two-layer convolutional neural network performed best across 25 train-test trials, achieving an accuracy of 99.2%. A three-layer MLP also performed well, achieving an accuracy of 98.8%, but the LSTM networks couldn't match either's performance with an accuracy of 97.9% — thus reflecting this heavily preprocessed and relatively simple dataset.

2. Literature Review

The earliest papers about diagnosing abnormal ECGs using neural networks managed to attain good accuracy through a simple one-layer multi-layer perceptron (MLP) neural network, focused on one specific condition. The experiments achieved 100% accuracy using eight nodes in the hidden layer and five expansion coefficients on a dataset of 56 patients, 24 of whom were ischemic [17].

To diagnose more subtle acute coronary syndromes, more sophisticated neural networks have subsequently been developed. Most approaches involve firstly selecting a method to extract the features, and then selecting a classification approach [11].

The main research developments throughout the 2000s focused on adding feature extraction layers before an MLP-type of the neural network. One potential feature extraction layer is to apply a Learning Vector Quantization (LVQ) algorithm to obtain the centers of clusters. This is done without using class information in the first instance and then the position of the centers of the clusters is refined, using the class information [3]. This was then subsequently extended to be applied to multi-class diagnoses, covering different types of arrhythmias [13].

Another potential feature extraction layer is to add a fuzzy self-organizing layer before the MLP, using c-means and Gustafson-Kessel algorithms for self-organization of the neural networks [14, 15]. It was also shown that the fuzzy clustering NN architecture (FCNN) can generalize better and learn better and faster than ordinary MLP architecture and LVQ.

Convolutional Neural Networks (CNN) were introduced in the 1990s, shortly after the introduction of MLP [10]. Their initial focus was image classification, but as the techniques developed and computing power increased, they came to be applied to other challenges, such as speech recognition and ECG diagnosis.

A1-dimensional CNN method was shown to classify ECG signals to 97.5% accuracy, using five hidden layers in addition to the input layer and the output layer (two convolution layers, two down sampling layers, and one full connection layer) [11].

CNN is also well-suited to multidimensional pattern or image recognition applications and this approach has been extended to other potential inputs, such as using two-dimensional ECG images or even a range of electrode measurements from different positions in the body [1, 2, 8].

The nature of ECG data means that one of the important pre-processing steps for the approaches described so far is noise reduction and baseline wander, for example, using wavelet filters [16].

Long Short-Term Memory (LSTM) networks have also been applied to this problem, extending to a deep recurrent neural network architecture in more recent years. One of the advantages of LSTM networks is that the ECG signal can be directly fed into the network without any elaborate preprocessing or knowledge of the abnormalities, and its results are better than other approaches [5].

More recently, the LSTM algorithm has been stacked with a CNN to further improve accuracy to 99.85% (with pre-processing) [14]. The CNN architecture consists of two convolution layers with two max-pooling layers, followed by three LSTM layers that further reduce the features, and the last layer is a fully-connected layer which makes the classification. The approaches described above are the most popular and the most applicable for ECG classification.

The approaches described above are the most popular and the most applicable for ECG classification, but there are many other potential NN variants that could also be used, such as Deep Belief Networks, Recurrent Neural Networks and Gated Recurrent Units [7]. There are also many potential approaches outside of NN, such as time-based methods (PCA/LDA) and non-linear methods [12].

3. Preliminary on Methodologies

3.1 Multi-Layer Perceptron (MLP)

Multi-layer perceptrons are one of the original and most fundamental neural networks, consisting of at least three layers of neurons: an input layer, one or more hidden layers, and an output layer. Each neuron has an activation function, a weight, and is fully connected to each of the neurons in the next layer of the network. The network is 'trained' by passing batches of data through the network, calculating the loss (error), and then back-propagating the error to update each of the neuron weights. This process is repeated for each batch of data and then repeated over a pre-defined number of epochs.

Key design parameters are the number of neurons, the activation function used in each neuron, the number of layers, the batch size, the loss function used to calculate the error and the 'optimizer' algorithm used to calculate the revised neuron weights in each back-propagation cycle.

The networks considered in this paper will use two of the most widely used activation functions. The output layer will use a sigmoid function, which will normalize the output between 0 and 1 to match the binary classification task. All other layers will use the Rectified Linear Unit (ReLU) function, as it is a nonlinear function that does not saturate at a limit of 1, it can prevent vanishing gradient, reduce overfitting and speed up the network training.

The loss function and optimizer together define the complex mathematics to recalculate the neuron weights with each back-propagation cycle. Mean Squared Error (MSE) is one of the simplest loss functions to usually combine with Stochastic Gradient Descent (or some variant). The Adam algorithm [9] is a widely-used variant of Stochastic Gradient Descent, based on adaptive estimates of lower-order moments, making it more effective and computationally efficient. In this work, we use different loss functions and optimizers during our experiments to find the best combination for the ECG-based Cardiological Diagnosis project.

3.2 Convolutional Neural Network (CNN)

Convolutional neural networks are often used when the problem involves time series or spatial features with varying positions, swith wide application to satellite image analysis [19], medical image analysis [20], behavior analysis [21], and facial modeling [22, 23]. Each convolutional layer consists of a bank of filters, it can change the input data as a new feature map by the convolutional kernel, which is a meaningful process to extract the obvious features from the input data.

Pooling layers are used to down-sample from the previous convolutional layer's feature map to compress the data and reduce over-fitting, with max-pooling most commonly used. Drop-out filters are also used to reduce over-fitting by systematically removing a proportion of the output nodes from the previous layer in each cycle. One or more fully connected layers are then usually used before the final output layer, to translate the convolutional feature maps into a useable output.

The result of this network design is that convolutional networks preserve spatial features, allowing them to be detected across different positions – something that would be useful for the ECG dataset, particularly if the phasing of each heartbeat hadn't been aligned in pre-processing.

3.3 Long Short Term Memory (LSTM)

Long Short Term Memory is a form of recurrent neural network, with connections between neighboring units within the LSTM layer, making it well-suited for time series data. The output of an RNN at each time step is based on the current input and the input at all previous time steps, using the back propagation through time algorithm for training.

The main design parameter for the Long Short Term Memory layer is the number of units. Similar to CNN, one or more fully connected layers are then usually used before the final output layer, to translate the LSTM output into a useable output.

4. Experimental Results

4.1 Experimental Setup

As previously discussed, the data had already been extensively pre-processed. It was normalized to values between 0 and 1 and randomly split into train/test in a ratio of 80:20.

The key parameters were explored for each neural network, repeating 25 train-test runs for each setting to ensure statistically robust results. The most important evaluation metrics for this ECG dataset were considered to be:

- *Accuracy*: A good overall measure, particularly given the data is balanced.
- *Precision*: The percentage of abnormal diagnoses that is true.
- *Recall*: The percentage of abnormal cases are caught.
- Training time.
- Testing time.

Accuracy and training time are the two main metrics. The models were built in Python, using the Keras libraries on Google Colab with GPUs enabled.

4.2 MLP Training

A range of different MLP designs was explored, with the results summarized in Table 1. A Rectified Linear Unit (ReLU) activation function was used in each dense layer. A 128 neurons layer performed best with the 148×1 input data and models with more than one hidden layer performed best when the subsequent layers also had 128 neurons. The two-layer MLP performed significantly better than a one-layer MLP; three layers performed slightly better than two, but there were no further gains from using a six-layer MLP. The optimal design was a three-layer MLP, with 128 neurons in each layer.

Table 1: Evaluation of MLP designs

MLP Design	Accuracy	Train Time
3 layers (3×128 neurons)	98.78 ± 0.56	8.52
Varying number of neurons in a 1-layer network:		
One layer with 32 neurons	97.91 ± 0.46	7.68
One layer with 64 neurons	98.24 ± 0.40	7.64
One layer with 128 neurons	98.40 ± 0.27	7.66
Exploring two layers models:		
Two layers (2×128 neurons)	98.66 ± 0.36	8.06
2 layers (128/16 neurons)	98.50 ± 0.25	8.04
Exploring alternative 3+ level deep networks:		
3 layers (128/16/4 neurons)	98.49 ± 0.65	8.43
6 layers (6×128 neurons)	98.78 ± 0.62	9.61

The MLP model was trained over 30 epochs and it can be seen in Fig. 1 that accuracy of >0.98 was achieved within five epochs and training was done over 30 epochs to maximize accuracy.

The remaining model parameters were then tuned, using the three-layer model, with 128 neurons in each layer. The impact of batch size on accuracy and training time is shown in Fig. 2. The right balance was judged to be a batch size of 64, which compromised < 0:01% of accuracy for a 75% reduction in training time.

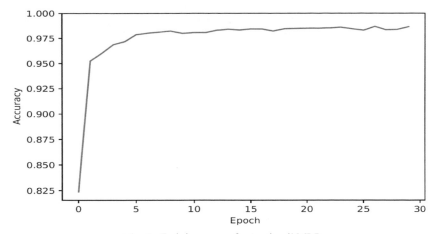

Fig. 1: Training curve for 'optimal' MLP

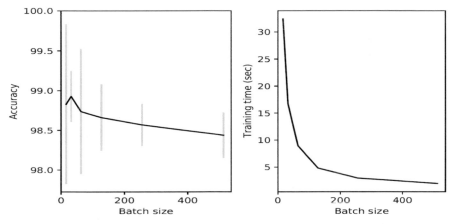

Fig. 2: Evaluation of MLP batch size

Different loss functions are evaluated in Fig. 3. The cosine similarity loss function did not successfully converge in training, given that it relies on dimensionality to generate vector angles and this is a one-dimensional dataset. The other four of the five loss functions performed similarly well, with accuracy ranging from 98.5-98.8; that is, well within the bounds of error. Given this similar performance, the simplest measure was used: Mean Squared Error.

Eight different 'optimizer' algorithms were evaluated to navigate the data and its gradients towards the optimal set of weights (shown in Fig. 4). Similar to the loss function evaluation, two of the optimizers did not successfully converge in training: Adadelta and Follow the Regularized Leader (FTRL). Adadelta's moving window of gradient updates is evidently not well suited to the ECG data and FTRL's feature vectorization is not well-suited to this one-dimensional data. Adamax performed best and so was selected.

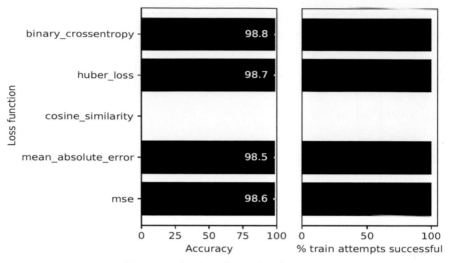

Fig. 3: Evaluation of MLP loss functions

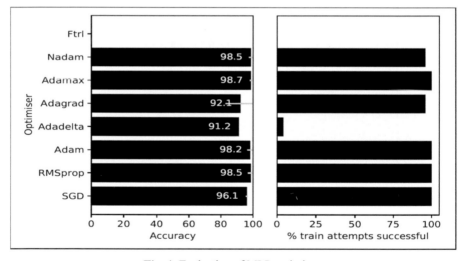

Fig. 4: Evaluation of MLP optimizers

4.3 CNN Training

A range of different CNN designs was also explored, with the results summarized in Table 2. The 'optimal' design for this data was a convolutional neural network with two convolutional layers, each with 64 filters, a Rectified Linear Unit (ReLU) activation function and a kernel size of 32. A drop-out of 0.2 was used between the two layers, to reduce over-fitting. Max pooling was used after each convolutional layer, with a pooling size of three. A fully connected dense layer of 32 neurons was used before the output layer, sized at one neuron with a sigmoid function to convert the out into a single binary classification. Additional convolutional layers were evaluated, but they did not significantly improve performance with this relatively simple dataset.

Table 2: Evaluation of CNN design parameters

CNN Design	Accuracy	Train Time
'Optimal' design	99.23 ± 0.29	7.49
Tuning number filters in convolutional layers:		
16 filters	99.04 ± 0.42	7.39
32 filters	99.21 ± 0.29	7.44
128 filters	99.27 ± 0.36	7.63
256 filters	99.28 ± 0.20	8.22
Tuning kernel size in convolutional layers:		
Kernel size 8	99.16 ± 0.48	7.26
Kernel size 16	99.20 ± 0.27	7.21
Kernel size 64	99.08 ± 0.50	7.32
Tuning dropout filter between layers:		
Dropout=0.2	99.29 ± 0.27	7.34
Dropout=0.4	99.33 ± 0.27	7.32
Dropout=0.6	99.31 ± 0.23	7.24
Dropout=0.8	99.23 ± 0.35	7.24
Tuning pooling between layers:		
Pooling size 2	99.28 ± 0.18	7.31
Pooling size 4	99.23 ± 0.28	7.19
Pooling size 7/5	99.16 ± 0.33	7.29
Pooling size 14/10	98.90 ± 0.42	7.24
Adding more hidden layers:		
Adding 1 more	99.12 ± 0.35	7.95
Adding 4 more	99.30 ± 0.28	9.85
Tuning dense layer size:		
16-neurons	99.21 ± 0.28	7.23
64-neurons	99.16 ± 0.43	7.16

The convolutional neural network was trained over 20 epochs, as the network 'learned' faster than the MLP, offsetting its longer training time per epoch (*see* Fig. 5).

The loss function and optimizer were set to be the same as the MLP: Means Square Error and Adamax. The evaluation of these in the MLP section showed that performance is primarily determined by the dataset, rather than the model and that there was little difference between the ones that 'worked' for the dataset. The impact of batch size on accuracy and training time is shown in Fig. 6. The right balance was judged to be a batch size of 64, which didn't compromise any accuracy for a 75% reduction in training time.

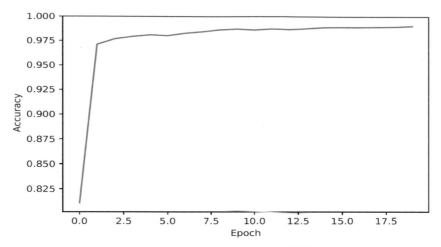

Fig. 5: Training curve for 'optimal' CNN

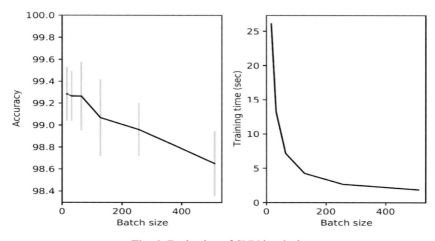

Fig. 6: Evaluation of CNN batch size

4.4 LSTM Training

A range of different LSTM designs was explored, with the results summarized in Table 3. The 'optimal' design for this data was an LSTM layer with 16 units, a drop-out of 0.2, a fully-connected dense layer of 16 neurons, before the one-neuron sigmoid output layer.

4.5 Test Results Compared

The optimized MLP, CNN, and LSTM models are compared across all of the key evaluation metrics in Table 4. The MLP set a high bar, with an accuracy of 98.78, which is also a reflection of the relatively simple and heavily pre-processed dataset. The CNN model slightly outperformed the MLP across nearly all the metrics, but it

Table 3: Evaluation of LSTM design parameters

LSTM Design	Accuracy	Train Time
'Optimal' design	97.89 ± 1.02	13.17
Tuning dropout filter between layers:		
Dropout=0	97.82 ± 1.19	13.24
Dropout=0.4	97.87 ± 0.93	13.16
Dropout=0.6	97.61 ± 1.09	13.18
Dropout=0.8	96.88 ± 1.44	13.20
Tuning the number of LSTM units:		
8 LSTM units	97.15 ± 1.62	13.43
32 LSTM units	98.01 ± 0.87	13.24
Tuning dense layer size:		
8-neurons	97.88 ± 1.29	13.37
32-neurons	97.87 ± 1.28	13.25

Table 4: Comparison of optimized models

	MLP	CNN	LSTM
Accuracy	98.78 ± 0.56	99.23 ± 0.29	97.89 ± 1.02
Precision	98.39	98.65	97.22
Recall	99.46	99.99	99.07
Training time	8.52	7.49	13.17
Testing time	0.41	0.46	0.77

should be noted that the gains are relatively small, for example, an accuracy of 99.23 versus 98.78. The LSTM model struggled to match even the MLP's performance, suggesting that the intensive pre-processing reduced the need for any further time alignment.

5. Discussion

The CNN outperformed the MLP and LSTM networks, which reflects it's a more sophisticated interconnected structure that enables the network to recognize patterns across different points in a time series, even if they are not aligned between data samples.

The evaluation of the various model parameters showed that tuning is more important than the type of neural network for this relatively simple dataset. There are many parameters that need to be tuned to the specific characteristics of each dataset and the more sophisticated models have more parameters. The MLP design was determined by the number of layers and number of neurons in each layer, whereas the CNN also needed to consider kernel sizes, drop-out filters and pooling size. This means that the more sophisticated models take longer to tune and so they should only be used if the application and dataset require the additional sophistication.

In this study, a basic one-layer MLP achieved accuracy of 98.4%, which provides limited scope for improvement and differentiation between more sophisticated models. A tuned three-layer MLP improves accuracy to 98.8% and a tuned two-layer CNN improves accuracy further to 99.2%. The tuned CNN had nearly 100% Recall, so the issue was False Negatives in the precision element. There may also be inaccuracies in the original classification, particularly given the very small levels of error. There are many more sophisticated models than those considered in this study, but the results suggest that this dataset does not warrant them. The LSTM model struggled to match any of the other models with an accuracy of 97.9%, perhaps because the data was heavily pre-processed.

Further studies could consider more realistic and complex ECG data as a continuous time series, with live simultaneous inputs from several different sensors. Similarly, more sophisticated multi-class classifications could be applied for different types of arrhythmia (e.g. non-ectopic, supraventricular ectopic, ventricular ectopic, fusion, and other unknown beats).

The performance gap between different neural networks is likely to be much greater when more challenging and realistic ECG data is used; more sophisticated data will require more sophisticated models (e.g. CNN-LSTM).

6. Conclusion

This paper compared three of the most popular and suitable types of neural networks for identifying abnormalities in ECG data.

The ECG dataset used was heavily pre-processed to align time dimensions across samples and only considered one metric. A two-layer convolutional neural network performed best across 25 train-test trials, achieving an accuracy of 99.2%. A three-layer MLP also performed well, achieving an accuracy of 98.8%, but the LSTM networks couldn't match either's performance with an accuracy of 97.9%.

Further studies could consider more realistic and complex ECG data as a continuous time series, with live simultaneous inputs from several different sensors. Similarly, more sophisticated multi-class classifications could be applied for different types of arrhythmia. These datasets are likely to require deeper and more sophisticated neural networks.

Acknowledgements

The dataset originates from the BIDMC Congestive Heart Failure Database via physionet.org and TensorFlow. The TensorFlow Keras libraries were used to implement the neural networks, helped by the associated manuals/tutorials.

References

1. Rajendra Acharya, U., Shu Lih Oh, Yuki Hagiwara, Jen Hong Tan, Muhammad Adam, Arkadiusz Gertych and Ru San Tan (2017). A deep convolutional neural network model to classify heartbeats, *Computers in Biology and Medicine*, 89: 389-396.

2. Avanzato, Roberta and Francesco Beritelli (2020). Automatic ECG diagnosis using convolutional neural network, *Electronics*, 9(6): 951.
3. Belgacem, N., Chikh, M.A. and Bereksi Reguig, F. (2003). Supervised classification of ECG using neural networks. Biomedical Engineering Laboratory, Department of Electronics, Science Engineering Faculty. Belkaid University.
4. Birnbaum, Yochai, James Michael Wilson, Miquel Fiol, Antonio Bayes de Luna, Markku Eskola and Kjell Nikus (2014). ECG diagnosis and classification of acute coronary syndromes, *Annals of Noninvasive Electrocardiology*, 19(1): 4-14.
5. Chauhan, Sucheta and Lovekesh Vig (2015). Anomaly detection in ECG time signals via deep long short-term memory networks. *In: 2015 IEEE International Conference on Data Science and Advanced Analytics (DSAA)*, pp. 1-7.
6. Corbett, Simon J., Saoussen Ftouh, Sedina Lewis and Kate Lovibond (2021). Acute coronary syndromes: Summary of updated nice guidance, *BMJ*, 372.
7. Ebrahimi, Zahra, Mohammad Loni, Masoud Daneshtalab and Arash Gharehbaghi (2020). A review on deep learning methods for ECG arrhythmia classification, *Expert Systems with Applications: X*, p. 100033.
8. Joon Jun, Tae, Hoang Minh Nguyen, Daeyoun Kang, Dohyeun Kim, Daeyoung Kim and Young-Hak Kim (2018). ECG arrhythmia classification using a 2-d convolutional neural network, arXiv preprint arXiv: 1804.06812.
9. Kingma, Diederik P. and Jimmy Ba (2014). Adam: A method for stochastic optimization, arXiv preprint arXiv: 1412.6980.
10. Yann, LeCun, Yoshua Bengio, *et al.* (1995). Convolutional networks for images, speech, and time series, *The Handbook of Brain Theory and Neural Networks*, 3361(10): 1995.
11. Li, Dan, Jianxin Zhang, Qiang Zhang and Xiaopeng Wei (2017). Classification of ECG signals based on 1d convolution neural network. *In: 2017 IEEE 19th International Conference on e-Health Networking, Applications and Services (Healthcom)*, pp. 1-6.
12. Joy Martis, Roshan, U. Rajendra Acharya and Hojjat Adeli (2014). Current methods in electrocardiogram characterization, *Computers in Biology and Medicine*, 48: 133-149.
13. Melin, Patricia, Jonathan Amezcua, Fevrier Valdez and Oscar Castillo (2014). A new neural network model based on the LVQ algorithm for multiclass classification of arrhythmias, *Information Sciences*, 279: 483-497.
14. Osowski, Stanislaw and Tran Hoai Linh (2001). ECG beat recognition using fuzzy hybrid neural network, *IEEE Transactions on Biomedical Engineering*, 48(11): 1265-1271.
15. Ozbay, Yuksel, Rahime Ceylan and Bekir Karlik (2006). A fuzzy clustering neural network architecture for classification of ECG arrhythmias, *Computers in Biology and Medicine*, 36(4): 376-388.
16. Singh, Brij N. and Tiwari, Arvind K. (2006). Optimal selection of wavelet basis function applied to ECG signal denoising, *Digital Signal Processing*, 16(3): 275-287.
17. Sun, Gang, Thomas, Cecil W., Jerome Liebman, Yoram Rudy, Yehuda Reich, Stilli, D. and Macchi, E. (1988). Classification of normal and ischemia from BSPM by neural network approach. *In: Proceedings of the Annual International Conference of the IEEE Engineering in Medicine and Biology Society*, pp. 1504-1505.
18. Tan, Jen Hong, Yuki Hagiwara, Winnie Pang, Ivy Lim, Shu Lih Oh, Muhammad Adam, Ru San Tan, Ming Chen and Rajendra Acharya, U. (2018). Application of stacked convolutional and long short-term memory network for accurate identification of cad ECG signals, *Computers in Biology and Medicine*, 94: 19-26.
19. Chiang, C., Barnes, C., Angelov, P. and Jiang, R. (2020). Deep Learning based Automated Forest Health Diagnosis from Aerial Images. IEEE Access.

20. Storey, G., Jiang, R., Bouridane, A. and Li, C.T. (2019). 3DPalsyNet: A Facial Palsy Grading and Motion Recognition Framework using Fully 3D Convolutional Neural Networks. IEEE Access.
21. Jiang, Z., Chazot, P.L., Celebi, M.E., Crookes, D. and Jiang, R. (2019). Social Behavioral Phenotyping of Drosophila with a 2D-3D Hybrid CNN Framework. IEEE Access.
22. Storey, G., Bouridane, A. and Jiang, R. (2018). Integrated Deep Model for Face Detection and Landmark Localisation from 'in the wild' Images. IEEE Access.
23. Storey, G., Jiang, R. and Bouridane, A. (2017). Role for 2D image generated 3D face models in the rehabilitation of facial palsy. *IET Healthcare Technology Letters*.

EEG-based Deep Emotional Diagnosis: A Comparative Study

Geyi Liu[1], Zhaonian Zhang[1], Richard Jiang[1], Danny Crookes[2] and Paul Chazot[3]

[1] School of Computing and Communication, Lancaster University, Lancaster, UK
[2] ECIT Institute, Queen's University Belfast, Belfast, UK
[3] Department of Biosciences, Durham University, Durham, UK

1. Introduction

Emotion plays an important role in human decision, interaction and cognition [1]. Due to the rapid development of society and the accelerated pace of life, people often feel pressure and anxiety. The persistence of this situation may lead to a variety of health problems or depression, thereby affecting people's daily life and self-development. Therefore, emotion recognition is gradually becoming a practical topic of researchers. Nowadays, emotion recognition is used in many fields, such as text, speech, expression, and posture. But these methods are subjective and cannot guarantee the authenticity of emotion. Physiological and psychological studies show that the changes of physiological signals are often closer to people's real emotions than facial expression, posture or voice [2]. Electroencephalogram (EEG) reflects all kinds of electrical activities and functional states of the brain, and contains the effective information of human emotional state. Using EEG signals for emotion detection has more advantages than other methods [3]. EEG based on emotion recognition will provide an accurate emotion in many fields.

In this paper, the performance of three deep learning technologies: CNN, LSTM, and DNN in binary classification is compared. The relevant parameters of the model are given and further comparative analysis is carried out. In addition, ensemble learning method, based on major voting, is proposed to further improve the accuracy of binary classification task. The median value 5 was used as the threshold to divide the rating values into two categories. Russell proposed a 2D model of emotion classification. In his model, valence is represented from negative to positive, whereas arousal is represented from not excited to aroused or dull to intense [15]. His model is commonly used previous studies. Arousal and valence are the target dimensions in this paper.

DEAP data set was used to conduct an emotion binary classification work to validate the efficacy of these deep learning techniques. The main purpose of the dataset is to establish a music video recommendation system based on the user's emotional state. The skin electrical response, plethysmograph, skin temperature, respiratory rate, EMG, EEG, and other physiological signals of 32 subjects were recorded [4, 5]. In this study, we mainly focused on the EEG signal in the dataset. The frequency domain features of the EEG signal were extracted and inputted into the model for binary classification work.

The rest of this paper is divided into six parts. The second section briefly reviews related works. The third section introduces the DEAP database and features of EEG signal used in this study. The fourth section introduces the basic principles of CNN, DNN, and LSTM. The fifth section presents the results of this study. Section sixth constitutes the conclusion.

2. Related Works

In recent years, many emotion recognition methods based on EEG signals have been proposed. Liu *et al.* [6] proposed a support vector machine (SVM) method for data classification using time-domain and frequency-domain features, where classification results are 70.3% and 72.6%. A supervised learning algorithm based on perceptron convergence algorithm and Bayesian weighted log a posteriori function was introduced by Yoon and Chung [7], and the accuracy of valence and arousal was 70.9% and 70.1%. Alhagry *et al.* [8] proposed an LSTM recurrent neural network model. The average accuracy of arousal and valence were 85.65% and 85.45% respectively. Zhan *et al.* [9] designed a shallow depth-wise parallel convolutional neural network (CNN). In DEAP dataset, the prediction accuracy of arousal and valence are 84.07% and 82.95%. Xing *et al.* [10] established a stacked automatic encoder (SAE) to decompose EEG signals and classify them through the LSTM model. The accuracy of valence and arousal observation were 81.1% and 74.38%. Bagzir *et al.* [11] established a valence/arousal emotion recognition system. Discrete wavelet transform was used to transform EEG signals, which are decomposed into alpha, beta, theta, and gamma bands, and the spectral features of each band are extracted. They proposed SVM, k nearest neighbor (KNN) and artificial neural network (ANN) for emotion recognition. The best accuracy on valence and arousal were 91.1% and 91.3%. A 3D convolutional neural network was invoked by Salama *et al.* for emotion recognition of multi-channel EEG signal [12]. The obtained accuracy of valence and arousal dimension were 87.44% and 88.49%. Chen *et al.* [13] proposed a new emotion-recognition method based on the ensemble learning method, AdaBoost. The best average accuracy could reach up to 88.70%.

In general, traditional machine learning, deep learning and ensemble learning methods have been used for emotion recognition of EEG signals. However, there is still a problem that EEG signal information cannot be well reflected in feature extraction. At the same time, the methods of feature extraction are also different, causing some difficulties in the comparison of model performance.

3. DEAP Dataset

In this paper, the DEAP database is applied for emotion recognition task. This database recorded the physiological signals of 32 subjects watching 40 minutes of music video (1 minute for each music video) and the psychological scales of valence, arousal, dominion, and liking of the subjects [14]. Each subject had 40 channels records. These channels consisted of 32 EEG channels and remaining eight channels were peripheral physiological signals, and the sampling frequency of these pre-processed signals was 128 Hz. The structure of the DEAP dataset is shown in Table 1.

Table 1: DEAP dataset description

Array Name	Array Shape	Array Contents
Data	40 × 40 × 8064	(Video/trial) × channels × data
Labels	40 × 4	(Video/trial) × label (valence, arousal, dominance, liking)

In this paper, all EEG channels are used. These 32 channels are denoted as Fp1, AF3, F3, F7, FC5, FC1, C3, T7, CP5, CP1, P3, P7, PO3, O1, Oz, Pz, Fp2, AF4, Fz, F4, F8, FC6, FC2, Cz, C4, T8, CP6, CP2, P4, P8, PO4 and O_2 respectively [14]. Figure 1 shows the placement position of the electrodes.

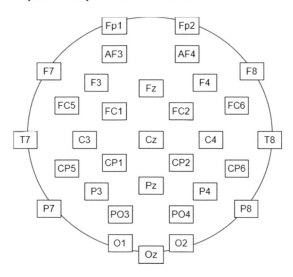

Fig. 1: Electrodes placement for EEG recording (32 electrodes)

People can observe brain waves in different frequency ranges through EEG and brain waves differ, depending on the information or instructions they convey [16]. For instance, high frequency brain waves can be observed when people are happy or excited. On the other hand, low frequency brain waves can be observed when people feel lazy or bored. According to the frequency range, EEG can be divided into five types as shown in Table 2 [17]. Gamma wave is closely related to learning, memory and information processing. Beta waves are associated with arousal. Too much stress and anxiety will be evident in the beta band. Alpha is the frequency range between beta and theta. This band is related to the state when we relax. Theta is involved in sleep. Too much θ activity may predispose people to depression. Delta is not our focus in this study because it is associated with the deepest levels of sleep and relaxation.

Table 2: EEG frequency bands

Name	Frequency Range
Gamma	40 Hz to 100 Hz
Beta	12 Hz to 40 Hz
Alpha	8 Hz to 12 Hz
Theta	4 Hz to 8 Hz
Delta	0 Hz to 4 Hz

In this paper, Fast Fourier Transform (FFT) was used to extract the specified frequency band features. We mainly extracted the information of five frequency bands, which were [4-8 Hz]: theta band, [8-12 Hz]: alpha band, [12-16 Hz]: low beta band, [16-25 Hz]: high beta band and [25-40 Hz].

4. Models

4.1 Convolutional Neural Network (CNN)

In recent years, convolutional neural network (CNN) has been widely used in satellite image analysis [18], medical image analysis [19], behavior analysis [20], and facial modelling [21, 22]. A CNN consists of an input and an output layer, as well as multiple hidden layers. Typically, hidden layers consist of three main feature layers: convolution layer, pool layer and fully connected layer [23]. The typical structure of CNN is shown in Fig. 2.

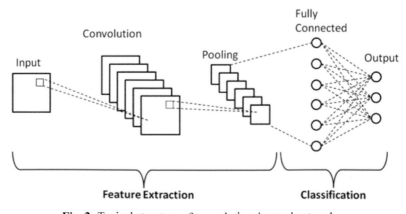

Fig. 2: Typical structure of convolutional neural network.

Convolution layer is the core of CNN operation. Convolution layer is composed of several feature maps and applies a convolution operation to the input, transferring the result to the next layer. And the Rectified Linear Unit (ReLu) aims to apply an activation function, such as sigmoid to the output produced by the previous layer [24]. The pooling layer aims to down sample in order to reduce the complexity for further layers and increase the robustness of feature extraction. After all of the features are generated by the neural network, they are passed to the fully connected Softmax layer. The function of fully connected layer in CNN is the same as that in standard ANN. It will produce probability distribution of all labels used for classification. In order to improve network performance, ReLu is usually used between these layers. Compared with traditional classifiers, CNNs have obvious advantages in analyzing high-dimensional data [23]. The convolution layers in CNN can control and reduce the number of parameters through parameter sharing scheme. And pooling layer could progressively reduce the spatial size of the representation and the number of parameters in the network, before solving the problem of overfitting to a certain extent.

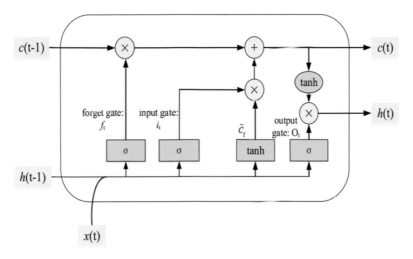

Fig. 3: LSTM block structure [25]

4.2 Long Short-Term Memory (LSTM)

The typical structure of an LSTM block is shown in Fig. 3 [25]. Compared with RNN, LSTM network performs better in dealing with explosion of long-term dependences and gradient vanishing. LSTM block includes three gate controllers, which are input gate, forget gate as well as output gate. And the input of LSTM block has its previous cell state $c_{(t-1)}$, previous hidden state $H_{(t-1)}$, and current input vector $x_{(t)}$.

The forget gate decides what information should be discarded or retained. Information from the previously hidden state and the current input are passed through the sigmoid function. The output value of sigmoid function is between 0 and 1. The closer the output value is to 0, the easier it will be forgotten, and the closer it is to 1, the more it will be retained.

The input gate is used to update the cell status. First, the previous hidden state and the current input will be passed to the sigmoid function. At the same time, the hidden state and the current input will be passed to the tanh function, compressing their values between −1 and 1 to help adjust the network. Finally, the tanh function output will be multiplied by the sigmoid function output, and sigmoid function output will determine which information is important.

The output gate determines what the next hidden state is. First, the previous hidden state and the current input are passed to the sigmoid function. The new cell state is then passed to the tanh function. And tanh output will be multiplied by the sigmoid output to determine the information to be carried in the hidden state. Finally, the new cell state and the new hidden state will be passed to the next time step.

The relevant calculation formula for three gates are as follows: b is the corresponding bias terms; σ is the nonlinear activation function like sigmoid function; W is the corresponding weights.

$$f_{(t)} = \sigma(W_{fx}x_{(t)} + W_{fh}h_{(t-1)} + b_f)$$

$$i_{(t)} = \sigma(W_{ix}x_{(t)} + W_{ih}h_{(t-1)} + b_i)$$

$$o_{(t)} = \sigma(W_{ox}x_{(t)} + W_{oh}h_{(t-1)} + b_o)$$

The memory cell $C_{(t)}$, intermediate state $\tilde{C}_{(t)}$ and hidden state of LSTM $h_{(t)}$ are updated as

$$\tilde{C}_{(t)} = \tanh\left(W_{cx}x_{(t)} + W_{ch}h_{(t-1)} + b_c\right)$$

$$C_{(t)} = f_{(t)} \cdot C_{(t-1)} + i_{(t)} \cdot \tilde{C}_{(t)}$$

$$h_{(t)} = O_{(t)} \cdot \tanh(C_{(t)})$$

"·" represents the point-wise multiplication of two vectors.

4.3 Deep Neural Network (DNN)

Deep neural network is a type of ANN which includes input layer, output layer, and several hidden layers [26]. The nodes of each layer are fully connected with the nodes of the next layer and all connections have their weighted values. At the same time, multiple hidden layers enhance the expression ability and complexity of the model. The structure of DNN model is shown in Fig. 4.

5. Methods

5.1 Pre-processing

As mentioned before, Fast Fourier Transform (FFT) was used to extract the specified frequency band features. There are five bands which were extracted: [4-8 Hz], [8-12 Hz], [12-16 Hz], [16-25 Hz] and [25-40 Hz]. The extracted features will be used as the input of the neural network model. In terms of data partition, the ratio of training set to test set is 8:2. In the division of classification labels, we set the label value less than 5 as negative (0), and the label value greater than 5 as positive (1).

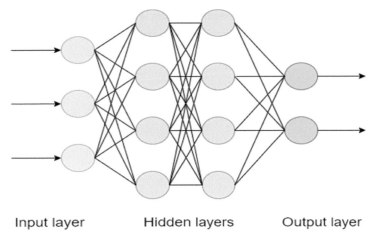

Input layer   Hidden layers   Output layer

Fig. 4: Description of deep neural network (DNN) model

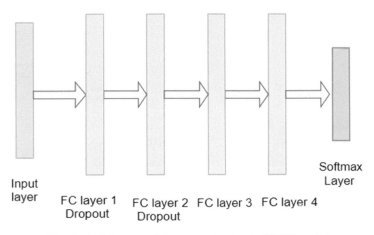

Fig. 5: Architecture of deep neural network (DNN) model

5.2 Model Architecture

First, the architecture of DNN will be introduced. The DNN model uses four fully connected, dense neural layers. And the output of one layer serves as the input for the next layer. The architecture of Deep Neural Network (DNN) model is shown in Fig. 5. Dropout is applied in the first two full connection layers with a probability of 0.2. Because dropout technology is added, it is very important to use high epoch. In the three models this time, more than 500 epochs are used.

In the selection of activation function, rectified linear units (ReLU) are applied to all hidden layers, and Softmax is used for the output layer. The neurons number in each layer are shown in Table 3.

The proposed CNN model is composed of three convolution layers, one Flatten layer, two fully connected (FC) layers, and an output layer. The specific model structure is shown in Fig. 6.

Table 3: Neurons number in each layer of DNN model

Layer	Neurons Number
Input layer	70
FC layer 1	512
FC layer 2	1024
FC layer 3	1024
FC layer 4	256

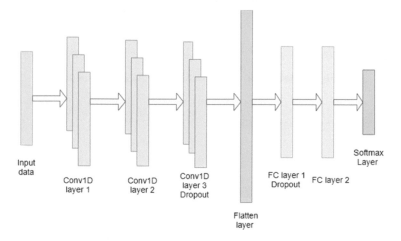

Fig. 6: Architecture of Convolutional Neural Network (CNN) model

In order to avoid over-fitting and improve network performance, batch normalization is added after the first two convolution layers, and dropout layer is added after the third convolution layer and the first fully connected (FC) layer. The drop probability of convolution layer is 0.5. And the fully connected (FC) layer has a dropout probability of 0.2. Furthermore, we deploy rectified linear units (ReLU) and Softmax as the non-linear activation function. In the selection of optimizer, we use the most common method, Adam. The kernels of the three convolution layers are 8, 3 and 3, respectively. The neurons of each layer are shown in Table 4.

Table 4: Neurons number in each layer of CNN model

Layer	Neurons Number
Convolution 1D layer 1	256
Convolution 1D layer 2	128
Convolution 1D layer 3	64
FC layer 1	64
FC layer 2	16

The third model is CNN-LSTM model. Many previous studies have achieved good results in CNN model and LSTM model. So this time we try to combine the two techniques to explore the feasibility of CNN-LSTM model in the classification task. The specific layer structure is shown in Fig. 7. Convolution layer is used to extract feature information, and batch normalization technology is added in the third convolution layer to avoid over-fitting. Compared with the previous CNN model, LSTM replaces the full connection layer, and the output of the convolution layer is then inputted into the LSTM layer.

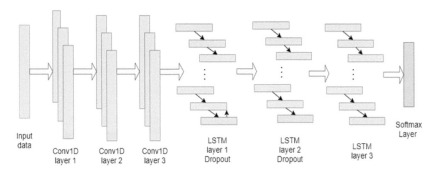

Fig. 7: Architecture of CNN-LSTM model

Table 5 records the neuron numbers in each layer of CNN-LSTM model. As in the previous models, the activation functions are ReLu and Softmax. The dropout probability of LSTM layer is set to 0.3. The kernel of convolution layer 1 is 15, the kernel of convolution layer is 2 as well as convolution layer 3 is 3. And the strides of convolution layer 1 and convolution layer 3 are all 2.

Table 5: Neurons number in each layer of CNN-LSTM model

Layer	Neurons Number
Convolution 1D layer 1	256
Convolution 1D layer 2	128
Convolution 1D layer 3	64
LSTM layer 1	512
LSTM layer 2	256
LSTM layer 3	128

6. Result

In this section, we compare the performance of the three models mentioned above, and compare and analyze the different settings of some hyperparameters of the models. Figure 8 shows the accuracy of the three models on the dimension of valence and arousal. From the figure, we can see that the average accuracy of the three models has reached more than 90%. Among them, the average accuracy of CNN model can reach about 95% on arousal dimension and about 94% on valence dimension.

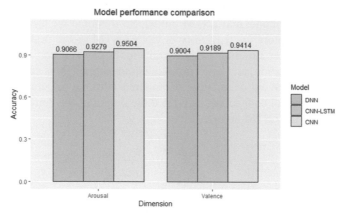

Fig. 8: Model comparison

Compared with some methods mentioned in previous articles, our method has made some progress, especially on CNN model. The comparison between the proposed method and some existing methods using DEAP data base is shown in Table 6.

Table 6: Comparison of the proposed method with some existing methods

Reference	Feature Domain	Class	Classifier	Valence (%)	Arousal (%)
Hwang *et al.* [27]	Frequency	Binary	SVM	64.90	64.90
Zhan *et al.* [9]	Frequency	Binary	CNN	82.95	84.07
Alhagry, S. [8]	Time	Binary	LSTM	85.65	85.45
Our model	Frequency	Binary	CNN-LSTM	92.79	91.89
Our model	Frequency	Binary	CNN	95.04	94.14

7. Conclusion

In conclusion, we examined three models and compared their performances on EEG-based mental diagnosis. In our experiments, it shows that our CNN model has a better performance than the other two models. From the results, the deep learning techniques exhibit high practical value and research potential. In our work, we used some specific frequency bands in feature extraction, where the performance of frequency bands is critical in the model. Further optimization of hyperparameters may help improve the performance, which can also be part of our future work.

References

1. Sreeshakthy, M. and Preethi, J. (2016). Classification of human emotion from Deap EEG signal using hybrid improved neural networks with cuckoo search, *BRAIN: Broad Research in Artificial Intelligence and Neuroscience*, 6(3-4): 60-73.

2. Chen, Y., Cui, Y. and Wang, S. (2017). Review of emotion recognition based on physiological signals, *System Simulation Technology*, 13(1): 1-5.
3. Li M., Chai, Q., Kaixiang, T., Wahab, A. and Abut, H. (2009) EEG Emotion recognition system. *In:* Takeda, K., Erdogan, H., Hansen, J.H.L. and Abut, H. (Eds.). *In-Vehicle Corpus and Signal Processing for Driver Behavior*. Springer, Boston, MA.
4. Kumar, D.K. and Nataraj, J.L. (2019). Analysis of EEG-based emotion detection of DEAP and SEED-IV databases using SVM, *SSRN Electron*, 15: 207-211.
5. Koelstra, S., Muhl, C., Soleymani, M., Lee, J.S., Yazdani, A., Ebrahimi, T., ... and Patras, I. (2011). Deap: A database for emotion analysis; using physiological signals, *IEEE Transactions on Affective Computing*, 3(1): 18-31.
6. Liu, J., Meng, H., Li, M., Zhang, F., Qin, R. and Nandi, A.K. (2018). Emotion detection from EEG recordings based on supervised and unsupervised dimension reduction, *Concurrency and Computation: Practice and Experience*, 30(23): e4446.
7. Yoon, H.J. and Chung, S.Y. (2013). EEG-based emotion estimation using Bayesian weighted-log-posterior function and perceptron convergence algorithm, *Computers in Biology and Medicine*, 43(12): 2230-2237.
8. Alhagry, S., Fahmy, A.A. and El-Khoribi, R.A. (2017). Emotion recognition based on EEG using LSTM recurrent neural network, *Emotion*, 8(10): 355-358.
9. Zhan, Y., Vai, M.I., Barma, S., Pun, S.H., Li, J.W. and Mak, P.U. (2019, June). A computation resource-friendly convolutional neural-network engine for EEG-based emotion recognition. *In: 2019 IEEE International Conference on Computational Intelligence and Virtual Environments for Measurement Systems and Applications (CIVEMSA)*, pp. 1-6, IEEE.
10. Xing, X., Li, Z., Xu, T., Shu, L., Hu, B. and Xu, X. (2019). SAE+LSTM: A new framework for emotion recognition from multi-channel EEG, *Frontiers In Neurorobotics*, 13: 37.
11. Bazgir, O., Mohammadi, Z. and Habibi, S.A.H. (2018, November). Emotion recognition with machine learning using EEG signals. *In: 2018 25th National and 3rd International Iranian Conference on Biomedical Engineering (ICBME)*, pp. 1-5, IEEE.
12. Salama, E.S., El-Khoribi, R.A., Shoman, M.E. and Shalaby, M.A.W. (2018). EEG-based emotion recognition using 3D convolutional neural networks, *Int. J. Adv. Comput. Sci. Appl.*, 9(8): 329-337.
13. Chen, Y., Chang, R. and Guo, J. (2021). Emotion recognition of EEG signals based on the ensemble learning method: AdaBoost, *Mathematical Problems in Engineering*, 2021: Article ID 8896062.
14. Dabas, H., Sethi, C., Dua, C., Dalawat, M. and Sethia, D. (2018, December). Emotion classification using EEG signals. *In: Proceedings of the 2018 2nd International Conference on Computer Science and Artificial Intelligence*, pp. 380-384.
15. Russell, J. A. and Snodgrass, J. (1987). Emotion and the environment. *Handb. Environ. Psychol.*, 1: 245-281.
16. Mühl, C., Allison, B., Nijholt, A. and Chanel, G. (2014). A survey of affective brain computer interfaces: Principles, state-of-the-art, and challenges, *Brain-Computer Interfaces*, 1(2): 66-84.
17. Five types of brain waves frequencies; https://mentalhealthdaily.com/2014/04/15/5-types-of-brain-waves-frequencies-gamma-beta-alpha-theta-delta/ [accessed on 11 July, 2018]
18. Chiang, C., Barnes, C., Angelov, P. and Jiang, R. (2020). Deep Learning based Automated Forest Health Diagnosis from Aerial Images. IEEE Access.
19. Storey, G., Jiang, R., Bouridane, A. and Li, C.T. (2019). 3DPalsyNet: A Facial Palsy Grading and Motion Recognition Framework using Fully 3D Convolutional Neural Networks. IEEE Access.

20. Jiang, Z., Chazot, P.L., Celebi, M.E., Crookes, D. and Jiang, R. (2019). Social Behavioral Phenotyping of Drosophila with a 2D-3D Hybrid CNN Framework. IEEE Access.
21. Storey, G., Bouridane, A. and Jiang, R. (2018). Integrated Deep Model for Face Detection and Landmark Localisation from 'in the wild' Images. IEEE Access.
22. Storey, G., Jiang, R. and Bouridane, A. (2017). Role for 2D image generated 3D face models in the rehabilitation of facial palsy. *IET Healthcare Technology Letters*.
23. O'Shea, K. and Nash, R. (2015). An introduction to convolutional neural networks. arXiv preprint arXiv:1511.08458
24. Albawi, S., Mohammed, T.A. and Al-Zawi, S. (2017, August). Understanding of a Convolutional Neural Network. *In: 2017 International Conference on Engineering and Technology (ICET)*, pp. 1-6, IEEE.
25. Yuan, X., Li, L. and Wang, Y. (2019). Nonlinear dynamic soft sensor modeling with supervised long short-term memory network, *IEEE Transactions on Industrial Informatics*, 16(5): 3168-3176.
26. Acharya, U.R., Oh, S.L., Hagiwara, Y., Tan, J.H. and Adeli, H. (2018). Deep convolutional neural network for the automated detection and diagnosis of seizure using EEG signals, *Computers in Biology and Medicine*, 100: 270-278.
27. Hwang, S., Ki, M., Hong, K. and Byun, H. (2020, February). Subject-independent EEG-based emotion recognition using adversarial learning. *In: 2020 8th International Winter Conference on Brain-Computer Interface (BCI)*, pp. 1-4, IEEE.

A Novel Motor Imagery EEG Classification Approach Based on Time-Frequency Analysis and Convolutional Neural Network

Qinghua Wang[1,2,*], Lina Wang[1,2,*] and Song Xu[1,2]

[1] National Key Laboratory of Science and Technology on Aerospace Intelligence Control, Beijing, China
[2] Beijing Aerospace Automatic Control Institute, Beijing, China
Email: violina@126.com

1. Introduction

Motor Imagery (MI) EEG signal is a kind of spontaneous EEG signal, which refers to the EEG signal generated by imaging the body movement without the actual body movement [23, 26]. The EEG provides excellent temporal resolution and abundant information in a non-invasive way. Therefore, the feature extraction techniques of EEG signals are fundamental to the automated classification of MI tasks. Indeed, considerable research efforts have been made to analyze and characterize EEG signals [3, 28, 37, 29]. Presently, there are three major types of motor imagery EEG signal feature extraction methods: spatial filtering, time frequency analysis, and deep learning.

Common Spatial Pattern (CSP) [24] is a spatial domain filtering feature extraction algorithm for two classification tasks, which can decompose each class's spatially distributed components from multi-channel brain-computer interface data by simultaneously diagonalizing the covariance matrix. Filter bank common space pattern (FBCSP) [2] is one of the algorithms derived from CSP. FBCSP computes the mutual information of CSP features from multiple sub-bands to select the most discriminative features. Experimental results show that using FBCSP outperforms CSP in processing motion imagery EEG signals under two classification tasks (right hand, right foot).

Time-frequency analysis approaches, such as the short-time Fourier transform (STFT), continuous wavelet transform (CWT), and parametric spectral estimation methods, are widely applied to processing nonlinear signals. In the work of [25],

Peker *et al.* analyzed EEG data by utilizing a dual-tree complex wavelet transformation (WT), and five statistic features derived from feature vectors were employed for classification. Li *et al.* [15] adopted a time-varying autoregressive model (TVAR) to model the EEG time series and obtained precise power spectral density (PSD) estimates based on time-varying coefficients, which substantially improved the time-frequency resolution compared with the classical power spectrum estimation. Wang *et al.* [30] further combined the TVAR model with an ultra-regularized orthogonal forward regression (named UROFR) algorithm to identify a parsimonious model and obtain detailed time-frequency images for epilepsy detection.

More recently, deep learning methods have received increasing interest in EEG motor imagery decoding [3, 35, 36]. By drawing on the successful experience of deep learning in speech recognition [7], it is practical to convert raw EEG data into suitable intermediate representations and then classify motor imagery tasks based on deep learning methods. Literature [27] proposed a classification model based on STFT and one-dimensional CNN, which uses a one-dimensional convolution kernel to process time-frequency images of MI EEG signals generated by STFT. In order to alleviate the limitation of STFT, Li *et al.* [16] adopted continuous wavelet transform to capture a highly informative EEG image, and a simplified convolutional neural network is then developed to both classify motor imagery tasks and reduce computation complexity. More details of the latest developments and progress in deep learning for motor imagery EEG-based classification can be found in reference [1].

This paper is dedicated to developing a useful scheme for processing motor imagery EEG signals, including modeling, analyzing, and recognizing EEG signals. Driving this aim, this study presented a flexible framework termed time-varying autoregressive model-based power spectral density estimation and squeeze-excitation convolutional neural network (TVAR-PSD-SECNN) for feature extraction and classification of motor imagery EEG signals. Fig. 1 demonstrates a brief overview of the proposed framework. Specifically, the time-varying autoregressive model (TVAR) of the MI-EEG data is first constructed, using the multi-wavelet basis function (MWBF) extension method. The orthogonal forward regression algorithm incorporated zeroth-order regularization named ROFR to produce a parsimonious model. Then, the PSD estimation with the high resolution is estimated, based upon the reduced model. The size of the original time-frequency distribution image is large, which may contain massive noise and redundant information. Besides, training an accurate neural network model based on these input data costs considerable time and computing resources. To address these issues, we down-sample and resize the initial full spectrum to form a refined time-frequency representation. Finally, a well-designed squeeze-excitation convolutional neural network is exploited to automatically learn the most discriminating features and employ multi-channel feature-based fusion for high-accuracy classification of motor imagery EEG signals. The primary novelty of this work is the effectiveness of the time-varying parametric spectrum estimation approach, which provides a more explicit time-frequency distribution. Evaluation results using real datasets demonstrate the effectiveness and applicability of the proposed scheme.

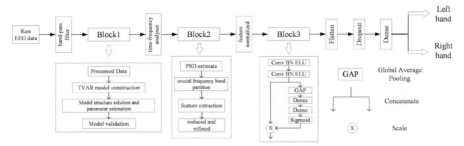

Fig. 1: The overall framework of the proposed TVAR-PSD-SECNN scheme

The remainder of this paper is arranged as follows; Section 2 describes the datasets used in our study and gives a detailed introduction to the proposed TVAR-PSD-SECNN framework. Section 3 presents the experiment details and results. Finally, the discussions and conclusions are given in Section 4.

2. Materials and Methods

2.1 Motor Imagery EEG Datasets

The dataset used in this paper is a widely used public BCI competition II dataset III available online [14]. The EEG data acquisition was recorded from a normal subject (female, 25y) during a feedback session. The experiment consists of seven runs with 40 trials each. All runs were performed on the same day with several minutes' break between tasks to reduce fatigue. The EEG data were sampled with 128 Hz measured over C3, Cz, and C4 three bipolar EEG channels, and filtered between 0.5 and 30 Hz. The experimental protocol was the same on every trial, as diagrammed below. The first 2s was quiet; at $t = 2s$, an acoustic stimulus indicates the beginning of the trial, and a blank screen with a fixation cross was displayed for 1s; then the left arrow or right arrow was displayed at 3s, the subject was instructed to imagine the movement of the respective hand. The dataset consisted of 280 trials of 9s length and was randomly divided into 140 training trials and 140 testing trials by the provider. In general, we selected MI-EEG signals from 3 to 6s for analysis and study.

2.2 Identification of TVAR Model Using Multi-wavelet Basis Function and Regularized Orthogonal Forward Regression

In this section, we will introduce the framework of multiscale wavelet-based TVAR model identification, including model structure determination and parameter estimation based on the ROFR algorithm.

2.2.1 Multiscale Wavelet-based TVAR Model

Numerous studies have shown that the TVAR model is more suitable for modeling and analyzing non-smooth signals, such as EEG signals [31, 32]. A k-order TVAR model can be written as:

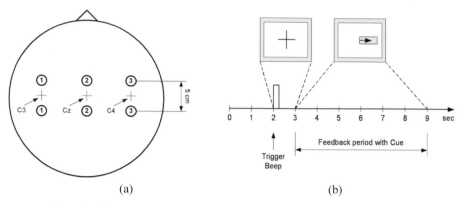

(a)   (b)

Fig. 2: Schematic diagram of electrode positions and experimental paradigm of the MI-EEG acquisition system. (a) Electrode positions; (b) experimental paradigm

$$y(t) = \sum_{i=1}^{k} \theta_i(t) y(t-i) + \varepsilon(t) \tag{1}$$

where $y(t)$ is the model output, k denotes the model order, $\theta_i(t)$ with $i = 1, 2, \ldots, k$ represents the i-th time-varying parameters, $\{y(t-i)\}_{i=1}^{k}$ represents the time-lag of signal y, and $\varepsilon(t)$ is a zero-mean white noise process with a constant variance σ_ε^2.

One effective way to identify the time-varying model 1 was to expand the time-varying parameters $\theta_i(t)$ using a group of basis functions [31]; thus, the complex time-varying system identification problem becomes a time-invariant system parameter estimation problem. In the specific procedure, we first expressed each time-varying parameter by a linear combination of suitable finite basis functions:

$$\theta_i(t) = \sum_{m=1}^{M} q_{i,m} \varphi_m(t), i = 1, 2, \ldots, k \tag{2}$$

where q is expansion coefficient of the basis function. Correspondingly, Equation (1) can be rewritten as:

$$y(t) = \sum_{i=1}^{k} \sum_{m=1}^{M} q_{i,m} y_m(t-i) + \varepsilon(t) \tag{3}$$

where $y_m(t-i)$ is defined as $\varphi_m(t) \cdot y(t-i)$. Following the simplification of time-varying system identification to a time-invariant system parameter estimation problem, the next important step is to select appropriate basis functions to approximate the time-varying coefficients. Multi-wavelet basis functions have proven quite effective for tracking both fast and slow parameter variations in time-varying processes in many research studies. Wavelet theory proves that the multi-resolution wavelet decomposition can approximate a square-integrable scalar signal f as follows [24, 8]:

$$f(x) = \sum_{n=-\infty}^{\infty} \alpha_{j_o,n} \phi_{j_o,n}(x) + \sum_{j \geq j_o}^{\infty} \sum_{n=-\infty}^{\infty} \beta_{j,n} \psi_{j,n}(x) \tag{4}$$

where $\phi_{j_o,n}(x) = 2^{j_o/2} \phi(2^{j_o}x - n)$ and $\psi_{j,n}(x) = 2^{j/2} \psi(2^j x - n)$ with $j_o, j, k \in Z$ (Z denotes the set of all integers) are the translated and dilated version of the scaling function $\phi(x)$ and the mother wavelet $\psi(x)$, $\alpha_{j_o,n}$ and $\beta_{j,n}$ denotes the wavelet expansion coefficients. In addition, when the resolution scale level of the scale functions $\phi_{j_o,n}(x)$

$= 2^{j_o/2} \, \phi(2^{j_o}x - n)$ is sufficiently large; namely, there exists an integer j, Equation (4) can be reduced to $f(x) = \sum_{n=-\infty}^{\infty} \alpha_{j,n} \, \phi_{j,n}(x)$.

Among the many commonly used basis functions, previous research efforts have revealed that multiple B-splines basis functions are more suitable for modeling dynamic systems due to their superior characteristics, such as compactly support and explicit analytical form [31, 33]. Therefore, we adopted the multiscale B-splines basis functions in this paper. The explicit recursive definition of cardinal B-splines function is formulated as follows:

$$B_m(x) = \frac{x}{m-1} B_{m-1}(x) + \frac{m-x}{m-1} B_{m-1}(x-1), \, m \geq 2 \tag{5}$$

where the first order of B-spline basis function $B_1(x) = 1$ with $x \in [0,1]$ is the famous Haar wavelet and the m-th order B-spline $B_m(x)$ is defined on $[0, m]$. Then $\phi_{j,n}(x)$ can be given as $\phi_{j,n}(x) = 2^{j/2} B_m(2^j x - n)$ when taking the cardinal B-splines as the basis function. Specifically, the dilation j and the translation indices n should satisfy $0 \leq 2^j x - n \leq m$. Assuming that the function $f(x)$ to be estimated with decompositions is defined within $[0,1]$, for any given dilation index j, the translation index n is restricted to the collection $\Gamma_m = \{n: -m \leq n \leq 2^j - 1\}$. The order m is usually chosen from 2 to 5, and a detailed derivation and discussion of B-splines characteristics can refer to [32] and [6]. Additionally, the value of scale factor j is usually a tradeoff between accuracy and computational efficiency. In practice, considerable prior research work has suggested that $j = 4$ is a suitable choice for most applications. Supplementary materials for determining an appropriate j can be found in literature [32].

Time-varying parameters $\{\theta_i(t)\}_{i=1}^k$ in Equation (1) can be estimated by utilizing a combination of B-splines basis functions from the families $\phi_{j,n}(x) = 2^{j/2} = B_m(2^j x - n)$ with $n \in \Gamma_m$, $m = 2 \sim 5$, which can be expressed as follows:

$$\theta_i(t) = \sum_{m=2}^{5} \sum_{n \in \Gamma_m} \lambda_{i,n}^m \phi_{i,n}^m \left(\frac{t}{M} \right) \tag{6}$$

where N is the number of sample observations for $t = 1, 2, ..., N$, and $\lambda_{i,n}^m$ are the time-invariant parameters to be estimate and will be used to recover the time-varying parameters $\theta_i(t)$ afterwards.

2.2.2 Model Structure Selection and Parameter Estimation Based on Regularized Orthogonal Forward Regression

Model structure selection and parameter estimation are two crucial steps in model identification. In this paper, the error reduction ratio (ERR) criterion developed by Billings *et al.* [4] is employed to define and measure the importance of each model term, and a regularized orthogonal forward regression algorithm combining the 0th-order regular term with the OFR algorithm is utilized to implement model structure detection and parameter estimation. Detailed analysis and procedure is presented below.

Firstly, a vector form of TVAR model (1) can be attained by adopting the above multi-wavelet basis function expansion method as:

$$y = GQ^T + Ey \tag{7}$$

where $G = [y(t-1)\varphi(t)^T, y(t-2)\varphi(t)^T, ..., y(t-k)\varphi(t)^T]$, $\varphi(t) = [\varphi_1(t), \varphi_2(t), ...$ $\varphi_m(t)]$, denotes the adaptive basis functions, and $Q = [q_{1,1}, q_{1,2}, ..., q_{1,M}, ..., q_{k,M}]$, the superscript T denotes the transpose of a matrix or vector. Considering the regression matrix G in Equation (7) to be of full column rank, the associated orthogonal decomposition can be defined as [31]:

$$G = [w_1, w_2, ...,w_N]\begin{bmatrix} 1 & \alpha_{1,2} & \cdots & \alpha_{1,N} \\ 0 & 1 & \ddots & \vdots \\ \vdots & \ddots & \ddots & \alpha_{N-1,N} \\ 0 & \cdots & 0 & 1 \end{bmatrix} = WA \tag{8}$$

where the orthogonal column satisfies $w_i^T w_{i,j} = 0, i \neq j$. Thus Equation (7) is equivalent to:

$$y = WH + E \tag{9}$$

where the weight vector $H = [h_1, h_2, ..., h_N]^T$ is an ancillary parameter and satisfies $AQ = H$. Note that E is the approximation error and is assumed to be independent of W; hence the vector H can be obtained via the orthogonal principle:

$$H = D^{-1}W^T Y \tag{10}$$

where $D = W^T W$. A more intuitive form can be written as:

$$\langle w_i, Y \rangle = h_i \langle w_i, w_i \rangle \Rightarrow h_i = \frac{\langle Y, w_i \rangle}{\langle w_i, w_i \rangle}, i = 1, 2, ..., M \tag{11}$$

Considering the 0-th order regularization term $\lambda H^T H$, the normalized error criterion is defined as [5]:

$$(E^T E + \lambda H^T H)\Big/ Y^T Y = 1 - \sum_{i=1}^{N} \frac{(w_i^T w_i + \lambda)h_i^2}{Y^T Y} = 1 - \sum_{i=1}^{N} \mathrm{rerr}_i \tag{12}$$

where λ denotes the penalty parameter and $\mathrm{rerr}_i (i = 1, 2, ... m)$ is termed the i-th regularized error reduction ratio, indicating how much of the approximation error can be reduced by the i-th vector. Note that $0 \leq \mathrm{rerr}_i \leq 1$, and $\Sigma \leq \mathrm{rerr}_i \leq 1$. Obviously, the regularized orthogonal forward regression process is to select the term that gives the largest rerr_i at each iteration. In summary, a general procedure for identifying multi-wavelets-based TVAR model using ROFR algorithm can be concluded as follows:

Step 1: Establish a suitable linear time-varying autoregressive model (such as Equation (1)) with real data.

Step 2: Extending all the time-varying parameters employing multiple B-splines basis functions and construct Equation (7).

Step 3: Calculate rerr_i index for each of regression vector $\{g_1, g_2, ..., g_m\}$:

$$\mathrm{rerr}_i = \frac{\left(\langle w_i, w_i \rangle + \lambda \right) h_i^2}{\langle Y, Y \rangle}, k = 1, 2, ..., m$$

Choose the vector that has the maximum 'rerr' as the first orthogonal vector.

Step 4: Orthogonalizing each of $\{g_1, g_2,..., g_m\}$ (except that selected in Step 3) and work out rerr value for each of the orthogonalized vectors, then choose the one with the maximum 'rerr' as the second orthogonal vector.

Step 5: Repeat the same process as in Step 4 until a satisfactory approximation is achieved.

Step 6: Estimate the correlation parameters from the selected model term and recover the initial time-varying coefficients by Equation (6).

2.3 High-resolution Time-Frequency Representation Construction and Refinement

The above-described OFR-RERR-based algorithm can produce a parsimonious model structure and accurate parameter estimations, while the multiple B-splines wavelets provide better time-frequency resolution. Thus, the time-dependent power spectrum estimation of the TVAR model is calculated by [17]:

$$\text{PSD}(t,f) = \frac{\hat{\delta}_\varepsilon^2}{\left|1 - \sum_{i=1}^k \hat{\theta}_i(t)e^{-j2\pi i f/f_s}\right|^2} \tag{13}$$

where f and f_s are the actual frequency and sampling frequency, respectively, $\hat{\delta}$ and $\hat{\theta}$ represent the estimated version of the variables in the original model. The time spectrum function in Equation (13) is continuous concerning the frequency f and can be applied to estimate the spectrum for any frequency point below the Nyquist frequency $f_s/2$. However, some neuronal activity energy changes in EEG signal are not pronounced in the initial full-spectrum representation. As mentioned in [17], the initial dimensions of the full-spectrum map were assumed to be $N_t \times N_f$, where $N_t = T \times f_s$ (T denotes the length of a trial in samples) and $N_f = F \times f$, F was the upper-frequency limit (considered 32 Hz here). This implies that the size of the original image may be immense (here 384×320). While higher resolution time-frequency images can aid physicians in visual examinations, training an effective convolutional neural network based on such dense images requires extensive computational time and resources. Thus, the following two steps are taken to solve the above problem:

Firstly, it is well documented that the Mu(8~14 Hz) and Beta(18~26 Hz) rhythms mainly reflect the activation in motor-related brain regions [23, 36, 1]. Therefore, we focused our subsequent analysis on these two frequency bands and down-sampled the image's size to avoid overfitting and maintain parsimony. Secondly, the two rhythms' effective frequency range is appropriately extended to 4~16 Hz(Mu) and 18~30Hz(Beta) [16]. We take an average of every four points in the time axis and every five points in the frequency domain separately. Thus, the aggregate features of size 96×48 are extracted more effectively. Note that the above procedure was carried out for multi-channel (i.e., C3, C_z, and C4 in this study). In this way, we can obtain a robust and refined representation $S_i \in R^{C \times H \times W}$ of the original MI-EEG signal, which is more conducive to training convolutional neural networks.

2.4 Squeeze-and-Excitation Convention Neural Networks Classifier

A standard convolutional neural network (CNN) operates in a manner inspired by the human visual cortex and consists of convolutional (feature-extraction), pooling (dimensionality-reduction), and activation (nonlinearity) layers [12]. Compared to traditional CNN models that make extensive use of 2D-kernel in computer vision, considering the specific characteristics (electrode positions and time-frequency components) in the time-frequency maps of MI-EEG, most scholars are more inclined to using a filter with 1D-kernel for feature extraction in this field [3, 27, 16, 11].

Based on these studies, this paper implemented a variation of the 1D CNN combined with a squeeze-and-excitation module termed Squeeze-Excitation Convolutional Neural Network (SECNN) as our machine learning classifiers. The overall flowchart of the proposed SECNN model is depicted in Fig. 3. The model comprises three central components: spectral feature extraction module (part 1), temporal feature extraction module (part 2), and squeeze-and-excitation module (part 3). In part 1, we employed F1 (here, 8) 1D filter of size $(1, N_{fc})$ along the frequency axis to extract spectral features, where N_{fc} (here, 24) is relevant to the width of the frequency band. Correspondingly, F2 (here, 16) convolution kernels of size $(N_{tc}, 1)$ were exploited to learn the temporal features in part 2 automatically, where N_{tc} (here, 32) was chosen empirically as one-third of the total sampling points. Note that batch normalization (BN) was added after each convolution operation to speed up convergence and prevent over-fitting. Subsequently, the average pooling layer and Exponential Linear Unit (ELU) activation function was adopted to reduce the dimensionality of the extracted features and enhance the nonlinear ability, respectively.

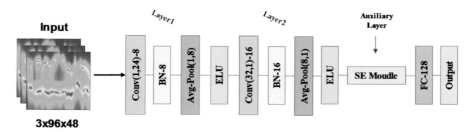

Fig. 3: Schematic illustration of the squeeze-excitation convolutional neural network model

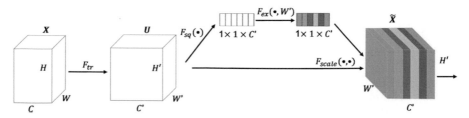

Fig. 4: Structure illustration of the squeeze-and-excitation block

Part 3 utilized the squeeze-and-excitation block to explicitly model the interdependencies between channels, which can better recalibrate the filter outputs [9]. This central idea can be further explained using the squeeze-and-excitation block image above. Specifically, the pipeline includes the following three main stages: The first stage, called squeeze, is to conduct a Global Average Pooling (GAP) operation in order to reduce the $C' \times H' \times W'$ feature map to $1 \times 1 \times C'$ and obtain a global statistic for each channel. Formally, the squeeze operation can be expressed as:

$$Z_c = F_{sq}(U) = \frac{1}{H' \times W'} \sum_{i=1}^{H'} \sum_{j=1}^{W'} U(i, j) \tag{14}$$

where F_{sq} denotes the squeeze operation, z_c is the compression vector formed by the GAP operation, and U represents the intermediate feature map (obtained from the input time-frequency image after part 1 and part 2 transformations).

The next stage is to generate the weight of each individual channel with the help of excitation operation, which is represented formally as follows:

$$\sigma = F_{ex}(z, W') = \sigma(g(z, W')) = \sigma(W_2 \delta(W_1 z)) \tag{15}$$

where σ denotes the sigmoid operation, δ refers to ReLU operation, z is the output from the squeeze stage, W_1 and W_2 denote two fully connected layers. The two FC layers form a bottleneck architecture; that is, the first W_1 layer is employed for dimensionality reduction by a ratio r, and the second W_2 layer is a dimensionality increasing layer returning to the channel dimension of U.

Since the sigmoid layer would return the channel weights between 0 and 1, thus the final output of the SE block is calculated by:

$$\tilde{X} = F_{scal}(u_c, s_c) = s_c \cdot u_c \tag{16}$$

Then $\tilde{X}$ is flattened and routed through a linear hidden layer and Ssoftmax layer to give the result predicted by the network.

3. Results

This section first presents results of MI-EEG representations obtained using the multiscale B-splines and ROFR-based modeling algorithm. Afterwards, we displayed and discussed the classification performance based on the TVAR-PSD-SECNN framework.

3.1 MI-EEG Data Modeling and Time-Frequency Analysis

The raw EEG signal is first processed by a 4~35 Hz FIR band-pass filter for filtering out noises at high and low frequencies along with eye-movement artifacts. The filtered EEG data is modeled adopting the multi-wavelets-based TVAR model as previously mentioned, while the order of B-spline is chosen as 2,3,4,5 and the scale factor j is taken to 4 as described in our previous study [30]. Then the OFR-RERR algorithm is utilized to determine a parsimonious model. To accurately assess the effect of the TVAR-ROFR-based approach, Fig. 5 displays the comparison of the filtered signal and the estimated signal. The following figure demonstrates that the generated TVAR

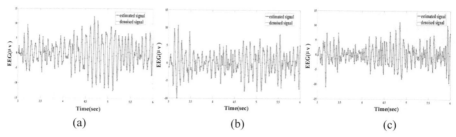

Fig. 5: Comparison between the reconstructed data by TVAR-ROFR and the filtered EEG sequence in various channels (C3,Cz,C4): (a) C3 channel; (b) Cz channel; (c) C4 channel.

model can reconstruct the original signals with high precision and accurately track changes in EEG activity, which contributes to further time-frequency analysis of EEG time series based on the resulting TVAR model.

For each instance in the dataset, the above preprocessing and modeling operations were carried out for the EEG data obtained from a single trial of motor imagery task (right and left hand). Then, the EEG signal energy distribution in the time-frequency plane of C3 and C4 channels under different tasks is obtained according to the time-spectrum function Equation (13), as depicted in Fig. 6.

It is evident from the diagram that EEG power spectra concentrated on the Mu (8–14 Hz) and Beta (18–26 Hz) frequency ranges in the motor imagery task. Specific to observing the energy change of the Mu rhythm, when imagining left-hand movement, it can be seen that the relative enhancement of the energy in C3 channel 3 seconds after the trial begins, that is, the ERS phenomenon occurs, while the energy in the contralateral C4 channel is declined, i.e. the ERD phenomenon appears. Correspondingly, when imagining right-hand movement, the ipsilateral C4 channel's energy increases (ERS), and the energy of the contralateral C3 channel diminishes (ERD).

Fig. 6 reflects the energy density distribution of the MI-EEG signal in time-frequency plane pretty well. To make it more intuitively clear, we compare the proposed algorithm with existing popular time-frequency analysis methods. Fig. 7 displays the time-frequency spectrograms generated using STFT, CWT, and the proposed TVAR-ROFR algorithms. It is clear from Fig. 7 that the proposed multi-scale B-splines and ROFR-based modeling method provides clearer resolution and higher quality of the time-varying spectral distribution. The window size of the traditional STFT algorithm is fixed; thus, the time-frequency resolution is poor. Although wavelet transform has the merits of multi-resolution decomposition and avoids the problem of window size selection, it cannot obtain high resolution in both low-and-high frequency regions simultaneously. Compared to non-parametric spectral estimation methods, the time-varying parametric spectral estimation approach presented in this paper substantially improved the frequency resolution. It enabled a more accurate estimate of the energy at each sampling time-point and at an arbitrary point on the frequency, which helps to mirror the subtle evolutions of the energy distribution versus time.

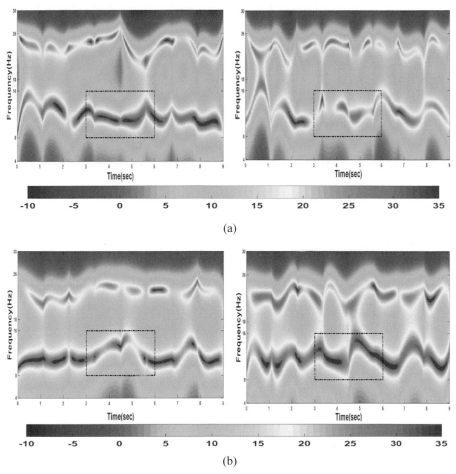

(a)

(b)

Fig. 6: Variation of time-frequency power spectrum of channels C3 and C4 during motor imagery: (a) imagining left-handed movements (left: C3; right: C4); (b) imagining right-handed movements (left: C3; right: C4)

3.2 Classification Result

As detailed in Section 2, we down-sampled and resized the full-spectrum image to better fit the CNN model. Fig. 8 illustrates the refined representation obtained after cropping the uninformative area from the raw time-frequency image. We retain the frequency range 4~16 Hz and 18~30 Hz, which mainly contains motor imagery information covering the Mu and Beta frequency bands. Notice that the input to the CNN model is an informative 3D tensor as illustrated below, where the dimensions are electrodes locations, time, and frequency, respectively.

The Z-score normalization method was standardized for the refined MI-EEG representations; then the input features were fed into the SECNN model for classification. During the training process, we used the Adam optimizer with a

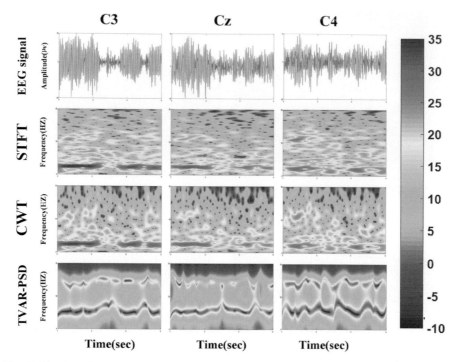

Fig. 7: The time-frequency power spectrum of MI-EEG signals using various time-frequency analysis approaches

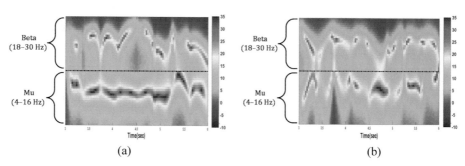

Fig. 8: The refined and highly informative time-frequency images as the input of the SECNN model: (a) C3 channel (imagining left-handed movements); (b) C4 channel (imagining left-handed movements)

learning rate of 0.001 to minimize the categorical cross-entropy loss. In particular, all training was run on an Nvidia GeForce RTX 2080Ti GPU with 11G memory.

In this article, we utilize accuracy and kappa values to evaluate the performance of the proposed scheme. Accuracy is defined as follows:

$$\text{Accuracy} = \frac{TP + TN}{TP + TN + FP + FN} \tag{17}$$

where TP, TN, FP, and FN stand for true positive, true negative, false positive, and false negative, respectively.

The formula for the calculation of kappa(κ) is:

$$
\begin{cases}
\kappa = \dfrac{ACC - p_e}{1 - p_e} \\
p_e = \dfrac{a_1 \times b_1 + a_2 \times b_2 + \ldots + a_c \times b_c}{n \times n}
\end{cases}
\tag{18}
$$

where n is the total sample size, $a_1, a_2, \ldots, a_c$ denote the actual number of samples in each class, and $b_1, b_2, \ldots, b_c$ represent the number of correctly classified samples for each category.

Table 1 presents the accuracy and kappa results of the BCI competition II datasets III using the STFT-CNN [27], CWT-CNN [13], TVAR-PSD-CNN, and TVAR-PSD-SECNN method, respectively. There are two significant findings to note in the table.

Table 1: Classification result using different time-frequency methods on real BCI dataset

Evaluation Metric	Method			
	STFT-CNN	CWT-CNN	TVAR-PSD-CNN	TVAR-PSD-SECNN
Accuracy (%)	89.3	91.4	92.8	**94.3**
Kappa	0.786	0.829	0.857	**0.886**

Firstly, by comparing the MI-EEG classification results obtained, based on different time-frequency analysis methods, it can be seen that higher accuracy and kappa values can be attained using the proposed TVAR-PSD approach. This is mainly attributed to the fact that the time-varying parametric spectrum estimation method produces an increased resolution time-frequency representation of the MI-EEG signal, which leads to more precise results. Another noteworthy outcome is that the CNN model incorporating the squeeze-and-excitation block improves the accuracy by about 1.5% as compared to a simplified CNN model. This is understandable since the squeeze-and-excitation block enables a better map of the channel (electrode position) dependency. Therefore, they can better recalibrate the filter outputs and thus, lead to performance gains.

Additionally, we analyze the impact of different frequency band features in recognition of motor imagery tasks. The results in Table 2 explain the accuracy of classifications based on single Mu-rhythm, Beta-rhythm, and hybrid Mu+Beta features. It is apparent from this table that adopting Mu-rhythm (extended range) alone was able to yield satisfactory accuracy, whereas using Beta-rhythm (extended range) alone, we obtain just accuracy of around 73%. These results were in accordance with the findings of a great deal of the previous work [18, 34] where Mu or alpha band is dominant for decoding MI-EEG signal and, most of the time, provides promising results. Indeed, this result can be intuitively understood by looking at the time-frequency spectrograms illustrated in Fig. 7; namely, the time-frequency energy change in Mu band is more apparent and more discriminative.

Table 2: Classification accuracy using features in different frequency bands

Method	Frequency Band		
	Mu + Beta	Mu (4~16 Hz)	Beta (18~30 Hz)
TVAR-PSD-CNN	**92.8%**	90.7%	71.4%
TVAR-PSD-SECNN	**94.3%**	93.6%	73.6%

To make an appropriate comparison, Table 3 shows the accuracy results of our proposed TVAR-PSD-SECNN framework and other related work on the same dataset. From Table 3, it can be observed that the proposed TVAR-PSD-SECNN approach achieved competitive results as compared to the state-of-the-art methods. Compared with earlier methods that mainly extracted single-dimensional features from raw data, this study can achieve a significant performance boost (about 10%) by combining the time-frequency-energy representation and electrode locations as input features. Furthermore, in comparison with recent related work, we still obtain a satisfactory 2~4% performance improvement. We ascribe this performance boost to the validity and reliability of the proposed time-varying parametric modeling approach, which produces a more detailed MI-EEG data representation. It is worth pointing out that although literature [10] reports a lightly higher classification accuracy, which was achieved by adopting a deep VGG19 network for feature extraction of the input images (of size of 224*224). This makes it time consuming for network model training. In other words, the slightly higher accuracy was achieved at the expense of time and resource consumption. Comparatively, the proposed SECNN model is compact and lightweight, which may facilitate fast processing of MI-EEG signals with more electrodes (multiple channels) and larger datasets.

Table 3: Performance comparison of different methods for MI-EEG classification

Authors	Year	Method Used	Accuracy (%)
Liu *et al.* [22]	2010	Common spatial pattern (CSP) and SVM classifier	84.29
Li *et al.* [19]	2013	Adaptive power projection and Bayesian classifier	90
Tabar *et al.* [27]	2016	STFT and convolutional neural network model	89.3
Xu *et al.* [34]	2018	Wavelet transform time-frequency image and deep learning	92.75
Lee *et al.* [13]	2019	Continuous wavelet transform (CWT) and 1D CNN	92.9
Kant *et al.* [10]	2020	CWT filter-bank, transfer learning and deep VGG19	**95.71***
This work		TVAR-PSD and squeeze-excitation CNN model	94.3

Note: [10] adopted a deep VGG19 network for feature extraction of the input images (of size of 224*224). As a consequence, the model training is time consuming.

4. Conclusion

In this paper, we propose a novel time-frequency analysis and convolutional neural network-based scheme termed TVAR-PSD-SECNN to process, characterize and identify MI-EEG signals. There are three main improvements in this study: (a) The time-varying parameter spectrum estimation method based on multiscale B-splines wavelet and TVAR model provides a more precise time-frequency energy distribution of MI-EEG signal; (b) Highly informative and refined MI-EEG representations were obtained by further extraction and collation of features in the specific frequency bands (Mu rhythm: 8-14 Hz; Beta rhythm: 18-26 Hz); (c) A well-designed SECNN model is utilized to automatically extract the high-level discriminant features, while the squeeze-and-excitation block is capable of learning a nonlinear interaction between channels to realize high-precision MI-EEG classification.

The proposed TVAR-PSD-SECNN scheme achieves competitive results on the publicly-available BCI competition data, suggesting that the combination of the new time-frequency analysis methods and deep learning help to increase the discrimination accuracy of motor imagery-based BCI systems. As detailed in Section 3, the time-varying parameter spectrum estimation approach can yield an improved and more efficient energy distribution than the non-parametric spectrum estimation method. This helps obtain the information of the specific frequency and the energy distribution of the frequency content varying over time along with deepening the understanding of the mechanisms of motor imagery brain activity. Moreover, a vital strength of the present scheme was the validity and applicability of the proposed multi-scale wavelet-based TVAR modeling framework, which is applicable for analyzing EEG signals in various states; for instance, a significant practical application is the time-varying nonlinear causality detection of MI-EEG signal [20, 21], which may be beneficial to uncover the causal information flow over the time course between various sensorimotor related channels. We will continue to exploit the proposed scheme for more application scenarios in future studies.

Authors' Contributions

Qinghua Wang and Lina Wang conceived of and wrote the paper; Qinghua Wang performed simulations and experiments; and Lina Wang and Song Xu offered useful suggestions for the paper preparation and writing. All authors have read and approved the final manuscript.

Funding

This research received no external funding.

Conflicts of Interest

The authors declare no conflict of interest.

References

1. Al-Saegh, A., Dawwd, S.A. and Abdul-Jabbar, J.M. (2021). Deep learning for motor imagery EEG-based classification: A review, *Biomedical Signal Processing and Control*, 63: 102172.
2. Ang, K.K., Chin, Z.Y., Zhang, H. and Guan, C. (2008). Filter Bank Common Spatial Pattern (FBCSP) in brain-computer interface. *In: Proceedings of 2008 IEEE International Joint Conference on Neural Networks (IEEE World Congress on Computational Intelligence)*, pp. 2390-2397.
3. Chen, J., Yu, Z., Gu, Z. and Li, Y. (2020). Deep temporal-spatial feature learning for motor imagery-based brain–computer interfaces, *IEEE Transactions on Neural Systems and Rehabilitation Engineering*, 28: 2356-2366.
4. Chen, S., Billings, S.A. and Luo, W. (1989). Orthogonal least squares methods and their application to non-linear system identification, *International Journal of Control*, 50: 1873-1896.
5. Chen, S., Chng, E. and Alkadhimi, K. (1996). Regularized orthogonal least squares algorithm for constructing radial basis function networks, *International Journal of Control*, 64: 829-837.
6. Chui, C.K. and Wang, J.Z. (1992). On compactly supported spline wavelets and a duality principle, *Transactions of the American Mathematical Society*, 330: 903-915.
7. Deng, L., Li, J., Huang, J.-T., Yao, K., Yu, D., Seide, F., Seltzer, M., Zweig, G., He, X. and Williams, J. (2013). Recent advances in deep learning for speech research at Microsoft. *In: Proceedings of 2013 IEEE International Conference on Acoustics, Speech and Signal Processing*, pp. 8604-8608.
8. Graps, A. (1995). An introduction to wavelets, *IEEE Computational Science and Engineering*, 2: 50-61.
9. Hu, J., Shen, L. and Sun, G. (2018). Squeeze-and-excitation networks. *In: Proceedings of Proceedings of the IEEE Conference on Computer Vision and Pattern Recognition*, pp. 7132-7141.
10. Kant, P., Laskar, S.H., Hazarika, J. and Mahamune, R. (2020). CWT based transfer learning for motor imagery classification for brain computer Interfaces, *Journal of Neuroscience Methods*, 345: 108886.
11. Lawhern, V.J., Solon, A.J., Waytowich, N.R., Gordon, S.M., Hung, C.P. and Lance, B.J. (2018). EEGNet: A compact convolutional neural network for EEG-based brain–computer interfaces, *Journal of Neural Engineering*, 15: 056013.
12. LeCun, Y., Bengio, Y. and Hinton, G. (2015). Deep learning, *Nature*, 521: 436-444.
13. Lee, H.K. and Choi, Y.-S. (2019). Application of continuous wavelet transform and convolutional neural network in decoding motor imagery brain-computer interface, *Entropy*, 21: 1199.
14. Lemm, S., Schafer, C. and Curio, G. (2004). BCI competition 2003-data set III: Probabilistic modeling of sensorimotor/spl mu/rhythms for classification of imaginary hand movements, *IEEE Transactions on Biomedical Engineering*, 51: 1077-1080.
15. Li, Y., Cui, W.-G., Huang, H.5, Guo, Y.-Z., Li, K. and Tan, T. (2019). Epileptic seizure detection in EEG signals using sparse multiscale radial basis function networks and the Fisher vector approach, *Knowledge-based Systems*, 164: 96-106.
16. Li, F., He, F., Wang, F., Zhang, D., Xia, Y. and Li, X. (2020). A novel simplified convolutional neural network classification algorithm of motor imagery EEG signals based on deep learning, *Applied Sciences*, 10: 1605.

17. Li, Y., Luo, M.-L. and Li, K. (2016). A multiwavelet-based time-varying model identification approach for time-frequency analysis of EEG signals, *Neurocomputing*, 193: 106-114.

18. Li, M., Wang, R. and Xu, D. (2019). An improved composite multiscale fuzzy entropy for feature extraction of MI-EEG, *Entropy*, 22: 1356.

19. Li, C.-Y., Liu, R., Wang, Y.-Y., Wang, Y.X. and Li, X. (2013). Adaptive power projection method for accumulative EEG classification. *In: Proceedings of 2013 35th Annual International Conference of the IEEE Engineering in Medicine and Biology Society (EMBC)*, pp. 7052-7055.

20. Li, Y., Lei, M.-Y., Guo, Y., Hu, Z. and Wei, H.-L. (2018). Time-varying nonlinear causality detection using regularized orthogonal least squares and multi-wavelets with applications to EEG, *IEEE Access*, 6: 17826-17840.

21. Li, Y., Lei, M., Cui, W., Guo, Y. and Wei, H.-L. (2019). A parametric time-frequency conditional Granger causality method using ultra-regularized orthogonal least squares and multiwavelets for dynamic connectivity analysis in EEGs, *IEEE Transactions on Biomedical Engineering*, 66: 3509-3525.

22. Liu, C., Zhao, H., Li, C. and Wang, H. (2010). CSP/SVM-based EEG classification of imagined hand movements, *Journal of Northeastern University (Natural Science)*, 31: 1098-1011.

23. Lotze, M. and Halsband, U. (2006). Motor imagery, *Journal of Physiology*, Paris, 99: 386-395.

24. Mallat, S.G. (1989). A theory for multiresolution signal decomposition: The wavelet representation, *IEEE Transactions on Pattern Analysis and Machine Intelligence*, 11: 674-693.

25. Peker, M., Sen, B. and Delen, D. (2015). A novel method for automated diagnosis of epilepsy using complex-valued classifiers, *IEEE Journal of Biomedical and Health Informatics*, 20: 108-118.

26. Stevens, J.A. and Stoykov, M.E.P. (2003). Using motor imagery in the rehabilitation of hemiparesis, *Archives of Physical Medicine and Rehabilitation*, 84: 1090-1092.

27. Tabar, Y.R. and Halici, U. (2016). A novel deep learning approach for classification of EEG motor imagery signals, *Journal of Neural Engineering*, 14: 016003.

28. Wang, Y., Gao, S. and Gao, X. (2005). Common spatial pattern method for channel selelction in motor imagery based brain-computer interface. *In: Proceedings of 2005 IEEE Engineering in Medicine and Biology, 27th Annual Conference*, pp. 5392-5395.

29. Wang, L., Xue, W., Li, Y., Luo, M., Huang, J., Cui, W. and Huang, C. (2017). Automatic epileptic seizure detection in EEG signals using multi-domain feature extraction and nonlinear analysis, *Entropy*, 19: 22.

30. Wang, Q., Wei, H.-L., Wang, L. and Xu, S. (2020). A novel time-varying modeling and signal processing approach for epileptic seizure detection and classification, *Neural Computing and Applications*, 1-17.

31. Wei, H.-L., Billings, S.A. and Liu, J. (2004). Term and variable selection for non-linear system identification, *International Journal of Control*, 77: 86-110.

32. Wei, H.L., Billings, S.A. and Liu, J. (2010). Time-varying parametric modelling and time-dependent spectral characterisation with applications to EEG signals using multiwavelets, *International Journal of Modeling, Identification and Control,* 9(3): 215-224.

33. Wei, H.-L., Billings, S.A. and Balikhin, M. (2004). Prediction of the DST index using multiresolution wavelet models, *Journal of Geophysical Research: Space Physics*, 109.

34. Xu, B., Zhang, L., Song, A., Wu, C., Li, W., Zhang, D., Xu, G., Li, H. and Zeng, H. (2018). Wavelet transform time-frequency image and convolutional network-based motor imagery EEG classification, *IEEE Access*, 7: 6084-6093.

35. Zhang, D., Chen, K., Jian, D. and Yao, L. (2020). Motor imagery classification via temporal attention cues of graph embedded EEG signals, *IEEE Journal of Biomedical and Health Informatics*, 24: 2570-2579.

36. Zhang, X., Yao, L., Wang, X., Monaghan, J., Mcalpine, D. and Zhang, Y. (2019). A survey on deep learning based brain computer interface: Recent advances and new frontiers, arXiv preprint arXiv:1905.04149

37. Ziyu, J., Youfang, L., Tianhang, L., Kaixin, Y., Xinwang, Z. and Jing, W. (2020). Motor imagery classification based on multiscale feature extraction and squeeze-excitation model, *Journal of Computer Research and Development*, 57: 2481.

Classification of EEG Signals for Brain-Computer Interfaces using a Bayesian-Fuzzy Extreme Learning Machine

Adrian Rubio-Solis[1], Carlos Beltran-Perez[2] and Hua-Liang Wei[3]

[1] Institute for Materials Discovery, University College London (UCL)
[2] Tecnologico de Monterrey, Campus Toluca, Mexico
[3] Automatic Control and Systems Engineering, University of Sheffield

1. Introduction

A Brain-Interface Computer (BCI) is an emanating technology that allows a direct communication between a human brain and a computer [1, 2, 3, 4]. BCI has been largely used as a technology to exploit the electrical activity in the brain for the diagnosis of neurological diseases in handicapped patients, who have lost their abilities and to understand psycho-physiological processes [5]. EEG is a non-invasive BCI system, in which brain activities are captured with a high temporal resolution, usability, portability and low setup cost [1]. In applications that involve handicapped patients, BCI is frequently used as a system to revive those elementary capabilities by creating an information pathway between the human brain and processing/computing devices. In other words, a BCI system utilizes the information from the brain activity in disabled people to assist them while mapping their sensory-motor functions [1, 2].

For the last two decades, the most adopted EEG patterns for BCI development include sensorimotor Mu rhythm [7, 8], and beta rhythms (15-30 Hz) [9, 10], recorded from the scalp over the sensorimotor cortex and widely used methods in BCI systems. Mu rhythm usually results from a power change of EEG frequency band between 8-18 Hz. It occurs as a simultaneous neural response at contralateral sensorimotor area during motor imagery (MI) tasks. Over the last decades, BCI systems have been increasingly developed towards the solution of several problems that involve the control of prosthesis, wheelchair navigation, writing and communication assistance in disabled people [3]. MI is one of the most popular methods in BCI applications that involves carrying out a motor task merely by thinking or imagining [1]. It can be just a simple shifting of hands or legs, and/or closing eyes. Accordingly, MI-based BCI systems have become a prominent solution to recognize the desired commands by classifying MI tasks for deprived people of their motor abilities and for rehabilitation [1, 2]. However, MI signals are highly non-stationary and inevitably contaminated with noise, meanwhile, they strongly depend on subjects [11]. Moreover, the classification of EEG is usually a complex and aperiodic time series, which is the sum of a large number of neuronal membrane potentials [1]. Therefore, a powerful pattern recognition model is required for the implementation of a MI-based BCI system with a high performance [13].

In literature, an extensive number of research efforts have been dedicated to improving EEG feature extraction and classification of MI tasks [1, 4]. Most of these efforts have suggested a two-step approach, where, first a process of feature extraction is implemented, and a second step is performed for the classification of the extracted features. Common Spatial Pattern (CSP) and Extreme Learning Machines (ELMs) have been successfully applied to the classification of MI tasks delivering a superior performance over traditional classification approaches, such as Support Vector Machine and Multilayer Perceptron neural networks [11]. On the one hand, CSP has been used as a robust spatial filter for multichannel optimization of EEG to the maximization of the variance of projected signal from one class while to minimize it from another class [12, 13]; on the other hand, ELM has been successfully applied to the classification of these projected signals [11]. For example in [13], a probabilistic framework based on Sparse Bayesian Extreme Learning Machine (SBELM) was suggested to improve the classification of traditional ELM [14]. SBELM was suggested as an improved ELM method that automatically controls model complexity, good generalization properties and exclude redundant hidden neurons by exploiting the advantages of ELM theory and Bayesian learning.

Similarly, other pattern recognition methods that are able to naturally deal with nonlinear and outlier characteristics of EEG signals have been suggested. For instance, in [14] a combined method based on wavelet transformation and Interval Type-2 Fuzzy Logic Systems (IT2-FLSs) called wavelet-IT2FLS was introduced. The proposed wavelet-IT2FLS is a higher order fuzzy system more capable of uncertainty handling than traditional ELM and SVM, in which the noisy, nonlinear and outlier-embedded nature of EEG signals can be modeled proficiently by type-2 fuzzy sets (FSs).

In this paper, a new Bayesian Fuzzy Extreme Learning Machine (BFELM) for EEG signal classification in BCI systems is presented. The proposed BFELM is a unified learning approach, in which a probabilistic method based on Bayesian learning and ELM theory is implemented for the training of Fuzzy Inference Systems (FISs) of Takagi-Sugeno-Kang (TSK). On the one hand, Bayesian learning incorporates a priori knowledge in the design of fuzzy rules while the confidence intervals of each consequent in the FIS are analytically determined. Within this context, a BFELM is a simultaneous learning method of antecedent and consequent parts of each fuzzy rule in an FIS, in which fuzziness and probability can work in a collaborative manner rather competitively. Moreover, the proposed BFELM inherits the capabilities of

FISs to naturally deal with uncertainty and noisy signals. To validate the performance of the proposed BFELM for EEG signal classification, two public datasets of BCI competitions are used. For feature extraction, CSP is implemented, and the extracted features are fed into the BFELM. To compare the BFELM with other benchmark techniques, traditional ELM, SVM, Multi Kernel ELM (MKELM), Fuzzy ELM (FELM) and Bayesian ELM (BELM) were also implemented.

The rest of this paper is structured as follows. In Section 2, basic concepts of Bayesian ELM and Fuzzy Inference Systems (FISs) is reviewed, as well as the proposed BFELM is described. In Section 3, experiments and results are presented, while in Section 4 the corresponding discussion is provided. Finally, in Section 5, conclusion and future work are drawn.

2. Background Material and Proposed Method

2.1 Extreme Learning Machine

Extreme Learning Machine (ELM) is a learning paradigm originally developed to train single-hidden-layer feedforward networks (SLFNs), in which parameters in the hidden neurons are initialized randomly and the output weights are optimized using the Moore-Penrose pseudoinverse. Given a number of 'P' distinct samples $D = (\mathbf{x}_p, t_p)$, with each $\mathbf{x}_p$ being a N dimensional vector and t_p as the target scalar output. Hence the goal of ELM is to find a relationship between $\mathbf{x}_p$ and $\mathbf{t}_p$. Standard SLFNs with M hidden nodes and activation $h(\cdot)$ function can be mathematically modeled by:

$$\sum_{k=1}^{M} \beta_k h_k(\mathbf{w}_k; \mathbf{x}_p) = \mathbf{h}(\mathbf{w}_k; \mathbf{x}_p)\,\mathbf{b} = y_p, \ 1 \le p \le P \tag{1}$$

in which $\mathbf{h}(\mathbf{w}_k; \mathbf{x}_p) = [h_1(\mathbf{w}_1; \mathbf{x}_1), \dots, h_M(\mathbf{w}_M; \mathbf{x}_M)]$ is the hidden feature mapping, $\mathbf{w}_k = [\mathbf{w}_1, \dots, \mathbf{w}_N]^T$ is the weight vector a randomly generated parameter of the hidden layer connecting the kth hidden node and the input nodes. The output weight $\beta_p = [\beta_{p1}, \dots, \beta_{p\bar{N}}]^T$ is the weight vector connecting the kth hidden node to the nth output. A SLFN with M hidden nodes and activation function $g(\mathbf{x})$ can approximate P samples with zero error means $\sum_{p=1}^{M}\|y_p - t_p\|$. Thus, a matrix representation of Eq. (1) is:

$$\mathbf{H} = \begin{pmatrix} h(\mathbf{w}_1; \mathbf{x}_1) & \dots & h(\mathbf{w}_k; \mathbf{x}_p) \\ \vdots & \vdots & \vdots \\ h(\mathbf{w}_k; \mathbf{x}_p) & \cdots & h(\mathbf{w}_k; \mathbf{x}_p) \end{pmatrix}_{p \times M} \tag{2}$$

Where $\mathbf{H}$ is the hidden matrix of an SLFN with respect to the inputs $\mathbf{x}_p$. The target vector is defined by $\mathbf{T} = [t_1, \dots, t_P]$. The minimum norm least-squares solution of the linear system $\mathbf{H}\beta = \mathbf{T}$ is unique and can be achieved by calculating the Moore-Penrose pseudo-inverse $H^\dagger$ as follows:

$$\hat{\beta} = \mathbf{H}^\dagger \mathbf{T} \tag{1.3}$$

in which, $\mathbf{H}^{\dagger}$ can be calculated using the orthogonal projection method: $\mathbf{H}^{\dagger} = (\mathbf{H}^T \mathbf{H})^{-1} \mathbf{H}^T$ when $\mathbf{H}^T \mathbf{H}$ is nonsingular, or $\mathbf{H}^{\dagger} = \mathbf{H}(\mathbf{HH}^T)^{-1} \mathbf{H}^T$ when $\mathbf{HH}^T$ is nonsingular. A penalty term can be added to the diagonal of $\mathbf{H}^T\mathbf{H}$ or $\mathbf{HH}^T$ for regularization purpose. However, the optimum value of this penalty is still subjected to minimization of the validation error. In many real-world applications, the number of hidden nodes is much smaller than the number of training samples $M \ll P$ [?]. Hence $\mathbf{H}$ is a non-square matrix, such that one specific value for $\hat{\mathbf{w}}_k, \hat{\mathbf{b}}_k$ and $\hat{\beta}_k$ needs to be determined.

2.2 Bayesian Extreme Learning Machine

Bayesian Extreme Learning Machine (BELM) was originally introduced to ELM theory to determine the output weight with Bayesian Inference Method (BIM). By using BELM, each observed t_p is assumed to have an independent noise component ϵ_p which is Gaussian distributed with zero mean and variance σ^2, that is $t_p = \mathbf{h}(\mathbf{w}; \mathbf{x}_p)$ $\beta + \epsilon_p$, where, $p(\varepsilon_p | \sigma^2) = \mathcal{N}(0, \sigma^2)$. The probabilistic model is then given by:

$$p(t_p | \mathbf{H}, \beta, \sigma^2) = \mathcal{N}(t_p | \mathbf{h}(\mathbf{w}; x_p)\beta, \sigma^2) \tag{4}$$

Using all the training sales, the likelihood function can be computed as:

$$p(\mathbf{T}|\mathbf{H}, \beta, \sigma^2) = \prod_{p=1}^{P} p(t_p | \mathbf{H}, (\beta), \sigma^2)$$

$$\prod_{p=1}^{P} \frac{1}{\sqrt{2\pi\sigma^2}} \exp\left[-\frac{(t_p - \mathbf{h}(\mathbf{w}; \mathbf{x}_p)\beta)^2}{2\sigma^2} \right] \tag{5}$$

To penalize large weights, a natural distribution is given by:

$$p(\beta|\alpha) = \mathcal{N}(\beta 0, \alpha^{-1}\mathbf{I}) = \left(\frac{\alpha}{2\pi}\right)^{P/2} \exp\left[-\frac{\alpha}{2}\beta^T\beta \right] \tag{6}$$

in which, α is a shared prior, and $\mathbf{I}$ is the identity matrix. As the prior and likelihood functions follow a Gaussian distribution, the posterior is also a Gaussian distribution defined by:

$$p(\beta|\mathbf{T}, \mathbf{H}, \alpha, \sigma^2) = \mathcal{N}(\beta|m, S) \tag{7}$$

where $\mathbf{m}$ and $\mathbf{S}$ is the mean and covariance respectively, which are be obtained by:

$$\mathbf{m} = \sigma^{-2} \cdot \mathbf{S} \cdot \mathbf{H}^T \cdot \mathbf{T} \tag{8}$$

$$\mathbf{S} = (\alpha\mathbf{I} + \sigma^{-2} \cdot \mathbf{H}^T \cdot \mathbf{H})^{-1} \tag{9}$$

in which, the posterior distribution of parameters α and σ^2 is $p(\alpha, \sigma^2 | \mathbf{T}, \mathbf{H}) \propto p(\mathbf{T}|\mathbf{H}, \alpha, \sigma^2)$. Therefore, the optimal values for the parameters α and σ^2 can be determined with type-II maximum likelihood (ML-II). Such process involves the maximization of the marginal likelihood $p(\mathbf{T}|\mathbf{H}, \alpha, \sigma^2)$ inferred from integral:

$$\int p(\mathbf{T} | \mathbf{H}, \beta, \sigma^2) p(\beta | \alpha) d\beta \tag{10}$$

By differentiating the marginal log-likelihood function log $p(\mathbf{T}|\mathbf{H}, \beta, \sigma_2)$ with respect to parameters α and σ^2, their optimal values can be computed by:

$$\alpha^{new} = \frac{M - \alpha \cdot trace[\mathbf{S}]}{\mathbf{m}^T \mathbf{m}} \tag{11}$$

$$\sigma^{2,new} = \frac{\sum_{p=1}^{P}(y_p - \mathbf{h}(\mathbf{w}; \mathbf{x}_p)\mathbf{m})^2}{P - M + \alpha \cdot trace[\mathbf{S}]} \tag{12}$$

Thus, by initializing α and σ^2, the terms $\mathbf{m}$ and $\mathbf{S}$, are updated iteratively with Eqs. (8) − (9) and (11) − (12) until convergence. The new $\mathbf{m}$ can be employed for computing the new output y_{new} when new input data $\mathbf{x}_{new}$ is presented following the distributions:

$$p(y_{new}|\mathbf{h}(\mathbf{w}; \mathbf{x}_{new}), \mathbf{m}, \alpha, \sigma^2) = \mathcal{N}(\mathbf{h}(\mathbf{w}; \mathbf{x}_{new})\mathbf{m}, \sigma^2(\mathbf{x}_{new})) \tag{13}$$

where

$$\sigma^2(\mathbf{x}_{new}) = \sigma^2 + \mathbf{h}(\mathbf{w}; \mathbf{x}_{new})^T \cdot \mathbf{S} \cdot \mathbf{h}(\mathbf{w}; \mathbf{x}_{new})$$

Since α is a natural consequence of a Gaussian process, compared to ELM, BELM does not require to include any regularization term. Hence BELM provides better generalization properties than traditional ELM.

2.3 Fuzzy Extreme Learning Machine

According to ELM theory, a Fuzzy Inference System (FIS) can be interpreted as a SLFN if for a given number of distinct training samples $D = (\mathbf{x}_p, \mathbf{t}_p)$, a model of the FIS with M fuzzy rules is given by [20]:

$$\mathbf{y}_p(\mathbf{x}_p) = \sum_{k}^{M} \beta_k G(x_p, c_k, a_k) = \mathbf{t}_p, \; p = 1, \ldots, P \tag{14}$$

An FIS either of Takagi-Sugeno-Kang (TSK) or Mamdani type can be defined by a number of fuzzy rules R^k of the form [?, ?]

$$R^k : \text{IF } x_{p1} \text{ is } A_{1k} \text{ AND } x_{p2} \text{ is } A_{2k} \text{ AND} \ldots$$

$$\text{IF } x_N \text{ is } A_{Nk} \text{ THEN } (y_p \text{ is } \beta_k) \tag{15}$$

where, A_{sk} ($s = 1, \ldots, N, k = 1, \ldots, M$) are the fuzzy sets that correspond to the *sth* input variable x_{ps} in the *kth* rule. When an FIS employs a TSK inference engine, β_k ($k = 1, \ldots, M$) is defined by a linear combination of input variables, i.e. $\beta_k = q_k, 0 + q_k, 0 x_1 + \ldots q_k, N xN$, otherwise if the FIS is of Mamdani type, β_k is a crisp value. In Fuzzy Logic System theory (FLS), the degree to which any given input x_{ps} satisfies the quantifier A_{sk} is specified by its Membership Function (MF) $\mu A_{ks}(c_{ks}, a_k)$, where usually a non-constant piece-wise continuous MF is used [19]. By using the symbol $\otimes$ for the repre-sentation of fuzzy logic AND operations, the firing strength of the *kth* fuzzy rule can be computed as [28]:

$$R^k(\mathbf{x}_p; \mathbf{c}_k, a_k) = \mu A_{k1}(x_{p1}, c_{k1}, a_k)$$
$$\otimes \mu A_{k2}(x_{p2}; c_{k2}, a_k) \otimes \ldots \otimes \mu A_{kN}(x_{pN}; c_{kN}, a_k) \qquad (16)$$

Each fuzzy rule R^K can be normalised as

$$G^k(\mathbf{x}_p; c_{ks}, c_k) = \frac{R^k(\mathbf{x}_p; c_k, a_k)}{\sum_{k=1}^{M} R^k(\mathbf{x}_p; c_k, a_k)} \qquad (17)$$

G^k is called fuzzy basis function, where for the *pth* input-output, each $\mathbf{y}_p$ is:

$$\mathbf{y}_p = \sum_{k=1}^{M} \beta_k G^k(x_p; c_{ks}, c_k) \qquad (18)$$

For an FIS with a TSK, consequent parameters are linear combinations of input parameters computed as:

$$\beta_k = \mathbf{x}_{p,e} \mathbf{q}_k^T, \ k = 1, \ldots, M \qquad (19)$$

For a TSK fuzzy model, $\mathbf{x}_{p,e} = [1 \ \mathbf{x}_p]$ is the extended version of $\mathbf{x}_p$ defined as $\mathbf{x}_{p,e} = [x_0, x_1, \ldots, x_N]$ and each coefficient $\mathbf{q}_k = [q_{k,0}, q_{k,1}, \ldots, q_{k,N}]$. The output weight of SLFN is defined by $\beta = [\beta_1, \ldots, \beta_M]$. For a Mamdani fuzzy model, $\mathbf{x}_{p,e} = [0, \mathbf{x}_p]$, and β_k is a single crisp value. In this work, a FIS with a TSK inference is considered. A compact representation for Eq. (14) is

$$\mathbf{H}_{TSK} \mathbf{Q} = \mathbf{T} \qquad (20)$$

in which, $\mathbf{Q}$ is the matrix of coefficients $q_{kj,s}$. If a TSK implication is employed, $\mathbf{H}_{TSK} = [\mathbf{h}_1(\mathbf{c}_k, \mathbf{a}_k; \mathbf{x}_1), \ldots, \mathbf{h}_P(\mathbf{c}_k, \mathbf{a}_k; \mathbf{x}_P)]^T$, where the optimal value for matrix $\hat{\mathbf{Q}}$ is obtained

$$\hat{\mathbf{Q}} = H_{TSK}^{\dagger} T \qquad (21)$$

where the vector of firing strengths is $\mathbf{h}_p = (h_{p1}, \ldots, h_{pM})$, and h_{pk} is:

$$h_{pk} = (G^1(\mathbf{x}_p, \mathbf{c}_1, a_1)x_0, \ldots, G^1(\mathbf{x}_p, \mathbf{c}_1, a_1)x_p, \ldots,$$
$$G^M(\mathbf{x}_p, \mathbf{c}_M, a_M)x_0, \ldots, G^M(\mathbf{x}_p, \mathbf{c}_M, a_M) x_p) \qquad (22)$$

Matrix $\hat{\mathbf{Q}} = [\hat{\mathbf{q}}_1, \ldots, \hat{\mathbf{q}}_M]^T$. In FELM, the parameters of each MF $(\mathbf{c}_k, a_k)$ are randomly generated. Based on this, the consequent parameters β_i are analytically estimated.

2.4 Bayesian-Fuzzy ELM for EEG Classification

Bayesian-Fuzzy Extreme Learning Machine (BFELM) is a unified learning framework based on Bayesian learning regression, Fuzzy Logic Systems (FLSs) theory and ELM to the training of Fuzzy Inference Systems (FISs) of either TSK or Mamdani type. In this study, BFELM is applied to the learning of the consequent parts of a FIS of Takagi-Sugeno-Kang (TSK) that can handle any bounded non-constant piecewise continuous membership function (MF). As illustrated in Fig. 1, given a set of training samples $(\mathbf{x}_p, t_p)$, $p = 1, \ldots, P$, where $\mathbf{x}_p = [x_{p1}, \ldots, x_{pN}]^T \in \mathbf{R}^N$ and the corresponding class labels $t_p = \{-1, 1\}$ denotes the class of MI task. Assume that

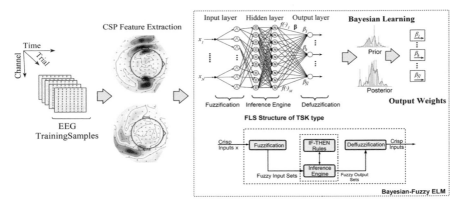

Fig. 1: Illustration of the BFELM for EEG signals classification

each of the training samples is a feature vector obtained by using Common Spatial Pattern (CSP) as detailed in [21]. BFELM aims at finding an optimal hyper-plane that maximizes the separating margin between the two classes. Given certain MF g and its parameters ($\mathbf{c}_k$, $\mathbf{a}_k$), and rule number M for EEG signal classification, BFELM is formulated as an iterative learning algorithm that is implemented in two steps as described in Algorithm 1.

(I) Inference of the posterior distribution of the consequent parameters $\mathbf{q}_k$ of the TSK FIS using a Gaussian function with mean $\mathbf{m}_{q,t}$ and covariance $\mathbf{S}_{q,t}$ (Eq. 23 and 24).

(II) With the evidence procedure, estimate the two hyperparameters α_t and σ_t^2 as described in Equations 25 and 26.

Thus, from algorithm 1, fuzzy consequent parts $\mathbf{q}_k$ are optimized iteratively by means of ML-II [23] or evidence procedure [24].

Algorithm 1: Bayesian-Fuzzy ELM for EEG Classification

Result: optimized α, σ^2, $\mathbf{S}_q$, $\mathbf{m}_q$

Initialize: α, σ^2, $\mathbf{c}_k$, $\mathbf{a}_k$, $\in$, M;

$t = 0$;

while *While* $\|m_{q,t+1} - m_{q,t}\| > \in$ **do**

    $t = t + 1$;

    Calculate;

$$\mathbf{m}_{q,t} = \sigma^{-2} \cdot \mathbf{S}_{q,t} \cdot \mathbf{H}_{TSK}^{T} \cdot \mathbf{T} \tag{23}$$

$$\mathbf{S}_{q,t} = (\alpha\mathbf{I} + \sigma^{-2} \cdot \mathbf{H}_{TSK}^{T} \cdot \mathbf{H}_{TSK})^{-1} \tag{24}$$

    Update;

$$\alpha_t = \frac{M - \alpha \cdot trace[\mathbf{S}_{q,t}]}{\mathbf{m}_{q,t}^{T}\mathbf{m}_{q,t}} \tag{25}$$

$$\sigma_t^2 = \frac{\sum_{p=1}^{P}(y_p - \mathbf{h}_p(\mathbf{c}_k, \mathbf{a}_k, \mathbf{x}_p)\mathbf{m}_{q,t})}{P - M + \alpha \cdot trace[\mathbf{S}_{q,t}]} \tag{26}$$

end

The iterative process for the proposed BFELM is stopped when the difference of the norm $\mathbf{m}_{q,t}$ between successive iterations has fallen below a given value $\in$ [22]. Therefore, relevant CSP features are automatically selected for the subsequent classification [13]. Given a test sample $\hat{\mathbf{x}}$, the prediction distribution of the consequent parts of an FIS can be defined as:

$$p(\hat{y} \mid \alpha, \sigma^2, \mathbf{h}(c_k, \hat{a}_k; \mathbf{x})) = \int (\hat{y} \mid \hat{\mathbf{q}}_k, \sigma^2, \mathbf{h}(c_k, a_k; \hat{\mathbf{x}}))p(c_k, a_k; \hat{\mathbf{q}}_k \mid \alpha, \sigma^2, y)d\hat{\mathbf{q}}_k \tag{27}$$

where $\mathbf{h}(c_k, a_k; \hat{\mathbf{x}}) = [h1 (c_k, a_k; \hat{\mathbf{x}}), \ldots, h_M (c_k, a_k; \hat{\mathbf{x}})]$ is again Gaussian with mean $\hat{\mathbf{m}}$ and variance $\hat{\sigma}^2$ defined as:

$$\hat{\mathbf{m}}_q = \mathbf{h}(c_k, \hat{a}_k; \mathbf{x})\mathbf{m}_q \tag{28}$$

$$\hat{\sigma}^2 = \sigma^2 + \mathbf{h}(c_k, a_k; \mathbf{x}) \mathbf{S}_q \mathbf{h}(c_k, \hat{a}_k; \mathbf{x})^T \tag{29}$$

The test sample is then classified using the criterion:

$$\hat{\mathbf{y}} = \begin{cases} -1, & \hat{\mathbf{m}} < 0 \\ +1 & \hat{\mathbf{m}} > 0 \end{cases} \tag{30}$$

Note: Compared to other learning approaches for the training of FISs, the hyperparameter α can be viewed as a regularization parameter in Eq. (24) that naturally results from the Gaussian process.

3. Experimental Study

In this section, the performance of the proposed BFELM on two public MI EEG datasets is evaluated and compared to other existing ELM-based techniques. Description of suggested datasets as well as experimental setup and evaluation is also provided in this section.

3.1 EEG Data Acquisition

In this work, two public data EEG datasets are used to study the proposed BFELM [6, 17, 18]. The first dataset is available from BCI competition IV dataset IIb. The EEG data was collected from nine different individuals using three bipolar channel ($C3$, Cz and $C4$) with a sampling rate of $250Hz$ for the discrimination of two classes (left-hand and right-hand motor imagery - (MI)). The electrooculogram (EOG) was recorded using three monopolar electrodes. They were recorded from each subject in five sessions [18]. For comparison purposes, in this study only the sets that correspond to B0103T, B0203T, …, B01903T are used for training the proposed BFELM. Each subject was required to complete 160 trials (half for each class of MI). For each trial, the subject received visual guidance to perform MI task for a period of time of 4.5 seconds.

Table 1: Classification accuracy (%) obtained by SVM, ELM, MKELM, FELM and BFELM, for competition IV, dataset IIb

Subject	SVM	ELM	MKELM	SBELM	FELM	BFELM[1]	BFELM[2]
B0103T	76.1	76.8	77.5	77.6	77.7	77.9	77.6
B0203T	56.9	59.8	65.4	61.8	62.3	63.1	63.2
B0303T	50.9	51.5	54.3	54.1	54.2	54.2	54.3
B0403T	98.7	98.7	99.3	99.0	99.1	99.1	99.0
B0503T	83.2	84.1	84.6	84.3	84.5	84.6	84.7
B0603T	67.6	68.3	69.5	69.2	69.3	69.5	69.3
B0703T	82.9	84.2	86.8	84.7	84.9	85.2	85.0
B0803T	87.0	87.5	89.9	89.0	89.3	89.6	89.3
B0903T	81.3	82.9	83.7	83.5	83.6	83.6	83.4
Average	76.0±15.3	77.0±14.8	79.0±14.0	78.1±14.3	78.3±14.1	78.5±14.1	78.4±14.2
Time (s)	16.4± 0.1	2.12±0.2	6.8±0.6	3.7±0.4	2.8±0.3	3.8±0.1	3.9±0.2

1 BFELM with Gaussian Membership Function
2 BFELM with Cauchy Membership Function

The second dataset corresponds to BCI competition III data set IVa. Such dataset was collected at a sampling rate of $100Hz$ from 188 electrodes from five different subjects, namely: "aa", "al", "av", "aw" and "ay". Each subject was required to complete 280 trials for the imagination of two tasks, i.e. right hand or foot movements. Each subject completed 140 trials for each MI task for a period of time of 3.5 seconds.

3.2 Experimental Setup and Evaluation

For comparison purposes, first data preprocessing was performed on the raw EEG data. For each trial, data was band-pass filtered between 4-40 Hz using a fifth-order Butterworth filter. Next, the dimension of EEG signals was reduced using Common Spatial Pattern (CSP), a widely used technique for feature selection in the classification of MI-based BCIs [21]. Finally, to discriminate the filtered EEG signals, a number of different techniques were implemented. To validate the performance of the proposed BFELM, a comparison study with other existing techniques was implemented. In this study, seven algorithms are compared:

(1) SVM: Support Vector Machine [5].
(2) ELM: Extreme Learning Machine [25].
(3) MLKELM: Multilayer Kernel Extreme Learning Machine [17].
(4) SBELM: Sparse Bayesian Extreme Learning Machine [13].
(5) FELM: Fuzzy Extreme Learning Machine [26, 27, 28].
(6) BFELM1: Bayesian-Fuzzy Extreme Learning Machine with Gaussian MFs.
(7) BFELM2: Bayesian-Fuzzy Extreme Learning Machine with Cauchy MFs.

For each algorithm, a 5×5 cross-validation was implemented. The experimental setup for each model involves the regularization parameter 'C' for SVM; number fuzzy rules and hidden neurons for FELM and the proposed BFELM and ELM respectively. For MKELM Gaussian and polynomial kernels are selected [17]. Tables 1 and 2 summarize the average classification performance of ten random experiments obtained by different learning algorithms for BCI competition IV data set IIb and BCI competition III data set IVa respectively. In both tables, the computational time was also compared among the seven methods under MATLAB R2016a on a laptop with 2.7 GHz CPU (16 GB RAM). As it can be observed from both tables, in general ELM-based techniques outperform the performance accuracy of SVM. In particular, MLKELM yields the highest accuracy for datasets IIb and IVa on almost all subjects.

For the classification of dataset IIb, the proposed BFELM with Gaussian MFs not only achieves a similar performance to that produced by a MKELM, but also improves the model accuracy of traditional SBELM. From Table 1, it can be seen that the implementation of an BFELM represents an accuracy improvement over ELM, FELM and SBELM on subjects $B0103T$, $B0203T$, $B0703T$ and $B0803T$. For subjects $B0103T$, $B0503T$, $B0603T$, the proposed BFELM improved the performance achieved by an MKELM. In general, from Table 1, the proposed BFELM either with Gaussian or Cauchy MSs achieves an accuracy of 78.5% and 78.4% correspondingly, an improvement of 1.02% over ELM and 1.03% over SVM.

For data IVa, the BFELM with a Gaussian and Cauchy MFs achieved a mean accuracy of 87.2% and 87.3%, an improvement of 0.1% with respect to ELM and SVM. As described in Table 2, BFELM provides the highest accuracy among BELM-based methods with a similar performance to that obtained by an MKELM. In terms of the computational load required to train each model, it can also be observed from Table 2, the incorporation of TSK inference engine does not implies a significant increase over traditional SBELM. Moreover, while the training time for an MKELM is about 10.1s to produce an accuracy of 87.5%, the training time of a BFELM is 7.3s to achieve an accuracy of 87.3%. This represents a decrease of 28% for the mean computational training. As illustrated in Table 2, the proposed BFELM approach represents an improvement on each subject over ELM, SBELM, FELM and SVM.

Finally, in Fig. 2, the effect of varying the number of fuzzy rules on the average accuracy of all subjects using 80% as training for the proposed BFELM models is presented. As illustrated in Fig. 2, an increase in the number of fuzzy rules does not necessarily imply an increase in the final accuracy. It can be observed, degraded accuracy of BFELM occurs when either using a large or small number of fuzzy rules. For data IIb, the optimal number of fuzzy rules for a BFELM either with Gaussian or Cauchy MFs is produced by using 30 fuzzy rules. In contrast, for data IVa, the optimal number of fuzzy rules using a BFELM with Gaussian MFs can be achieved using between 40-50 rules, while for a BFELM with Cauchy MFs, the highest accuracy can be obtained with 30 or more fuzzy rules.

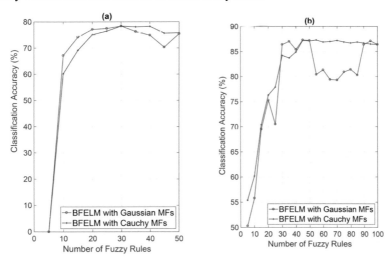

Fig. 2: Average accuracy of ten experiments for the BFELM using a different number of fuzzy rules for the classification of (a) competition IV, data IIb and (b) competition III, dataset IVa

4. Discussion

This section provides a performance analysis as well as its pros and cons for MI classification between the proposed BFELM, and a number of literature techniques such as MKELM, SVM, BELM, FELM and ELM. From Tables 1 and 2, it can be observed that the proposed BFELM either with Gaussian MFs or Cauchy MFs

Table 2: Classification accuracy (%) obtained by SVM, ELM, MKELM, FELM and BFELM, for competition III, dataset IVa.

Subject	SVM	ELM	MKELM	SBELM	FELM	BFELM[1]	BFELM[2]
aa	80.6	82.6	83.3	82.9	82.8	83.0	83.1
al	97.7	97.9	98.5	98.2	98.1	98.3	98.3
av	69.3	70.0	71.4	70.6	70.7	70.9	71.1
aw	89.7	90.2	91.3	90.7	90.9	91.0	91.1
ay	91.7	92.4	93.0	92.6	92.5	92.9	92.8
Average	85.8±10.7	86.6±11.3	87.5±10.8	87.0±10.9	87.0±11.1	87.2±10.7	87.3±10.6
Time (s)	26.3± 0.1	3.1±0.1	10.1±0.2	7.1±0.2	3.2±0.1	7.3±0.1	7.2±0.1

[1]BFELM with Gaussian Membership Function
[2]BFELM with Cauchy Membership Function

produces comparable results to that obtained by using a Multi-Kelnel ELM (MK-ELM) and in general it is more efficient than SVM, BELM, ELM and FELM. Generally speaking, similar to all ELM-based approaches, the suggested BFELM presents universal approximation properties for continuous functions [17, 20]. Moreover, the proposed BFELM also inherits the ability of ELM learning for tuning-free of the hidden layer weights (consequent part in the design of FISs) while delivering an improved generalization performance. The results presented in this study demonstrate the following:

(a) The aforementioned BFELM is a probabilistic method that provides an effective approach to build the confidence intervals of an FIS of TSK type while eliminating the need to incorporate a regularization parameter.

(b) Since the hyperparameters α and σ_i^2 are a consequent of a Gaussian process, the training of a FIS using the proposed BFELM does not require a regularization term. Therefore, by applying BFELM, the FIS naturally provides an improved generalization performance compared to traditional ELMs.

(c) The proposed BFELM is a probabilistic model for TSK FISs, in which fuzziness and probability work well for the classification EEG signals in a collaborative manner. Moreover, new advances in either theory may be implemented under appropriate conditions.

Table 3: Pros and cons of SVM, ELM, MKELM, FELM and the proposed BFELM

Method	Pros
SVM	• Good generalization properties • Can deal with high dimensional data
ELM	• Tuning-free for the hidden layer weights • High efficiency and good generalization performance
BELM	• Not need to include a regularization parameter
MKELM	• Not need to determine the number of hidden units • Can explore nonlinear features • Can combine multiple kernels
BFELM	• Unifies the concept of fuzziness and probability • Good generalization performance • New advances in Fuzzy theory and Bayesian learning may be implemented • Less number of hidden units than MKELM to provide a similar performance • Consequent part is a linear combination of each hidden activation.

Method	Cons
SVM	• Parameter selection is data dependent
ELM	• Need to determine the number of hidden units • Need to determine a regularization parameter to improve generalization
BELM	• Iterative method
MKELM	• Need to specify balance between kernels • Computationally burden increases as data size increases • Can combine multiple kernels
BFELM	• Iterative method, hence its training may prove more expensive than ELM • Number of fuzzy rules needs to be tuned

Finally, the main pros and cons for the proposed BFELM and other techniques for the classification of the datasets IIb and IVa are described in Table 3.

5. Conclusions

In this paper, a Bayesian probabilistic method based on ELM theory for the construction of of TSK Fuzzy Inference Systems (FISs), in which fuzziness and probability can work in a collaborative manner is presented. The proposed method called Bayesian Fuzzy Extreme Learning Machine (BFELM) is a unified learning algorithm based on Bayesian learning regression and Fuzzy Inference Systems theory. On the one hand, Bayesian learning allows the introduction of a priori knowledge while the confidence intervals of each consequent in the Fuzzy Inference System is analytically determined. On the other hand, simultaneous learning of antecedent and consequent part is achieved, where fuzziness and probability can work complementary rather than competitively for the classification of MI EEG tasks.

To evaluate the performance of the proposed BFELM, two public datasets about BCI competition IV dataset IIb and BCI competition III, dataset IVa are suggested. To compare the performance of a BFELM, other techniques such as traditional ELM and SVM as well as Fuzzy ELM (FELM), Bayesian ELM (BELM) and Multi Kernel ELM (MKELM) have been also implemented. As described in our results, it was demonstrated that the proposed BFELM shows similar performance to that provided by a MKELM and better than traditional SVM, BFELM, FELM and ELM. It can also be concluded that the associated computational training time is approximately 28% less expensive than MKELM.

It can also be concluded that the proposed learning framework inherits the capability of extreme learning machines for universal approximation of continuous functions as well as to randomly select the parameters of antecedent of each fuzzy rule in the FIS, and analytically determine their consequent. Future work includes the incorporation of new advances not only from Bayesian theory, but also from the design of higher FISs for the classification of MI EGG signals.

References

1. Al-Saegh, A., Dawwd, S.A. and Abdul-Jabbar, J.M. (2021). Deep learning for motor imagery EEG-based classification: A review, *Biomedical Signal Processing and Control*, 63: 102172.
2. Al-Saegh, A., Dawwd, S.A. and Abdul-Jabbar, J.M. (2021). Deep learning for motor imagery EEG-based classification: A review, *Biomedical Signal Processing and Control*, 63: 102172.
3. Berger, J.O. (2013). Statistical decision theory and Bayesian analysis. Springer Science & Business Media.
4. Chaudhary, S., Taran, S., Bajaj, V. and Sengur, A. (2019). Convolutional neural network based approach towards motor imagery tasks EEG signals classification, *IEEE Sensors Journal*, 19(12): 4494 4500.

5. Guler, I. and Ubeyli, E.D. (2007). Multiclass support vector machines for EEG-signals classification. *IEEE Transactions on Information Technology in Biomedicine*, 11(2): 117-126.

6. Gu, X. and Wang, S. (2018). Bayesian Takagi-Sugeno-Kang fuzzy model and its joint learning of structure identification and parameter estimation, *IEEE Transactions on Industrial Informatics*, 14(12): 5327-5337.

7. Hernandez-Hernandez, R.A., Martinez-Hernandez, U. and Rubio-Solis, A. (2020, July). Multilayer Fuzzy Extreme Learning Machine Applied to Active classification and Transport of objects using an Unmanned Aerial Vehicle. *In:* 2020 IEEE International Conference on Fuzzy Systems (FUZZ-IEEE), pp. 1-8.

8. Huang, G.B., Zhu, Q.Y. and Siew, C.K. (2006). Extreme learning machine: theory and applications. *Neurocomputing*, 70(1-3): 489-501.

9. Jin, Z., Zhou, G., Gao, D. and Zhang, Y. (2020). EEG classification using sparse Bayesian extreme learning machine for brain-computer interface. *Neural Computing and Applications*, 32(11): 6601-6609.

10. Khosla, A., Khandnor, P. and Chand, T. (2020). A comparative analysis of signal processing and classification methods for different applications based on EEG signals, *Biocybernetics and Biomedical Engineering*, 40(2): 649-690.

11. Liang, N.Y., Saratchandran, P., Huang, G.B. and Sundararajan, N. (2006). Classification of mental tasks from EEG signals using extreme learning machine, *International Journal of Neural Systems,* 16(01): 29-38.

12. MacKay, D.J. (1995). Probable networks and plausible predictions – a review of practical Bayesian methods for supervised neural networks. *Network: Computation in Neural Systems*, 6(3): 469.

13. Martinez-Hernandez, U., Rubio-Solis, A., Panoutsos, G. and Dehghani-Sanij, A.A. (2017, July). A combined adaptive neuro-fuzzy and Bayesian strategy for recognition and prediction of gait events using wearable sensors. *In*: 2017 IEEE International Conference on Fuzzy Systems (FUZZ-IEEE), pp. 1-6.

14. Nguyen, T., Khosravi, A., Creighton, D. and Nahavandi, S. (2015). EEG signal classification for BCI applications by wavelets and interval type-2 fuzzy logic systems. *Expert Systems with Applications*, 42(9): 4370-4380.

15. Park, C., Looney, D., Ur Rehman, N., Ahrabian, A. and Mandic, D.P. (2012). Classification of motor imagery BCI using multivariate empirical mode decomposition, *IEEE Transactions on Neural Systems and Rehabilitation Engineering*, 21(1): 10-22.

16. Pfurtscheller, G. and Neuper, C. (2001). Motor imagery and direct brain-computer communication, *Proceedings of the IEEE*, 89(7): 1123-1134.

17. Ren, W. and Han, M. (2019). Classification of EEG signals using hybrid feature extraction and ensemble extreme learning machine, *Neural Processing Letters*, 50(2): 1281-1301.

18. Rubio-Solis, A., Panoutsos, G., Beltran-Perez, C. and Martinez-Hernandez, U. (2020). A multilayer interval type-2 fuzzy extreme learning machine for the recognition of walking activities and gait events using wearable sensors, *Neurocomputing*, 389: 42-55.

19. Rubio-Solis, A. and Panoutsos, G. (2014). Interval type-2 radial basis function neural network: a modeling framework. *IEEE Transactions on Fuzzy Systems*, 23(2): 457-473.

20. Rubio-Solis, A., Martinez-Hernandez, U. and Panoutsos, G. (2018, July). Evolutionary extreme learning machine for the interval type-2 radial basis function neural network: A fuzzy modelling approach. *In:* 2018 IEEE International Conference on Fuzzy Systems (FUZZ-IEEE), pp. 1-8.

21. She, Q., Chen, K., Ma, Y., Nguyen, T. and Zhang, Y. (2018). Sparse representation-based extreme learning machine for motor imagery EEG classification, *Computational intelligence and Neuroscience*, 2018.

22. Soria-Olivas, E., Gomez Sanchis, J., Martin, J.D., Vila-Frances, J., Martinez, M., Magdalena, J.R. and Serrano, A.J. (2011). BELM: Bayesian extreme learning machine. *IEEE Transactions on Neural Networks*, 22(3): 505-509.

23. Zhang, Yangsong, *et al.* (2018). Two-stage frequency recognition method based on correlated component analysis for SSVEP-based BCI. *IEEE Transactions on Neural Systems and Rehabilitation Engineering* 26.7, pp. 1314-1323.

24. Tangermann, M., Mu¨ller, K.R., Aertsen, A., Birbaumer, N., Braun, C., Brunner, C., Leeb, R., Mehring, C., Miller, K.J., Mueller-Putz, G. and Nolte, G. (2012). Review of the BCI competition IV. *Frontiers in Neuroscience*, 6: 55.

25. Zhang, Yu, *et al.* (2013). Spatial-temporal discriminant analysis for ERP-based brain-computer interface. *IEEE Transactions on Neural Systems and Rehabilitation Engineering* 21.2, pp. 233-243.

26. Wang, Y., Gao, S. and Gao, X. (2006, January). Common spatial pattern method for channel selelction in motor imagery based brain-computer interface. *In:* 2005 IEEE Engineering in Medicine and Biology 27th Annual Conference, pp. 5392-5395.

27. Zhang, Y., Guo, D., Li, F., Yin, E., Zhang, Y., Li, P., Zhao, Q., Tanaka, T., Yao, D. and Xu, P. (2018). Correlated component analysis for enhancing the performance of SSVEP-based brain-computer interface, *IEEE Transactions on Neural Systems and Rehabilitation Engineering*, 26(5): 948-956.

28. Zhang, Y., Wang, Y., Zhou, G., Jin, J., Wang, B., Wang, X. and Cichocki, A. (2018). Multi-kernel extreme learning machine for EEG classification in brain-computer interfaces. *Expert Systems with Applications*, 96: 302-310.

Index

A

Acute lymphoblastic leukaemia, 123
AI in medicine, 135
Animal biometrics, 110
Anomaly detection, 281-284, 286-290, 292, 293, 295-301
Artificial intelligence, 64, 81, 111, 113, 115, 116, 135-140, 142, 173, 269

B

Bayesian learning, 348, 352, 359, 360
BCI system, 268-270, 276
Behavioral psychology, 110
Bi-directional LSTM, 30
Bioimpedance, 253-257, 259, 262
Biomedical informatics, 157
Blood biomarkers, 98-100, 105
Blood vessel segmentation, 157-159, 162, 164, 169, 170
Body mass index, 253, 259
Brain age prediction, 81
Brain sMRI, 102, 103, 105
Brain, 215, 218, 235
Brain-computer interface, 329, 347

C

Classification, 159, 160, 163, 164, 170
Clinic decision support, 81
Closed-loop motor training, 268
CNN, 305-307, 310-314
Common spatial pattern, 348, 353, 356
Contrastive learning, 238, 239, 241, 242, 249
Convolutional neural network, 329, 330, 335, 336, 342, 343
COVID diagnosis, 29, 32, 33, 38, 63, 72
COVID-19 detection, 11

COVID-19, 13-15, 17, 22, 41- 43, 49, 53-57, 238, 239, 244
CT Scan, 62, 63, 65, 69, 72

D

Deep learning, 10, 11, 29, 30, 36, 41, 42, 44, 56, 62-66, 68-71, 74, 76, 77, 111, 113-115, 117-119, 317, 318, 326
Diagnosis, 135-142
Dynamic mode decomposition, 176, 198
Dynamic time warping, 281, 282, 290

E

ECG, 281-283, 286-290, 292-301, 304-309, 314
EEG classification, 329, 341-343
EEG signals, 347, 348, 353, 356, 359
EEG, 317-320, 326
Emotion, 146-148
Emotional diagnosis, 317
Explainable AI, 81, 82
Explainable deep learning, 81
Extreme learning machine, 347-352, 356, 360

F

Facial expression synthesis, 146
Facial images, 104, 105
Feature selection, 123-132
Forecasting, 41-44, 47-49, 54, 56, 57
Freezing of gait, 173, 174, 176, 180, 188, 204, 207-208
Fuzzy C-mean clustering, 123, 158
Fuzzy logic, 348, 351, 352

G

Generative adversarial networks, 146

Genetic algorithm, 123
Group-LASSO, 216, 217, 220, 221, 228, 235

H

Healthcare, 1, 22
Human aging, 96
Hybrid differential evolution, 157

I

India, 41-43, 49, 53-57
Individual differences, 276
Infection rate, 42
Information fusion, 190, 197
Interpretable machine learning method, 239, 249

L

Long short-term memory, 29
LSTM networks, 43-45, 47, 48, 53
LSTM, 305-307, 312-314
Lung computerized tomography (CT) scan, 1

M

Machine learning, 13, 14, 19, 22, 96, 124, 125, 135, 137-141, 175
Medical imaging, 62, 72
Medical prediction and diagnosis, 135
MLP, 305, 306, 308, 309-314
Motor imagery, 329-331, 338, 339, 341, 343
MRI scan, 82, 91
Multimodal information, 188, 190, 197
Multi-wavelet basis functions, 217, 222, 235

N

NARMAX method, 241, 248, 249
NARMAX model, 16, 18, 21
NARMAX, 238-245, 248, 249
Neuro-decoding methods, 269

Neurodegenerative, 81, 92, 100, 173
Nonlinear, 215-221, 225, 227, 228, 234
Non-self match distance, 281, 282, 284, 290, 292-294, 296, 297, 299-301

O

Orthogonal matching pursuit, 216

P

Parkinson's disease, 173, 174, 188
Prediction, 135-137, 140-142
Proxy measurement, 176, 189, 196

R

Rehabilitation paradigms, 269
Respiratory anomaly, 29, 31, 38, 39
Retinal images, 157, 159, 160, 164, 170

S

SEIR model, 13, 15
SIR model, 13
Social behavioral biometrics, 110
Sparsity, 216, 217, 220, 221, 228, 235
Spike train, 216-219, 221, 225, 228
Support vector machines, 4

T

Takagi-Sugeno fuzzy logic, 257
Texture analyses, 1
Time-frequency analysis, 329, 337, 338, 340, 341, 343
Time-varying, 215-223, 225, 227-229, 234, 235
Trauma, 135, 137-142

W

Well-being monitoring, 146

X

X-Ray, 63, 66, 70-72, 77